Biochar

Emerging applications

Biochar

Emerging applications

Edited by
Alberto Tagliaferro, Carlo Rosso and Mauro Giorcelli
Politecnico di Torino, Turin, Italy

IOP Publishing, Bristol, UK

ISBN 978-0-7503-2660-5 (ebook)
ISBN 978-0-7503-2658-2 (print)
ISBN 978-0-7503-2661-2 (myPrint)
ISBN 978-0-7503-2659-9 (mobi)

DOI 10.1088/978-0-7503-2660-5

Version: 20201201

IOP ebooks

British Library Cataloguing-in-Publication Data: A catalogue record for this book is available from the British Library.

Published by IOP Publishing, wholly owned by The Institute of Physics, London

IOP Publishing, Temple Circus, Temple Way, Bristol, BS1 6HG, UK

US Office: IOP Publishing, Inc., 190 North Independence Mall West, Suite 601, Philadelphia, PA 19106, USA

To all our friends that make possible this eBook.

—Alberto, Carlo and Mauro

Contents

Preface

This eBook is aimed to highlight the perspectives of biochar as a substitute for oil derived carbon materials in advanced applications. In this eBook the most renowned research team from all over the world have brought their experience to assess the viability and perspectives of several innovative biochar applications. Many of the applications discussed in this eBook have been already proposed/tested/developed using other environment unfriendly carbon materials like carbon black, carbon nanotubes and graphene. We will show in this eBook that the environment friendly biochar is a viable alternative to them, its widespread use eventually leading to an eco-friendly era for carbon materials.

A quick look to the index readily shows that biochar has interesting perspectives in various field, some traditional, some innovative. A specific attention is dedicated to the use of biocahar as a filler in composite materials, where it can represent a viable alternative to existing fillers for large scale and low cost applications.

We really hope that you enjoy to read this eBook discovering new biochar applications or deepening your understanding in a particular application where you have not yet think that biochar could be a key material.

We are sure that after reading the eBook you'll share our view: biochar will be among the leadres of a new eco-friendly carbon era.

Enjoy!

Acknowledgements

Alberto, Carlo and Mauro would be grateful to all the authors that with their friendship and professional efforts have given their outstanding contribution that brought this eBook form the realm of wishful thinking to reality. IOP staff support in all steps of the editorial process is also gratefully acknowledged.

Editor biographies

Alberto Tagliaferro

Alberto Tagliaferro is an associate professor of solid-state physics at Politecnico di Torino, Italy, where he is head of the Carbon Group, and an adjunct professor at the University of Ontario Institute of Technology, Canada. He has been active in the field of carbon materials, their properties and applications for almost 30 years and has co-authored over 190 publications.

Carlo Rosso

Carlo Rosso holds a PhD in machine design and construction from Politecnico di Torino (2005) and he has been an Associate Professor in machine design at the Department of Mechanical and Aerospace Engineering of Politecnico di Torino since 2016. His main research topics focus on the dynamics of mechanical components with particular attention on gears and metal replacements in the automotive industries. In particular, he focuses on the usage of nanofiller for improving performance in composite materials and the usage of thermoplastic reinforced materials for structural applications. He is the author of four patents and the founder of two start-ups, one of which is Spin-Off of Politecnico di Torino. He is the (co-)author of 70+ publications on machine design topics. He has a good relationship with the industrial framework of the Piedmont region and he has signed industrial research agreements valuing more than €920 000.

Mauro Giorcelli

Mauro Giorcelli is an electronic engineer with PhD in physics. He is a co-founder of the Carbon Group of Politecnico di Torino (Italy) and his career is dedicated to carbon materials. In particular, he is interested in the properties that carbon materials could impart to composite materials. He started to work in the biochar field over five years ago and his collaborations are worldwide, from Canada to Asia and the European Union. He has published over 80 articles which have garnered over 900 citations (Scopus).

List of contributors

Justice Asomaning
Department of Agricultural, Food and Nutritional Science, University of Alberta, Edmonton T6G 2P5, AB, Canada

Hanieh Bamdad
Institute for Chemicals and Fuels from Alternative Resources, Department of Chemical and Biochemical Engineering, Western University, London, ON, Canada

Mattia Bartoli
Department of Applied Science and Technology, Politecnico di Torino, Corso Duca degli Abruzzi 24, 10129 Turin, Italy

Franco Berruti
Institute for Chemicals and Fuels from Alternative Resources, Department of Chemical and Biochemical Engineering, Western University, London, ON, Canada

Anna Bogush
Centre for Agroecology, Water and Resilience, Coventry University, Coventry, UK

David C Bresslerb
Department of Agricultural, Food and Nutritional Science, University of Alberta, Edmonton T6G 2P5, AB, Canada

Johnathon N Caguiat
Department of Chemical Engineering and Applied Chemistry, University of Toronto, Toronto M5S 3E5, ON, Canada

Michael Chae
Department of Agricultural, Food and Nutritional Science, University of Alberta, Edmonton T6G 2P5, AB, Canada

Jamal Chaouki
Mohammed VI Polytechnic University, Ben Guerir, Morocco

David Chiaramonti
POLITO/RE-CORD, Via del Campo di Marte 9, 50137 Florence, Italy

Oisik Das
Material Science Division, Department of Engineering Sciences and Mathematics, Luleå University of Technology, 97187 Luleå, Sweden

Sherif Farag
Department of Chemical Engineering, Polytechnique Montreal, Montréal, QC, Canada

Elisabetta Di Francia
Politecnico di Torino, Department of Applied Science and Technology, Corso Duca degli Abruzzi 24, 10129 Turin, Italy

Paola Giudicianni
Istituto di Ricerche sulla Combustione, CNR, Naples, Italy

Souradeep Gupta
School of Civil and Environmental Engineering, The University of New South Wales, Sydney, Australia

Kelly Hawboldt
Department of Engineering and Applied Science, Memorial University, St John's, NL, Canada

Anjali Jayakumar
UK Biochar Research Centre, School of GeoSciences, University of Edinburgh, Edinburgh, UK

Charles Q Jia
Department of Chemical Engineering and Applied Chemistry, University of Toronto, Toronto M5S 3E5, ON, Canada

Donald W Kirk
Department of Chemical Engineering and Applied Chemistry, University of Toronto, Toronto M5S 3E5, ON, Canada

Naomi B Klinghoffer
Institute for Chemicals and Fuels from Alternative Resources, Department of Chemical and Biochemical Engineering, Western University, London, ON, Canada

Sek Teng Koh
Department of Building, National University of Singapore, Singapore

Harn Wei Kua
Department of Building, National University of Singapore, Singapore

Tao Li
Department of Chemical Engineering and Applied Chemistry, University of Toronto, Toronto M5S 3E5, ON, Canada

Griffin Loebsack
Institute for Chemicals and Fuels from Alternative Resources, Department of Chemical and Biochemical Engineering, Western University, London, ON, Canada

Giulio Malucelli
Politecnico di Torino, Department of Applied Science and Technology and Local INSTM Unit, Viale Teresa Michel 5, 15121 Alessandria, Italy

Ondřej Mašek
UK Biochar Research Centre, School of GeoSciences, University of Edinburgh, Edinburgh, UK

Samuele Matta
Politecnico di Torino, Department of Applied Science and Technology and Local INSTM Unit, Viale Teresa Michel 5, 15121 Alessandria, Italy

Thomas R Miles
Executive Director, United States Biochar Initiative, Golden, CO, USA

Edoardo Miliotti
RE-CORD, Piazza degli Innocenti 2b, 59100, Florence, Italy

Aldrich Ngan
Department of Chemical Engineering and Applied Chemistry, University of Toronto, Toronto M5S 3E5, ON, Canada

Clare Peters
UK Biochar Research Centre, School of GeoSciences, University of Edinburgh, Edinburgh, UK

Raffaele Ragucci
Istituto di Ricerche sulla Combustione, CNR, Naples, Italy

Carlo Rosso
Department of Mechanical and Aerospace Engineering, Politecnico di Torino, Corso Duca degli Abruzzi 24, 10129 Turin, Italy

Patrizia Savi
Politecnico di Torino, Department of Electronics and Telecommunications, Corso Duca degli Abruzzi 24, 10129 Turin, Italy

Jean-Marc Tulliani
Politecnico di Torino, Department of Applied Science and Technology, Corso Duca degli Abruzzi 24, 10129 Turin, Italy

Muhammad Yasir
Politecnico di Torino, Department of Electronics and Telecommunications, Corso Duca degli Abruzzi 24, 10129 Turin, Italy

Ken Yeung
Departments of Chemistry and Biochemistry, Western University, London, ON, Canada

Daniele Ziegler
Politecnico di Torino, Department of Applied Science and Technology, Corso Duca degli Abruzzi 24, 10129 Turin, Italy

Part I

Biochar: feedstocks, production, and characterization

IOP Publishing

Biochar

Emerging applications

Alberto Tagliaferro, Carlo Rosso and Mauro Giorcelli

Chapter 1

Introduction to the biochar world with a focus on new possible applications

Thomas R Miles

1.1 Introduction

The exploration of biochars has evolved since their discovery as components of Amazonian dark earths in the last century. The process has been international, and individuals, businesses, and research organizations have developed biochar products to meet the demands of farming, markets, and climate change. Biochars are generally viewed as soil modifiers which can sequester carbon and improve soil health by retaining nutrients and water, however, their physical, chemical, and biological properties make biochars versatile materials which can be used in a variety of applications. Industry and academia are expanding the applications of biochars in agriculture and for the environment. They are being adapted to a variety of soils, crops, and uses. More efficient forms are being developed for use in large scale agriculture. New products are being developed for green infrastructure, water quality, and construction. Increased production and the development of new technologies and new analytical tools have led to new markets and uses. This chapter will provide an overview of the current use of biochars, biochar production, and emerging biochar products with examples from agriculture and industry.

1.2 Biochar properties

The physical properties of biochars, such as their porosity, are important for their current uses. When biomass is heated a skeleton of carbon is left as fine-grained, highly porous charcoal. Porous biochar can modify the soil texture to reduce soil density which helps to relieve compaction. Biochars have been used in growing media as renewable substitutes for manufactured media such as vermiculite. The low density of biochars has enabled the City of Stockholm to combine biochars made from urban wood waste with rock to create structured soils to relieve compaction and thus to revive urban trees. This use has been adopted in other Scandinavian

doi:10.1088/978-0-7503-2660-5ch1

cities (Gustafson 2019). Biochars have been useful additions to compost where the low density and porosity improves aeration, water infiltration, and water retention. Biochars make more water available to plants which extends the growing season and helps plant survival. The pore sizes and surface areas of biochar are suitable for capturing the gases and nutrients essential to plant growth. The biological properties have also been important. The discovery of the electrochemical properties of biochars, which affect biological functions, has led to new products and applications. Biochars have been described as 'biogeobatteries' which store energy and nutrients, provide habitats for organisms, and help plants and organisms in the microbiome cycle carbon, nutrients, and energy through the soil.

The capacity of biochars to help plant growth varies with the composition of ash and carbon. Research has helped to explain how some of the interactions of these components may lead to more efficient commercial products. For example, some biochars can increase the soil E_h, or redox potential, which results in greater plant nutrient content (Chew *et al* 2020, Chacon *et al* 2020). High ash biochars can enhance the cation exchange capacity (Domingues *et al* 2020). These studies should lead to the creation of products which can be more efficiently placed in the rhizosphere to reduce fertilizer inputs and improve plant growth.

The large specific surface area, porous structure, and the content of the non-carbon organic fractions provide a high variability in surface functional groups which enables the sorption of pollutants and nutrients. These properties enable biochar to capture pollutants which is why they have been increasingly used in soil remediation and stormwater filtration. New methods to modify the surface functional groups have improved the efficiency of biochars in nutrient and pollutant capture and made them suited for new uses. These properties have helped to establish viable uses of biochars in developing and developed economies.

1.3 Products and markets

Farmers have been slow to adopt biochars even though such materials are used in a wide variety of soils and crops and at all levels of agricultural production. The high cost of processing and the logistics of biochars have limited the broader use of biochars in agriculture. Consequently, many biochars are used in small quantities to enhance soil improvement or in higher value urban applications. The 62 biochar suppliers who are members of the National Alliance of Biochar Science and Technology Innovation in China have demonstrated that small quantities of biochar can be used in biochar-based fertilizers to improve the economic feasibility for general adoption. Their biochar-based fertilizers incorporate mineral rich biochars made from crop residues. Farmers are provided incentives to remove straw and deliver it to fertilizer companies to avoid open burning. The straw is carbonized and processed into fertilizers which are granulated and sold back to the farmers. Large scale production of biochar-based fertilizers has provided opportunities for the long term evaluation of biochar use (Meng *et al* 2019). The collaboration of universities and industry has led to the further discovery of nanocarbons in liquids from biochars and wood vinegars, which may lead to other ways in which the carbon from biochar production can be used more broadly in agriculture (Pan 2018).

In North America, most biochars are made from residues of the wood and energy industries. Future supplies are expected to come from forest residues. The innovative testing of biochar in forestry includes reforestation, range improvement, growing media, and revegetation. Wood derived biochars are used by a few progressive farmers who use conservation and regenerative agricultural techniques such as no-till and cover crops. For example, a corn and soybean farmer increased soil carbon, crop quality, and yield by adding a small amount of biochar to an organic fertilizer mix each year using strip tillage. His yield and quality increased the first year and has continued to increase through the last five years. In his view biochar combined with other amendments has helped improve the soil health and productivity. Another farmer who applied a manure-coated wood-based biochar in strip tillage declared that 'biochar is a change agent'. These farmers will adopt biochar-based fertilizers if they can be handled safely using existing equipment (Terrachar).

Some new products have emerged which farmers are more likely to use, including biopesticides and biofertilizers, where micronized biochars are used as carriers for nutrients, micro-organisms, or liquids. The powdered fertilizers can then be applied using air seeders, liquid suspensions, electrostatic sprayers, or irrigation systems (Greenquest). Formulated biotic soil amendments which combine biologicals, biochars, minerals, and nutrients are now produced in Europe and North America where they are used in high value horticultural and landscape applications (Carbon Gold).

Biochars are increasingly being used as ingredients instead of as bulk soil amendments even in developing countries. Smallholders and mid-sized farmers in some Asian countries have traditions of using biochars. They recognize that biochars can improve soil and animal health where carbon stocks are low and sources of nutrition are scarce. These biochars vary in composition but are often used as ingredients with nutrient sources such as manures. Biochars from grasses and agricultural residues usually contain important minerals, such as silica, potassium, and calcium which are often scarce. The benefits of these minerals can be seen in smallholder plots on poor soils where small amounts of biochars from crop residues combined with manures improve plant health and disease resistance. These uses continue to grow as methods improve to make biochars which are suited to smallholders in developing countries (Warm Heart, B4SS).

Mineral modified biochars mimic the man-made, or anthropogenic, terra preta soils in the Amazons. Minerals such as magnetite or expanding clays such as bentonite are added to wood during carbonization. The intermediate biochar is then baked at lower temperatures with additional minerals, such as manures. The results are mineral modified biochars (CPMTP). These biochar-based fertilizers improve soil properties, such as an enhanced redox potential. Less biochar is required for improved plant growth. The development of new enhancements to increase biochar functions can provide products for both agricultural and environmental applications which will improve the economic feasibility of producing biochars.

Biochar markets continue to grow in Europe and North America, creating new opportunities to convert wastes and residues into sustainable carbon while restoring soil health and improving water quality. Biochar-based products are useful amendments in poor quality urban soils. Urban landscaping, including gardens, turf and trees, and remediation are high value markets. Agricultural use is increasing as growers find ways to use high cost biochars strategically in field crop production. New products and management techniques are being developed such as biochars with compost, co-composted biochar with manures and green wastes, biochar-based fertilizers, biotic soil amendments, and micro and nano-carbons. Biotic soil amendments combine biochar, nutrients and biologicals and are delivered in granulated and liquid forms (Lesco). Public agencies have developed cost share programs to promote the addition of biochars and compost to increase soil carbon in agriculture. Biochar suppliers are learning to provide biochars of consistent quality to soil blenders and formulators, in particular for use in horticulture. Arborists include biochars in media to establish or repair urban trees and to ensure plant survival (Bartlett). Biochars are used in structured soils, rain gardens, and bioswales to improve the growing conditions and reduce pollutants. These biochars are also used in green roofs and other green infrastructure.

Environmental markets include the use of biochars to sorb pollutants in stormwater filters, wetlands, and bioretention facilities. Biochars combined with organic amendments and nutrients are used to reclaim mine land and remediate oilfields. They are also used to filter water and reduce organic pollutants (Boehm *et al* 2020).

1.4 Forms of biochar

Biochars can be produced at scale as co-products in existing bioenergy, wood products, and carbon producing facilities for delivery in bulk to farms and compost facilities. Biochars from industrial processes are often fine or can be screened to less than 6 mm. Wood chars usually contain more than 60% carbon. Sometimes the fine ash is removed to increase carbon content to 80%. Finer chars are used in growing media for forestry, arboriculture, horticulture, landscaping, erosion control, liquid applications, hydroseeding, biotic soil amendments, and even animal feed. These biochars can be used in spreaders and seeders. They are often combined with other ingredients. The physical properties of fine biochars from gasifiers and boilers make them effective replacements for vermiculite and perlite in nurseries. Increased root growth is often seen when biochars are used in growing media. In hydroseeding, crushed biochars are added to fiber, seed, and fertilizer to improve plant establishment on poor soils. Chip-sized chars are preferred in filters, vegetative strips, and in bioretention filters. They are increasingly used in bioswales and rain gardens to ensure plant survival and to prevent metals such as copper, zinc, and cadmium from entering waterways. Chip-sized biochars are used to filter *E. coli* from stormwater (Mohanty and Boehm 2014).

1.5 Methods to apply biochar

Biochars can be delivered in 1.5 m^3 bulk bags or in bulk trucks. The dry weight of a low ash biochar is often 150 kg m^{-3} or about 7 m^3 per dry tonne. A 10 tonne load could be used to amend 1–15 ha depending on how it is applied. Biochars are usually wetted to about 30% to reduce dusting and fire hazards in storage and transport.

Spreaders have been developed for broadcasting crushed biochars for revegetation. Applications can be 20 tonnes per hectare for range improvement or to establish grasses and forbs for mine reclamation, or less than 100 kg ha^{-1} for biochar-based fertilizers.

Methods of incorporating biochars are evolving as farmers learn to blend biochars with organic fertilizers such as poultry litter. A 'subsurfer' has been developed to incorporate poultry litter which would be an effective method to apply coarser biochars, manures, or composted biochars to increase moisture under cover crops (Leytem *et al* 2009). Broad agriculture is increasingly liquid based so new biochar products must be adapted for application in granules or liquids. Farmers who practice conservation and regenerative techniques have been the most receptive to using biochar. Biochars are readily applied with no-till equipment in organic fertilizer blends. No-till applications can also include biochar-based fertilizer or liquids. Fine biochars added to compost teas improve foliar applications.

Biochars are combined with green waste to make composted soil amendments. Five to ten percent biochar added to organics reduces odor, retains nitrogen, improves the quality of compost and makes biochars easier to handle (Sanchez-Monedero 2018). Biochars composted with organics increases water retention in vineyards. Biochars can capture and make nutrients available to plants which suggests that biochar could reduce the leaching of nitrates from compost. Poultry farmers have found that adding biochar to litter for in-house composting reduces odor, improves bird health, and provides low cost nitrogen to their farm customers. Improved soil amendments are also made by vermicomposting poultry litter and biochar. Vermicomposting with biochar has been proposed for treating municipal biosolids. An Australian avocado farmer has increased fruit quality and yield by composting wood chips and poultry manure in orchard rows on top of a clayey soil amended with biochar (Joseph *et al* 2020).

Biotic soil amendments, which combine biochars with fertilizers, minerals, and microorganisms, help accelerate growth (Lesco). Biochars in liquid suspensions are being developed for use in no-till to offset acidity from over fertilization and to change root zone dynamics. They are convenient to apply and can be effective in plant establishment and growth. Biotic soil amendments made from biochar can be applied in a granular or liquid form. Biochar is used in formulated seed balls for aerial seeding by drones and fixed wing aircraft in Africa, Australia, and America (Airseed, Droneseed, Seedballs Kenya).

1.6 Production

Mobile and stationary systems are used to produce biochars. Systems with capacities up to 2 tonnes per hour are usually mobile or modular. Capacities of 2–6 tonnes per hour are usually processed with industrial pyrolysis or gasification technologies.

Biochar is made by heating wood in the absence of air. In a simple kiln, a high-quality biochar can be made by limiting air at the base of the fire. As the wood is heated, gases evolve which mix with air to burn above the kiln. Radiant heat from this 'flame cap' heats the wood to convert it to pyrolysis gas and biochar. Wood is charred as it sits in a pool of pyrolysis gas. A 'flame cap' kiln principal has been applied to mobile kilns for carbonizing small diameter wood. These kilns produce good quality biochars while reducing emissions. A 1 m^3 kiln will produce a cubic meter of biochar from about 5 m^3 of wood in 4 h. Several kilns can be tended at once. Flame cap kilns such as the 'Kon Tiki' are used in many parts of the world (Ithaka Institute). Clays and manures can be added in these simple kilns to enrich the biochars.

A pit version of the flame cap kiln can be made in the ground with local tools. It is covered with metal to quench the char. Biochar is mixed with manure at planting. Smallholders in Ethiopia, Malawi, Ghana, Kenya, and Peru make biochar in smokeless 'flame cap' pit kilns. Smallholders in Kenya are saying that 'Biochar is a farmer's best friend'. The fuels are residues from their corn and sorghum crops (Warm Heart).

The flame cap principle can be used to make char and reduce emissions when burning piles of wood in small 'conservation burns'. The technique prevents high temperatures at the ground which otherwise kill organisms and causes scars from open burning. The char is then spread in place. The conservation burn does not kill soil organisms and the biochar facilitates regrowth. Conservation burns are often used for piles of 2–5 m^3. Many foresters, contractors, and fire crews have been trained in this technique (Wilson). A mobile 'flame cap' carbonizer is under development with public funding in which a curtain of air ensures complete combustion of the gases for reduced emissions. The combustion air is adjusted to increase the size of the pool of pyrolysis gases which carbonize the wood. Biochar is removed as it is formed. The systems in development are intended to recover more carbon than existing mobile devices. Scaling up to larger capacities will be determined by commercial partners (Air Burners Inc). Mobile carbonizers can process wood at landings or distributed sites at capacities of up to 100 dry tonnes per day, producing 2–5 tonnes of biochar. There are about 30 of these systems in operation. The air quality is acceptable to most authorities. Permitting is based on the amount of material to be burned in a season. They are currently used for residues with tipping fees such as land clearing and orchard removal. These chars can be used directly or modified for value added markets (Tigercat).

Modular pyrolysis systems are emerging which are suited to distributed biochar production and reprocessing. A typical modular pyrolysis system can process 1.5 tonnes of fuel per hour to produce up to 10 tonnes of biochar per day or 3000 tonnes per year (Artichar, Earth Systems). They can process materials with high annual growth such as branches, needles, and grasses. Stationary pyrolysis provides additional process control for value added biochars. They are a good fit for integrated wood utilization facilities. An array of modular batch kilns is used to convert shredded logs and logging debris to biochar. The batch kilns can be used at distributed locations. Biochars are crushed and sized to chips, medium, small, and powder for different purposes such as direct placement, air seeding, irrigation or water spray, soil injection, biofilters, bio-fillers, or carriers for microbes (BiocharNow).

Small scale European gasifiers and boilers can generate heat and biochar. Systems can consume 250 kg of fuel per hour to produce 400 kW heat and 60 kg of biochar. They will produce about 500 tonnes of biochar per year (Biomacon). Systems in greenhouses convert wood to heat, power, and biochar. At one installation fuel is delivered to the greenhouse in exchange for biochar. A self-unloading trailer meters the fuel to a gasifier which converts the fuel to gas and char. Part of the gas is fed to an electric generator. The rest goes to a boiler which provides heat for the greenhouse. The biochar is retrieved by the fuel supplier who uses it to enhance food waste compost (Rainbow Beeeater).

Most biomass energy plants are designed to re-inject and reburn flyash with a high carbon content, but some plants can recover it for use as biochar. A 30 MWe plant can produce about 8 m^3 of biochar per hour or 8000 tonnes, per year. The chars are often post-processed to remove fines depending on the market.

The commercial production of biochars and markets for biochar-based products have been slow to develop. Large scale industrial biochar production is limited. Most production facilities are small. China leads production with an estimated 500 000 tonnes of biochar-based fertilizers per year followed by smaller production in Southeast Asia, America, and Europe. Most plants use conventional pyrolysis or gasification technology. New pyrolysis technologics for liquid fuels are not yet producing commercial biochars. Some companies which produce charcoal for barbeque, non-fossil carbon metallurgical grade charcoals, and other remediation or heating products in large quantities have started to make biochars. They are investigating postprocessing techniques such as activation to increase product value.

Animals process biochars in important ways. Feeding biochars improves animal health, meat, and milk production. Cows, pigs, chickens, goats, and sheep will eat biochar given a free choice. Young cows are particularly fond of biochar as it facilitates feed conversion in their developing rumens. In Australia, cows in a 250 head dairy consumed small quantities of biochar each day which increased fluid milk earning the farmer an additional \$50 000 in a year. Biochar inoculated manure was consumed by dung beetles which further inoculated and mixed the nutrient laden char in the pasture improving the soil and forage quality (Rebbeck 2020). This has led to further dung beetle research and application. There is evidence that earth worms that ingest char can produce inoculated nanoparticles of char and iron which improve the use of nutrients such as phosphorous.

The use of biochars in animal production has been embryonic. Smallholders who use biochar for crops discover that they are also effective in improving disease resistance in animals. Reduced disease appears to improve egg, milk, and meat production (Warm Heart). A European feed standard has been developed for wood derived biochars (EBC). Producers report that biochar reduces the need for pharmaceuticals in large scale poultry operations (Charline). Increasing awareness of feed benefits has led to research and demonstration in Australia and North America. The importance of cascading biochar through animal feed to soil amendments is just beginning to be appreciated. As described above beef and dairy herds in Australia are discovering that manures distributed by the cattle are converted in the soil by dung beetles which improves the pasture soils and forage quality (Rebbeck 2020). Analysis of mechanisms in digesters and rumens has led to the benefits of the electrochemical properties of biochars, in particular the impact of biochars on increasing the redox potential of soils. Biochars added to anaerobic digestors are now used commercially to stabilize the microbial environment for renewable gas production. Biochar solids are incorporated in the digestate and retain nutrients when composted or applied as a soil amendment. Even countries which restrict chars in animal feed have been conducting research to justify their use (Winders *et al* 2019).

1.7 New applications

Biochars are finding their way into non-soil applications where carbon sequestration is still an objective. Initial uses include urban environmental applications for pollutant control and waste management, construction materials, and other commercial products. These uses often have higher values which can pay for processing or postprocessing.

Urban markets including landscaping, stormwater, and green infrastructure. New products are developed for landscaping and erosion control. Biochars are increasingly used in bioswales and rain gardens to prevent metals such as copper, zinc, and cadmium from entering waterways. Biochars are modified or amended to improve nitrogen and phosphorous removal. Biochars have been effective filters for *E. coli* in stormwater (Mohanty and Boehm 2014). They have been used to reduce chemical oxygen demand (COD) in leachates. Filter media assemblies are constructed using biochars for roof drains and stormwater outfalls (Stormwaterbiochar.com). General specifications are being developed for practices to use biochar to filter bacteria and organics from stormwater and water prepared for recharge in aquifers and constructed wetlands (Boehm). Biochars have been used to absorb mercury from a legacy chemical plant (Wang and Ptacek 2021). Functionalized biochars are likely to find their way into these filtration applications. The benefits and appropriate use of biochars in these applications has been demonstrated in science and in practice. Biochars are effective tools but more landscape architects and engineers in public agencies must write specifications and approve biochar-based products for demand to grow.

Urban waste management opportunities include carbonizing biosolids and urban wood waste. Some pyrolysis and staged gasification systems have been installed to process biosolids to reduce treatment costs, reduce fossil carbon consumption, and destroy pharmaceuticals such as polyfluorinated compounds (PFAS) which are not removed in existing processes (Bioforcetech, Ecoremedy, Pyreg). Hog manure has been pyrolyzed in China to destroy pharmaceuticals. Combinations of pyrolysis with anaerobic digestion have demonstrated the synergy of adding pyrolysis gases and biochars to the digesters, which can stabilize and increase gas production while producing renewable natural gas (RNG) (Anaergia).

Large volumes of urban wood waste are available for conversion to biochar. Mobile carbonizers are used to reduce the volume of urban tree removal (Tigercat). Construction and demolition material recovery facilities produce clean wood which can be converted to biochar (Construction and Demolition Recycling). Gypsum board, which is also called drywall, plasterboard, sheetrock, or wallboard, is common in demolition materials. It contains a paper face and a gypsum, or calcium sulfate, core. The paper is separated and recycled as animal bedding. Clean gypsum is combined with biochar and granulated for agricultural use. Dirty gypsum contains sulfur which creates odor problems at landfills when it is hydrated. New processes seek to combine the dirty gypsum with biochar to reduce odors and bind heavy metals from construction.

Biochars which are not suitable for soil application can be incorporated into road building and construction materials. An Australian company has demonstrated a unique cold formed asphalt which incorporates biochar. It is a promising method to sequester carbon in roads and pathways while providing weather and traffic resilient surfaces (Carboncor). It is likely that changing existing regulations to be able to use biochar amended asphalt may be the biggest barrier to its use.

Bates and Draper (2019) included other composites for biochar which are under development or in limited production. These include the addition of biochars to cement, plasters, gypsum board, plastics, building blocks, masonry, and several consumer products. New uses in building materials such as wallboard and cement increase the potential to sequester carbon with biochar. Biochars in wallboard could provide additional cooling and reduce the electromagnetic transmission from house wiring. Biochar improves the properties of plastic composites for building and decorative uses. At least one company is nearing commercial production of plastic-biochar composites for non-structural uses (American Cierra). Biochars are added to lighten masonry for easier construction. These masonry products have passed rigorous tests for flame resistance, safety, and strength. They should be approved for construction but they are not yet in commercial production (Biogreenstone).

New high value uses for biochar will depend on biochar modifications. Some of the modifications used today to enhance carbon performance include: surface area through physical activation (with CO_2, steam); char oxidation with the formation of surface carboxyl and carbonyl functional groups in contact with strong acids or oxidants (O_2, O_3, H_2O_2); and N-doping (reaction with NH_3 or co-processing with N sources). Other functionalization strategies such as co-composting and the addition of metals and enzymes are also used. These and other modifications are described by Wang *et al* (2020). Because the number of applications and products is so large, it is necessary to organize the creation of carbon companies specialized in targeted products and markets.

1.8 Summary

The increased production, research, and demonstration of biochars in a wide variety of applications have led to new products and uses, and to opportunities for the long term evaluation and discovery of the benefits of biochar. Biochars and processed biochar-based products are used strategically to improve the effectiveness and reduce costs. Higher value uses for agriculture, urban landscaping, waste management, and the environment have helped to build and sustain biochar businesses. Research will lead to new production processes and new products, and enhance the usefulness and economic feasibility of using biochars.

Organizations

Air Burners Inc., Palm City, FL, USA https://airburners.com
Airseed Technologies, Sydney, Australia; Capetown, South Africa https://airseedtech.com
American Cierra, Auburn, NY, USA http://www.cierraindustries.com/
Anaergia, Burlington, CA, USA https://www.anaergia.com/
Artichar, Prairie City, IA, USA https://arti.com
Australia New Zealand Biochar Industry Group, New South Wales, Australia https://anzbig.org/
Bartlett Tree Experts, Charlotte, NC, USA https://www.bartlett.com/
Biochar For Sustainable Soils (B4SS) https://biochar.international/
BiocharNow, Berthoud, CO, USA http://biocharnow.com
Bioforcetech, Redwood City, CA, USA https://www.bioforcetech.com/
Biogreenstone, KS, USA https://biogreenstone.com
Biomacon, Rehburg, Germany https://biomacon.com
CarbonCor, Perth, Western Australia, Australia https://carboncorproducts.com
Carbon Gold, Bristol, UK https://carbongold.com/
Charline, Riedlingsdorf, Austria https://char-line.com
Construction and Demolition Recycling Association, Chicago, IL, USA https://cdrecycling.org/
Droneseed, Seattle, WA, USA https://www.droneseed.com/
Carbon Powdered Mineral Technology and Products, Hallsville, New South Wales, Australia https://www.carbonpoweredmineral.com.au/
Earth Systems, Kew, Victoria, Australia https://earthsystems.com.au
Ecoremedy, Mechanicsburg, PA, USA http://ecoremedyllc.com/
European Biochar Certificate (EBC) https://www.european-biochar.org/en
European Biochar Industry Consortium http://www.biochar-industry.com/
International Biochar Initiative (IBI), Washington, DC, USA http://www.biochar-international.org
Ithaka Institute, Arbaz, Switzerland http://www.ithaka-institut.org/en/home
Greenquest, Richfield, WI, USA https://onagreenquest.net/
LESCO, Cleveland, OH, USA https://www.lesco.com/products/carbonpro
Pyreg, Dörth, Germany http://www.pyreg.de
Rainbow Beeeater, Melbourne, Victoria, Australia http://www.rainbowbeeeater.com.au
Seedballs Kenya, Nairobi, Kenya http://www.seedballskenya.com
Stormwaterbiochar.com, Portland, OR, USA http://Stormwaterbiochar.com
TerraChar, Colombia, MO, USA, http://www.Terra-char.com

Tigercat, Brantford, ON, Canada https://www.tigercat.com/products/material-processing/
Warm Heart Worldwide, Chain Mai, Thailand http://www.warmheartworldwide.org
Wilson Biochar Associates, Cave Junction, OR, USA http://www.wilsonbiochar.com
United States Biochar Initiative, Golden, CO, USA http://www.biochar-us.org

References

Bates A and Draper K 2019 *Burn: Igniting a New Carbon Drawdown Economy to End the Climate Crisis* (New York: Chelsea Green)

Boehm A B *et al* 2020 Biochar-augmented biofilters to improve pollutant removal from stormwater—can they improve receiving water quality? *Environ. Sci.: Water Res. Technol.* **6** 1520–37

Chacon F J *et al* 2020 Enhancing biochar redox properties through feedstock selection, metal preloading and post-pyrolysis treatments *Chem. Eng. J.* **395** 125100

Chew J *et al* 2020 Biochar-based fertilizer: supercharging root membrane potential and biomass yield of rice *Sci. Total Environ.* **713** 136431

Domingues R R, Sánchez-Monedero M A, Spokas K A, Melo L C A, Trugilho P F, Valenciano M N and Silva C A 2020 Enhancing cation exchange capacity of weathered soils using biochar: feedstock, pyrolysis conditions and addition rate *Agronomy* **10** 824

Gustafson M 2019 Stockholm biochar project *Biochar and Bioenergy 2019 (Fort Collins, CO)* https://biochar-us.org/presentation/stockholm-biochar-project-1

Joseph S *et al* 2020 Biochar increases soil organic carbon, avocado yields and economic return over 4 years of cultivation *Sci. Total Environ.* **724** 138153

Leytem A B *et al* 2009 Case study: on-farm evaluation of liquid dairy manure application methods to reduce ammonia losses *Prof. Anim. Sci.* **25** 93–8 https://ars.usda.gov/ARSUserFiles/np212/LivestockGRACEnet/ManureInjection.pdf

Meng J *et al* 2019 Development of the straw biochar returning concept in China *Biochar* **1** 139–49

Mohanty S K and Boehm A B 2014 *Escherichia coli* removal in biochar-augmented biofilter: effect of infiltration rate, initial bacterial concentration, biochar particle size, and presence of compost *Environ. Sci. Technol.* **48** 11535–42

Pan G 2018 China's biochar story: from crop residues to biomass industry *Biochar* (Wilmington, DE: US Biochar Initiative) https://biochar-us.org/presentation/china-biochar-story-crop-straw-biomass-industry

Rebbeck M 2020 Feeding Biochar to Dairy Cows, *Presented to the Australia New Zealand Biochar Conf. Study Tour 2020* https://anzbc.org.au

Sanchez-Monedero M A *et al* 2018 Role of biochar as an additive in organic waste composting *Bioresour. Technol.* **247** 1155–64

Wang A O and Ptacek C J 2021 Impact of multiple drying and rewetting events on biochar amendments for Hg stabilization in floodplain soil from South River *VA Chemosphere* **262** 127794

Wang L *et al* 2020 New trends in biochar pyrolysis and modification strategies: feedstock, pyrolysis conditions, sustainability concerns and implications for soil amendment *Soil Use Manage* **36** 358–86

Winders T M *et al* 2019 Evaluation of the effects of biochar on diet digestibility and methane production from growing and finishing steers *Translational Animal Sci.* **3** 775–83

Chapter 2

Controlling the conversion of biomass to biochar

Paola Giudicianni, Raffaele Ragucci and Ondřej Mašek

Pyrolysis is being explored as a possible thermal treatment capable of producing, in the absence of molecular oxygen, a solid residue (char) suitable for applications in several fields (as a fertilizer, activated carbon, etc) and liquid (bio-oil) and gaseous products that can be exploited for energy applications. Depending on the thermal conditions and the vapor residence time applied in the process, different types of pyrolysis exist: slow, fast, and flash pyrolysis. A lower process temperature and heating rate and longer vapor residence times (slow pyrolysis) favor the production of char, whereas a moderate temperature and short vapor residence time (fast and flash pyrolysis) generally produce bio-oil.

Many possibilities exist for improving the potential of char for diverse applications, such as use as a fuel in traditional and advanced power generation facilities, a fertilizer and carbon sink, a contaminant adsorbent in wastewater and soil, an adsorbent or catalyst for gas cleaning, a catalyst for syngas conversion to liquid hydrocarbons and biodiesel production, a raw material for supercapacitors, and a filler in wood and polymer composites.

Pyrolysis chars vary greatly in structure and chemistry due, in part, to the large degree of chemical heterogeneity of the feedstock. Previous works related to biomass pyrolysis in an inert atmosphere considered temperature, heating rate, pressure, gas residence time, and feedstock as the main operating variables affecting the char yield and characteristics. Moreover, the role of the pyrolyzing agent has relevant effects on char yield and morphology as well. Variations in pyrolysis operating conditions lead to products with a wide range of values of pH, specific surface area, pore volume, CEC, volatile matter, ash, and carbon content. A high pyrolysis temperature promotes the production of biochar with a strongly developed specific surface area, and high porosity, pH, and ash and carbon content, but with low values of CEC and volatile matter.

Therefore, to optimize the pyrolysis process, an overall analysis of the effects of the abovementioned variables on both the yield of char and its chemical and

doi:10.1088/978-0-7503-2660-5ch2

physical characteristics is required. In addition to the process conditions, the feedstock pre-processing (pelleting, milling, use of additives, etc) can also play an important role in determining the biochar properties and its performance in different applications.

2.1 Thermal decomposition of biomass undergoing pyrolysis

Biomass pyrolysis is a thermochemical decomposition conducted at a high temperature in the absence of molecular oxygen. Under these conditions the biomass undergoes complex decomposition reactions whose products can be grouped into three categories: condensable vapors, permanent gases, and carbon-rich solid residue (referred to as condensables, gas, and biochar in the following). By varying the experimental conditions of the process, three different types of pyrolysis can be performed:

- *Slow pyrolysis*, characterized by a low heating rate (~1–100 °C min^{-1}), a long residence time of the hot vapors in the reaction environment (~10–30 s), and a holding time at the final pyrolysis temperature varying from a few minutes to a few hours.
- *Fast pyrolysis*, characterized by a high heating rate (~100–1000 °C s^{-1}), a very short residence time of hot vapors (~1 s), and a temperature between 450 °C and 550 °C.
- *Conventional pyrolysis*, characterized by intermediate conditions between slow and fast pyrolysis.

Fast pyrolysis has been studied for many years and it is currently under investigation for the production of high yields of condensables that make up a liquid phase called bio-oil, considered as a possible substitute for fossil fuels [1]. The conditions of slow and conventional pyrolysis, on the other hand, are better suited to the production of biochar, also used in the past as a fuel [2] and of great interest in recent years for its chemical and physical characteristics. In-depth discussion of the chemical pathways involved in biomass pyrolysis is not within the scope of this chapter. The interested reader can consult several reviews dealing with both the process and technological aspects of pyrolysis [3, 4]. This section provides a brief description of the main phenomena that occur during pyrolysis in order to identify the role played by the main process variables in determining some characteristics of the biochars of interest for the applications presented in this book.

Figures 2.1(a)–(c) show a qualitative trend of the weight loss of lignocellulosic biomass, algae, and sludges of different origin. Biochar yields from different slow pyrolysis experiments are also reported [5–15]. It is clear that the temperature and the type of biomass greatly affect biochar yield. For each type of feedstock, biochar yield decreases as the pyrolysis temperature increases. However, although the characteristic temperatures of the decomposition process are different depending on the feedstock, some phenomena occurring during pyrolysis are common to all biomass types. As the temperature increases, the primary degradation of biomass

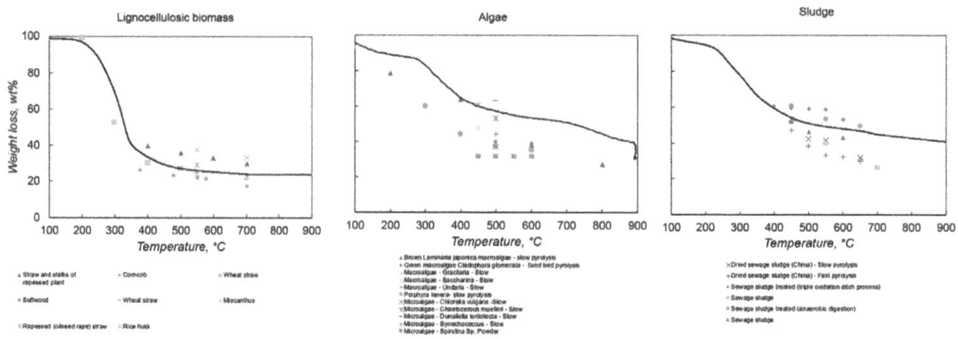

Figure 2.1. Biochar yields from slow pyrolysis of lignocellulosic biomass, algae, and sludges [5–15].

causes the formation of condensables at the expense of biochar, while the formation of gas (mainly CO, CO_2, CH_4) is limited; the condensables yield increases with temperature up to about 500 °C. At this temperature most of the volatile species are released and the weight loss rate decreases rapidly. In the meantime, condensables can undergo secondary degradation with a release of gas (mostly CO, CO_2, CH_4, and H_2) and low molecular weight condensables. In conditions where the secondary reactions are relevant (prolonged vapor residence times), the yield of condensables reaches a maximum, while the gas yield increases significantly. In the meantime, the vapors still entrapped in the porous biochar matrix undergo polymerization reactions and the biochar yield decreases more slowly due to the production of secondary char whose structure is much more compact than that of the primary char.

Under slow pyrolysis conditions lignocellulosic biomass typically exhibits a biochar yield decreasing from about 40 to 20 wt.% as temperature increases from 400 °C to 700 °C, whereas the decay curve is shifted to higher values for sludges and macroalgae due to the high ash content of this kind of biomass [16]. Microalgae are targeted as a source of lipids for algal oil including the production of biofuels [17], therefore their use for biochar production is less attractive. A greater data scattering can be observed for sludge and algae derived biochars due to the higher feedstock variability [16, 18].

At a high heating rate the biochar yields decreases in favor of gas and condensables. For temperatures lower than 500 °C the high heating rates correspond, on average, to a higher temperature at which primary decomposition occurs. In these conditions devolatilization is promoted at the expense of biochar production. This typically occurs under fast pyrolysis conditions where, due to the reduced particle size of the feedstock, the intra-particle thermal gradient is lower and devolatilization is favored over biochar forming reactions [19]. If a high heating rate is applied at temperatures exceeding 500 °C the activity of secondary reactions is enhanced, thus promoting gas production. Two opposing driving forces interact in determining the biochar yield: higher average temperatures inside the particles enhance the activity of the repolymerization reactions forming secondary chars;

conversely, the disruptive devolatilization induced by the higher heating rate creates a more open biochar structure, thus favoring the easier escape of volatiles. The reduced residence time of the volatiles inside the biochar pores affects negatively the formation of secondary char.

The effect of pressure on the product yield distribution is not easily predictable because of the formation of tars that prevent the attainment of thermodynamic equilibrium of pyrolysis reactions [20]. Limited data are presented in the literature on the effect of pressure on biochar yield [20–25]. Previous studies on cellulose pyrolysis have demonstrated that an increase in pressure, if associated with a high gas residence time in the reaction environment, generates higher biochar yields and produces lighter volatiles [20]. High pressure accompanied by a low gas flow rate (which implies a high residence time) limits the mass transport favoring the decomposition and further polymerization of primary pyrolysis products rather than their evaporation, thus promoting secondary biochar production. Conversely, low pressure and a high gas flow rate reduce the biochar yield, while if low pressure is associated with a long gas residence time some volatiles can undergo secondary cracking which leads to the formation of gas and light condensables, as observed for cellulose by Mok and Antal [24]. The mechanism observed for cellulose pyrolysis can be extended somewhat to most biomass types, however, it should be stressed that the biochar yield is a result of the contribution of all the biomass components. For lignocellulosic biomass, for example, the lignin content is closely related to the biochar yield. Also for lignin, an increase in pressure in the presence of steam has been found to promote the formation of non-volatile polyhydric phenols (e.g. hydroquinone) and therefore biochar [20]. Table 2.1 shows the experimental data concerning the yield of biochar from different lignocellulosic sources at different pressures. The datasets are not comparable with each other because they were obtained from very different processes, both in terms of the thermal (temperature and heating rate) and fluid-dynamic (gas flow rate) conditions. The data, however, confirm the positive correlation between pressure and biochar. Moreover, comparing the results obtained by Antal and Mok [24] for cellulose, it can be seen that at the same pressure an increase in the gas flow rate determines a reduction in the biochar yield. The results obtained by Pindoria *et al* [25] confirm the positive correlation of the residence time and the biochar yield. Indeed, the biochar yields obtained at $P = 1$ and 2 MPa are not very different because, despite of the high pressure variation, the gas flow rate had been set in such a way as to favor the removal of the condensables from the hot zone.

Opposite results were obtained by Mahinpey *et al* [23] for straw pyrolysis conducted in the pressure range $0.07 < P < 0.27$ MPa. In this case, the biochar yield remained constant as pressure increased. It is likely that, since the pyrolysis experiments were conducted at 500 °C, the temperature was too low to allow consistent advancement of the reactions producing secondary char whose role is decisive for the yield of biochar at high pressure. Moreover, the pressure range examined is rather limited to appreciate a significant increase in biochar yield under their temperature conditions.

Table 2.1. The effect of pressure on the yield of biochar from different feedstocks.

P (MPa)	Biochar yield (wt.%)						
	[24]			[22]	[25]	[21]	[20]
	Carrier gas flow rate = 1 $cm^3\ min^{-1}$	Carrier gas flow rate = 5 $cm^3\ min^{-1}$	Carrier gas flow rate = 20 $cm^3\ min^{-1}$				
0.001	12.1		10.3	–	–	–	–
0.1	12.1	11.5	10.3	6	–	–	28.9–33
0.2	14.5	13	12	–	–	–	–
0.4	–	–	–	–	–	40.5	–
0.5	17.4	14.2	13.2	–	–	–	–
0.7	–	–	–	–	–	40.2	–
1.0	18.4	16.8	14.8	–	13.8	44.4	34.6–37.5
1.1	–	–	–	–	–	50.8	–
2.0	–	–	–	–	14.5	–	–
2.5	22	18.5	18.5	–	–	–	–
3.3	–	–	–	–	–	51	–
4.0	–	–	–	15	–	–	–

2.2 Pyrolysis operating conditions affecting the electrical, mechanical, and adsorption properties of biochar

In the last decade the continuous pressure towards establishing a bio-based economy has in the last few decades motivated research on biochar as substitute for high cost and fossil-carbon-based carbonaceous materials. The physical and chemical characteristics of biochar suggest a wide variety of potential applications, such as use as a fuel in traditional and advanced power generation facilities [26], a fertilizer and carbon sink [27], a contaminant adsorbent in wastewater and soil [28], an adsorbent or catalyst for gas cleaning [29], a catalyst for syngas conversion into liquid hydrocarbons and biodiesel production [30], a raw material for supercapacitors [30], and a filler in wood and polymer composites [31]. The chemical, physical, mechanical, and adsorption properties play a pivotal role in selecting the appropriate biochar application. The biomass decomposition mechanism presented in section 2.1 suggests that the operational conditions of the pyrolysis process affect the biochar chemistry and structure greatly, which are responsible for the changes in biochar properties. In this section, a brief overview of some of the most relevant properties for the application of biochar as a filler in composites and an adsorbent in environmental remediation will be given in relation to the chemical and structural biochar modifications to which they are correlated. Then the changes in biochar chemistry and structure as result of variations in the pyrolysis process conditions will be discussed.

2.2.1 Physical, chemical, and mechanical properties of biochar as a filler in composites

Many applications of biochar as a filler in building materials and polymeric composites benefit from certain biochar properties, such as low density, low thermal conductivity, thermal stability, electrical conductivity, and intrinsic mechanical properties (hardness and elastic modulus). These are greatly affected by some chemical and structural biochar characteristics, such as elemental composition, volatiles, fixed carbon and ash content, aromatization degree, and porosity.

Density and thermal conductivity

The bulk density and thermal conductivity depend on the conversion of the low-density disordered carbon into high-density graphene sheets and on the arrangement of the graphene sheets in turbostratic crystallites. This means that they can be negatively correlated to the porosity caused by devolatilization and biochar shrinkage and to the H/C content as a measure of the advancement of the graphitization process [32, 33]. Ash should also be taken into account since they contribute to an increase in bulk density.

Thermal stability

The extent of aromatization and graphitization affect the thermal stability of biochar positively. The reactivity of biochar toward oxidation is depressed by the change in the biochar carbon structure induced by thermal annealing [34]. However, the content of inorganics, such as alkali metals, that are in such a form as to be catalytically active for oxidation are relevant in the determination of biochar reactivity [35]. The content of volatiles promotes the flammability of biochar, whereas the content and the chemical form of inorganics can act as a flame promoter or retardant [36].

Electrical conductivity

The process of electronic conduction within a carbon matrix is strongly dependent on the presence of sp^2 hybridized orbitals of the carbon atom associated with an electron delocalized in a p orbital perpendicular to the sp^2 orbitals' plane. It follows that the processes of aromatization and graphitization that occur at a molecular level in a biomass subjected to pyrolysis play a fundamental role in determining the electrical conductivity of biochar. Graphite represents a sort of asymptote towards which biochar tends as the process of aromatization progresses. Its crystallographic structure, organized in parallel layers, has electrons delocalized on the whole plane, thus determining an anisotropic electrical conductivity with the maximum in the plane direction equal to 0.33×10^6 s m^{-1}.

The electrical conductivity of complex micro-structured carbon materials has been explained through the two-phase composite model going through a percolation transition [37]. Kercher and Nagle [32], based on XRD spectra of carbonized monolithic fiberboards, adopted the so-called quasipercolation model to explain the microstructural evolution of woody material during carbonization and showed that the model is able to account for most of the changes to biochar's physical properties, such as electric conductivity. The presence of defects such as pores and inorganic elements is typically detrimental for the electrical conductivity and should be taken into account for interpreting some phenomena that occur in carbonaceous materials [38, 39]. Moreover, when the biochar structure has a low degree of aromatization, the polar functionalities on its surface play a pivotal role in the determination of the electrical conductivity [38].

Hardness and elastic modulus

Biochar chemistry exhibits a strong correlation with the mechanical properties. More specifically, the hardness/elastic modulus is positively correlated with the carbon content and the occurrence of covalent bonds between carbon atoms in aromatic structures associated with the loss of oxygenated functional groups. The evolution of charred material towards a turbostatic structure is also beneficial for the hardness/elastic modulus [40]. On the other hand, in some biomass sources such as sludge and poultry litter the order and crystallinity of the biochar structure is mostly contributed by the high amount of impurities present in form of crystal phases. In this case, biochar crystallinity cannot be correlated with its respective hardness/elastic modulus [41]. This means that in some cases the ash content and chemical form could also affect the biochar's mechanical properties.

2.2.2 The physical and chemical properties of biochar involved in adsorption mechanisms

Different mechanisms are involved in inorganic anion, cation, and organic compound adsorption. Usually, the different mechanisms interfere with each other, further complicating adsorption phenomena.

Porosity

The characteristics of the biochar porous structure, namely surface area, pore volume, and pore size distribution, are fundamental in adsorption processes and depend both on the original structure of the pyrolyzed biomass and its thermal history. The pore size distribution is a very important aspect since it is related to the mass transfer phenomena occurring during the adsorption process. A proper distribution of micro-, meso- and macropores is needed to provide a sufficient adsorption surface and to facilitate mass transfer into the smaller pores, in relation to the specific adsorbate.

Surface chemistry

Surface functional groups, aromaticity, and inorganic content and speciation determine the adsorption efficiency and selectivity of biochar with respect to different types of molecules or ions.

Adsorption in the pores and partition are both based on diffusion—organic molecules can diffuse and be adsorbed both inside the pores of the biochar and on the surface of the matrix not completely carbonized according to a mechanism called 'partition'. According to Han *et al* [42], the adsorption in the pores is closely related to the total volume of micropores and mesopores. It is often difficult to draw a clear boundary between the two adsorption mechanisms, although in general a higher pyrolysis temperature leads to the formation of biochar with a higher carbonized fraction, making adsorption in the pores predominant compared to 'partition' [43]. Depending on the other characteristics of biochar, these adsorption mechanisms may involve organic compounds of different nature and may occur in conjunction with other mechanisms.

Hydrophobic species tend, due to their nature, to minimize the contact surface with water by approaching each other and creating agglomerates. Biochar can have a hydrophobic or hydrophilic character according to its chemical composition, in particular according to the O/C ratio—the lower the ratio, the more hydrophobic the biochar.

Adsorption through the formation of hydrogen bonds occurs when the adsorbate presents some acceptor groups; the hydrogen atom that participates in the binding may be present on the biochar or in some surface functional group or in a water molecule bound to the surface of the biochar. This adsorption mechanism is influenced by the pH of the solution.

Another mechanism involved in biochar adsorption is the so-called electron donor acceptor (EDA). The electrons that are shared during the formation of this bond belong to a n orbital π. Therefore the groups that mainly contribute to this adsorption mechanism are the aromatic rings, which can behave both as acceptors and donors. The higher the degree of graphitization of biochar, the higher the role of this mechanism in the adsorption [44].

Surface complexation is an adsorption mechanism that involves the formation of complex structures, usually due to interactions between a metal ion and a ligand [45, 46]. Therefore, this mechanism is mainly involved in the adsorption of heavy metal species, but it can also involve organic species that can interact with the metals contained in biochar. Metals can also be adsorbed through a cation exchange with the biochar surface.

Biochar has a non-zero net charge distributed on its surface and this characteristic is involved in two possible mechanisms of adsorption, namely ion exchange and electrostatic interaction. The presence of oxygenated functional groups is positively correlated with the cation exchange capacity of biochar and this mechanism is involved in heavy metals adsorption [47]. Typically, biochar has a negative charge, making possible the adsorption by electrostatic attraction of positively charged compounds [48]. Some authors report that the presence of acidic functional groups such as phenolic and carboxylic groups on the surface promotes ammonium adsorption [49]. However, it is possible, through proper modifications, to obtain a biochar suitable for the adsorption of anionic species such as nitrates and phosphorus [50]. The electrostatic interaction is particularly affected by the pH of the solution, which is closely related to the surface charge of the biochar. Electrostatic attraction is one of the most common mechanisms in the adsorption of organic compounds as they often have a certain polarity (i.e. a charge accumulation) which allows the interaction with biochar.

2.2.3 The effect of operating variables on lignocellulosic biomass derived biochar

Biochar is a multiphase system composed of a carbon and a mineral phase associated with a liquid phase occluding partly the porosity of the solid matrix. The relative content and composition of these phases varies with the process operating variables, namely temperature, heating rate, pressure, and gas and solid residence times.

Temperature

At low temperatures (T < 250 °C) the biochar resulting from the pyrolysis of biomass is the result of dehydration and depolymerization reactions of holocellulose and is therefore mainly composed of oligosaccharides. Under these conditions, a condensed phase partially occludes the pores of the char. At higher temperatures (250 °C < T < 350 °C) the char is mainly composed of amorphous carbon. However, as the temperature increases it tends to form aromatic molecules, thus increasing the aromaticity of the structure. The char formed at this temperature has a high concentration of free radicals and is therefore very reactive in oxidative environments. This is the result of the competition between the mechanism of devolatilization and the reactions of dehydration, and the hemolytic cleavage of the sugars of cellulose. At temperatures higher than about 700 °C, charring reactions occur, determining the growth of graphene layers and the removal of etheroatoms.

Proximate analysis of a huge number of biochars obtained under slow heating conditions shows that an increase in temperature corresponds to an increasing content of fixed carbon due to the progress of the aromatization process. Fixed carbon yield is almost independent of the temperature and this is consistent with the ASTM procedure for measuring the fixed carbon, which is a sort of slow pyrolysis up to 950 °C of a material partially charred at lower temperatures [51]. The content of volatiles entrapped in the biochars structure decreases with the temperature due to the higher extent of devolatilization. At about 700 °C the contents of fixed carbon and volatiles vary to a lower extent as the temperature rises. The ash yield is independent of temperature, while its content increases almost linearly due to the devolatilization of the organic fraction [52].

O/C and H/C carbon ratios decrease according to the progress of the devolatilization of oxygenated compounds and the charring reactions. Most of the oxygen and hydrogen is removed during the devolatilization of the organic fraction at temperatures lower than 500 °C, whereas a slower reduction is observed at higher temperatures. It is worth noting that at high temperatures the H/C ratio tends to comparable values independent of the parent biomass. Ronsse *et al* [51] found that under the same pyrolysis conditions at 750 °C, wood, straw, green waste, and algae produced biochars with H/C approximately equal to 0.18. The H/C ratio is typically considered a measure of the graphitization process occurring as temperature increases during pyrolysis.

The temperature has a considerable influence on the type and concentration of the surface functional groups. Spectroscopic measurements (FTIR, XPS, and EELS) provide information about the types of functional groups on the biochar surface.

Inorganics are present as heteroatoms in the surface functional groups of the carbon phase which, depending on the specific etheroatoms, can be classified as oxygen, nitrogen, or sulfur containing functional groups. Oxygen containing groups are mainly OH, OR, O(C=O)R, (C=O)OH, and (C=O)H, whereas nitrogen containing groups are mostly NH_2. Sulfur is less abundant in lignocellulosic biomass and, in its organic form, it is mainly present in proteins. According to the Lewis definition, both nucleophile and electrophile groups are dislocated on the biochar surface and determine its acid or basic nature. More specifically, groups such as OH, NH_2, OR, and O(C=O)R are classified as nucleophile, whereas the (C=O)OH and (C=O)H groups are classified as electrophile. From an overall perspective, as the pyrolysis temperature increases, all the functional groups such as hydroxyl, carbonyl, carboxyl, ether, and lactone, are gradually reduced. As the temperature increases large amounts of N compounds are released so that their basic nature can be expressed only in biochars produced at low temperature. The N groups retained in the biochar are involved in a series of complex transformations [53, 54]. It was found that tobacco waste derived biochars exhibited three types of N-containing structures evolving with the temperature: pyridinic N, pyrrolic/pyridone N, and quaternary N. The pyrrolic/pyridone group is dominant in the biochar produced at 350 °C and is converted to pyridine which condenses to form mostly quaternary N structures in biochars produced between 750 °C and 950 °C.

Hydroxyl, carboxyl, carbonyl, and methoxyl groups are still present on the biochar surface at low temperatures (up to about 350 °C). These thermally labile functional groups are involved in the dehydration, acid production, and decarboxylation reactions of holocellulose, thus exhibiting weaker and weaker signals in the spectroscopic analyses up to the complete falling off at temperatures in the range 600 °C–700 °C. Low temperature biochars are characterized by a high content of aromatics arranged to form an amorphous structure. As the pyrolysis temperature increases, aromatics tend to form aromatic carbon sheets arranged in a turbostatic structure [28]. Evidence of this trend can be found in many lignocellulosic species such as white ash [55], pinewood [56], safflower seed press cake [57], walnut shells [58], and palm kernel shells [59].

The removal of acid groups at low temperature determines a strong increase of pH from mild acid to basic values in the temperature range between 200 °C and 400 °C. However, the further increase in pH towards alkaline values at high temperatures is mainly due to the concentration of inorganics in the alkaline form. The release of the OH group is correlated to the cation exchange capacity (CEC) of biochar whose non-monotonous trend with temperature reflects the concentration and the further removal of most of the OH groups on the biochar surface [60].

Inorganics are present in biomass mainly as mineral phase (K, Cl, N, Ca, Si, Mg, P, S, Mn, and Fe in the form of salts, oxides, and hydroxides). The mineral phase has acid, basic, and amphoteric groups (mainly oxides) on the surface whose concentration is highly dependent on the pyrolysis temperature. Cl is released to a large extent at very low temperatures. Harvey *et al* [60] reported a release of ~85% and ~99%, respectively, during combustion of bark and fiberboards at 500 °C. At such low temperatures, this anion is protonated during the decomposition of the organic matrix and released as HCl, whereas at higher temperatures the remaining Cl is vaporized. Inorganic S, mainly in the form of sulfate, is almost retained in the biochar even at high temperatures, but a transformation of S from soluble sulfate to insoluble S (CaS, K_2S, the addition of S to unsaturated sites on the biochar surface, and substitution of oxygen in oxides) is observed [53]. The alkalis (K and Na) associated with the organic matrix (typically a very low amount) are released at temperatures below 400 °C. When they are present in the form of carbonates, their decomposition and release can occur at very high temperatures [61] or they could be retained in a stable form if incorporated in silicates. Ca and Si, located in the cell wall, and Mg, bound with ionic and covalent bonds to organic molecules, are released at high temperatures. Fe and Mn are present in both the organic and inorganic forms and are mostly retained during pyrolysis. P is also retained during pyrolysis, but the increase in temperature turns it into more stable forms.

The morphological characteristics of biochar evolve as the pyrolysis temperature increases. There are two important parameters associated with biochar porosity:

- the specific surface measured as the BET surface, generally high in the presence of high microporosity (pore diameter lower than 2 nm);
- the pore volume, essentially determined by meso- (pore diameter between 2 and 50 nm) and macropores (pore diameter greater than 50 nm).

Depending on the use of biochar, the relative importance of the surface area and volume must be evaluated. In any case, for all the applications involving fluid/biochar interactions, the specific surface area that is really accessible is the most relevant. The macropores act mainly as access routes to the micropores, therefore in the presence of a low macroporosity there are diffusive limitations within the micropores that are not easily accessible. During pyrolysis, the increase of the crystalline zones and the turbostatic organization of the graphene planes above 700 °C evolve toward the graphite structure. The distance between the graphene planes is greater than that typical of graphite (0.335 nm) because the presence of heteroatoms prevents a further closing determining the formation of micropores. Moreover, as the temperature increases, the devolatilization of the lighter condensables is favored, thus increasing the total volume of the pores and the specific surface area. If the thermal conditions are too severe, devolatilization leads to thin walls of the micropores, which can coalesce to form larger pores. This leads to an increase in the volume of the pores and to a reduced microporous fraction in favor of the mesoporous and macroporous fractions (reduction of the specific surface area). In addition, at high temperatures (above 700 °C) condensables undergo repolymerization reactions producing heavy condensables that remain in the pores, reducing their total volume and specific surface area [52, 62]. These phenomena explain why the specific surface area of biochar has a non-monotonous relationship with the temperature with the maximum varying depending on the original biomass type. Moreover, fused intermediates can be produced by some inorganic species. In the presence of high amounts of alkaline metals and silicates, sintering phenomena can occur due to the formation of low melting point salts. Consistently, higher specific surface areas were observed for woody biochar characterized by low ash content compared to herbaceous plant derived biochar [51].

Heating rate

As the heating rate increases, devolatilization is promoted at the expense of biochar forming reactions. Consequently, biochars show a slightly reduced content of volatiles corresponding to an increase in fixed carbon and ash content. In any case the variations are very limited when the heating rate increases within the range of slow pyrolysis. Consistently, negligible variation of elemental composition, surface functional groups, and pH has been observed [64]. When fast and slow pyrolysis biochars are compared at the typical temperatures of fast pyrolysis, larger differences have been observed in the biochar characterizations obtained from both proximate and elemental analyses. In this case a higher volatile content, O/C, and H/C, and a lower fixed carbon and ash content have been measured for fast pyrolysis biochars. Accordingly, lower pH values have been measured and many oxygenated functional groups were observed on the biochar surfaces whose FTIR spectra still resembled those of the parent material. Moreover, the aromaticity of slow pyrolysis biochar is higher than that of fast pyrolysis, whereas the heating rate does not have relevant effects on the degree of aromatic condensation [...]. Tentatively, these results can be attributed to the long residence time of the solid matrix in slow pyrolysis, compared to that of fast pyrolysis, which results from the detection of unreacted components or less reacted biochar in the fast pyrolysis biochars [66, 67]. When high heating rates are associated with a sufficient holding time of the biochar at the pyrolysis temperature only slight differences have been noticed [68].

The heating rate affects the volatilization rate. At a low heating rate, volatilization is a slow process which does not alter the original structure of the biomass, thus allowing the development of a consistent microporous fraction. As the heating rate increases a very violent volatilization and a rapid shrinking of the solid phase occur that alter the structure of the biomass, causing the coalescence of micropores and the formation of larger pores and cracks inside the solid matrix. Therefore, given the creation of a more open structure, mass transfer through the pores is facilitated, thus reducing the possibility of repolymerization of condensable products and the consequent occlusion of micropores. The combined effect of a high heating rate and temperature and prolonged solid residence time can be detrimental for the BET surface and the microporous volume since the activity of the repolymerization reactions is more relevant at high temperatures [62].

Pressure

Among the pyrolysis operating variables, pressure has been less investigated, except for its effect on biochar yield and porosity. Antal and Grønli [2] explained the mechanisms contributing to the positive correlation between biochar yield and pressure as reported in section 2.1. A similar correlation was also observed for the fixed carbon yield since the reduced specific volume of volatiles induced by the high pressure determines an increase of the residence time of the volatiles inside the biochar matrix that, associated with the biochar catalytic activity, enhance the polymerization reactions leading to coke formation. This phenomenon is favored if a low carrier gas flow rate is applied and affects biochar porosity negatively. Cetin *et al* [69] observed a 20% reduction in the BET surface area of biochar obtained by fast pyrolysis for a pressure increase from 0.1 to 2 MPa. The formation of molten intermediates on the surface of the biochar and the agglomeration of adjacent biochar particles forming more compact structures also contributed to the lower BET surface.

Residence time
The residence times of both volatiles (governed by the devolatilization rate and carrier gas flow rate) and solids affect the chemistry and structure of the biochar. The effect of prolonged gas residence time can be correlated to a greater extent of polymerization reactions as discussed earlier, promoting the formation of a more aromatic and compact secondary char [70]. It should be noted that the carrier gas flow rate, in addition to affecting the residence time of the condensables in the particles and in the reaction environment, has significant effects on the fluid dynamics inside the pyrolyzer by altering the heat transfer resistance between the solid particles and the surrounding gas phase. A higher carrier gas flow rate promotes heat transfer, thus increasing the particle heating rate.

The effect of solid residence time can be connected to the effects of pyrolysis temperature. The longer the solid is held in the reaction environment at the highest final temperature, the more severe will be the pyrolysis process [25, 62].

2.3 The effect of feedstock composition on biochar properties

The feasibility of a biochar production chain is strictly dependent on the possibility of ensuring an adequate flow of incoming material. However, in order to achieve this objective it is often necessary to resort to the use of different biomass sources which, depending on the geographical context and territorial development, may have very different chemical and physical characteristics.

A wide variety of feedstocks can be used for biochar production, comprising agricultural or woody residues, aquatic biomass, and municipal, livestock, or industrial waste and by-products, etc. This implies that in order to produce high quality biochars with reliable and consistent properties, the composition of the biomass must be properly controlled. Therefore, it is essential to know the relationships between the composition and structure of the biochar produced under different pyrolysis conditions and biomass components.

In this section, the role of both organic and inorganic biomass constituents in pyrolysis will be discussed for different types of biomass, namely lignocellulosic biomass, algae, and residues from biological and biochemical treatments, in relation to specific biochar chemical and physical characteristics. Special emphasis has been placed on the toxicity that some of these constituents may induce in biochar during the production process.

2.3.1 Biochar from raw vegetal biomass

Vegetal biomass can be seen as a composite of many components acting together to determine the biochar properties. Organic components and inorganic elements evolve differently during the thermal treatment and they can interact with each other producing a nonlinear combination of effect [71, 72].

The main components of the lignocellulosic cell wall are cellulose, hemicellulose, and lignin. Cellulose is a linear macromolecular polysaccharide with a high degree of polymerization (an average DP comprising between 9000 and 10 000 [73]) that consists of a long chain of cellobiose units linked by β-1,4-glycosidic bonds. Hemicellulose is composed of short branched polysaccharide chains (500–3000 sugar units) and can differ in the side chain types, and the localization and/or types and distribution of glycosidic linkages in the main polysaccharide chain. Hemicelluloses are usually divided into four general groups: xylans, mannans, xyloglucans, and mixed-linkage β-glucans [74]. The 4-O-methyl-D-glucurono-D-xylan represents the most abundant (more than 90%) hemicellulose type in hardwoods and herbal plants while the dominant form in annual plants, such as straw and grass, is the (glucurono)-arabino-xylan [74]. Finally, lignin is a racemic heteropolymer consisting of three hydroxycinnamyl alcohol monomers differing in their degree of methoxylation: *p*-coumaryl, coniferyl, and sinapyl alcohols. The content of p-coumaryl, coniferyl, and sinapyl alcohol in lignin is dependent on the source feedstock: softwoods are characterized by a higher lignin content than hardwoods (24%–33% wt. versus 16%–24% wt.) [75] and the lignin of hardwoods consists mainly of coniferyl and synapilic alcohol units, while in softwood, coniferilic units prevail [76]. Usually, cellulose, hemicellulose, and lignin account for more than 90% of the entire lignocellulosic biomass; however, other species should be considered such as extractives and ash. A huge amount of different extractives can be identified so that they are typically classified as hydrophilic or hydrophobic extractives depending on their solubility in polar or non-polar solvents. Polyphenols are the most abundant hydrophilic extractives, whereas hydrophobic extractives (more abundant in softwoods) are mainly composed of terpenes and fatty acids. The biochemical composition of a large variety of lignocellulosic biomass is provided in [77]. In addition to the carbon fraction, biomass often contains high levels of mineral (ash) components. The main inorganic elements found in biomass, in order of abundance, are Ca, K, Si, Mg, Al, S, Fe, P, Cl, Na, Mn, and Ti. Additionally, trace elements (As, Ba, Co, Cr, Cu, Mo, Ni, Se, Sn, U, V, and Zn) are also found whose presence can create environmental issues when their concentration in biochar reaches critical values. Alkali metals are typically present in the form of salts and to a lesser extent organic salts, conversely, earth alkali metals are typically bound to the organic matrix. Sulfur is bonded both in the organic and organic form differently to phosphorus, whose inorganic form dominates its speciation in vegetal biomass [78].

Thermogravimetric analyses and pyrolysis experiments conducted on reference compounds representative of the main biomass components evidenced their different thermal stability and the consequent different tendencies to form biochar [79]. Lignin is characterized by a high content of crosslinking between the aromatic macromolecules of its complex structure, which are responsible for its high thermal stability. This leads higher biochar mass and energy yields than cellulose and hemicellulose.

According to the different temperature regions of maximum weight loss, the O/C ratio drops rapidly before 250 °C in the case of hemicellulose and between 250 °C and 350 °C in the case of cellulose. For both components, the removal of oxygenated gases and condensable species can be considered practically complete for temperatures above 500 °C. In contrast, lignin biochar shows a fairly homogeneous reduction of the O/C ratio up to 650 °C, above which no further noticeable reduction can be observed [80].

FTIR spectra of the three components of biochar show three main regions [81]: a region between 3100 and 3600 cm^{-1} containing signals of exchangeable protons (from alcohol, phenol, amine, amide, and carboxylic acid groups); a region between 2850 and 3050 cm^{-1} containing signals due to the stretchings of aromatic and aliphatic C–H groups; and a region between 1800 and 500 cm^{-1} containing overlapped signals of stretching and bending absorptions of many different functional groups (C=O of carbonylic and carboxylic groups, C–OH, C–H, C=C, and C–C). Functional groups such as –COOH and –OH contribute to lower biochar pH whereas pironic groups move pH towards alkaline values. They are distributed on the biochar surface not only in dependence on the pyrolysis temperature, as discussed in section 2.2.3, but also on the relative content of the main biomass components. The FTIR spectra of hemicellulose and cellulose, as expected, were very similar, both being polysaccharides characterized by β-(1→4)-glycosidic linkages [81]. The signal characteristics of the glycosidic linkages disappear and stretching vibration of the O–H bonds and C–H and CH_2 groups inside the sugar rings almost disappear in biochars produced at temperatures higher than 250 °C and 350 °C, respectively, for hemicellulose and cellulose. C–OH on carboxylic and phenolic groups and C=O on carbonylic groups remain on the surface of cellulose biochars up to about 550 °C. Hemicellulose biochars produced at low temperatures do not show the presence of O– in acetyl units due to the high thermal liability of this group and differently to cellulose biochars they preserve the carboxyl and carbonyl groups also at high temperature, although in a low concentration (at least up to 850 °C). As for lignin biochars, they exhibit the OH group in aromatic and aliphatic structures up to high temperatures even though their presence in biochar produced above 450 °C. In contrast, CH_2 and CH_3 in the aromatic methoxyl groups and in the methyl and methylene groups of the side chains and guaiacyl and syringyl units disappear at very low temperatures (about 350 °C), as well as carbonyl and carboxyl C=O. Even though the lignin precursor is characterized by a higher portion of aryl-C with respect to the carbohydrate fraction, both cellulose and hemicellulose show a high tendency to aromatization as evidenced by the high content of aryl-C detected in NMR spectra of the corresponding biochars [80]. At high temperatures the lower values of H/C ratios with respect to lignin biochars and symptoms of graphite patterns in XRD spectra show that the holocellulose component may contribute to biochar graphitization more than lignin that exhibits a relevant fraction of o-alkyl-C even in biochars produced at high temperatures [80].

In addition to the organic components, mineral matter contributes to the properties of the biochars both directly through its intrinsic characteristics and indirectly acting as a catalyst in the competing reactions involving the organic components [82]. The ash content in biochars is correlated to its content in the parent feedstock. Even though some inorganics are devolatilized during biochar production, mainly in the form of chlorides [83], they are mostly retained in the biochar structure (see section 2.2.3). De-ashing pre-treatments were found to reduce the O/C ratio even though this could be due to changes in the original organic composition induced by the acid solutions used for the pre-treatments [84], since the reduced decarboxylation, decarbonilation, and dehydration reactions promoted by the reduction of alkali and earth alkali metals should lead to the opposite result [82].

The role of these alkali and earth alkali metals in catalyzing the removal of light oxygenates corresponds to a reduction of acidic groups (COOH and OH) on the biochar surface, thus determining the increase of biochar pH, at least in the low temperature biochars. At higher temperatures, as discussed in section 2.2.3 the concentration of alkali and earth alkali metals in the form of salts is responsible of the alkaline nature of biochars and their solubility is positively correlated with the CEC [85].

The intrinsic structure of cellulose, hemicellulose, and lignin induces different transformations during the thermal treatment. Cellulose fibers preserve their original rigid structures as a result of the slow release of volatiles at a low heating rate, although their diameter undergoes a reduction of approximately 40% [86]. The roughness of the fiber surface does not disappear during charring reactions. Porosity development is mainly due to the voids left from the devolatilization of thermal labile fragments and the cracks formed during the charring process determining a reduction of fiber walls and the formation of hollow core fibers [86]. Thermal treatment of hemicellulose reference compounds induces an aggregation of the raw particles and a smoothing of the surface due to the formation of a melt phase during pyrolysis. The devolatilization determines the formation of an irregular porosity with very large pores [84]. Finally, lignin undergoes drastic changes due to its viscoelastic nature. Its melting produces a compact biochar with a shiny and smooth surface [84]. The tendency to graphitization of the three components would lead, at high temperatures, to form a graphite-like compact structure, reducing its internal porosity drastically. However, the arrangement of the different biomass components inside the cell wall, the presence of extractives and, to a large extent, of inorganic impurities deeply affects the determination of the final porosity of the biochar. The influence of extractives is often neglected, however, this can be a reasonable assumption for woody residues, and not for some agricultural residues where their concentration can be higher than 10 wt.% [77]. Some of them, in particular resins and waxes, can play a role in the mass transfer mechanism of volatiles evolving from the other main biomass components, affecting to a certain extent the development of biochar porosity [77]. However, this topic is only marginally discussed in the literature.

The mechanisms of thermal decomposition of the three components and charring reactions are altered by the presence of the melting layer on cellulose fibers and, to a large extent, by the presence of inorganic elements. Reactive fragments from hemicellulose and lignin decomposition were found to be stabilized by the H-rich products evolving from cellulose to form volatile compounds at the expense of the repolymerization reaction leading to the formation of biochar. The release of the formed volatiles contributes to a better development of the biochar porosity. Mineral matter negatively affects biochar porosity through different chemical and physical mechanisms. It may fuse and block the pores inside the biochar matrix and at the same time it favors the cellulose decomposition through dehydration and ring opening reactions at the expense of depolymerization and devolatilization of heavier condensables.

Non-lignocellulosic biomass comprises materials containing proteins and lipids in great amounts other than carbohydrates and small amounts of lignocellulosic components. Algae belongs to this category and have been the subject of increasing interest in recent decades for their potential in carbon sequestration and the fight against eutrophication [87]. The feasibility of their exploitation is greatly dependent on the possibilities for their valorization. Whereas microalgae are being investigated as source of lipids for production of biofuels through oil extraction and trans-esterification, the economic valorization of macroalgae is still challenging. Details on macroalgae and microalgae characterization are provided in the pertinent literature [87]. Briefly, their composition differs greatly from the lignocellulosic biomass and among the two algae categories, thus determining different properties of the derived biochars. A systematic approach to the correlation between algae composition and biochar characteristics is still lacking in the literature. However, most of the mechanisms involved in lignocellulosic biomass pyrolysis still occur also in algae pyrolysis. Typically, microalgae have a very variable content of lipids and proteins reaching in some cases very high values (up to 51 wt.% of lipids for *Schizochytrium limacinum* and 65 wt.% of proteins for *Spirulina platensis* [87]) and in general greatly higher than macroalgae. For this reason it would be advisable to consider for biochar production the residue after oil extraction. Other constituents, more abundant in macroalgae, are carbohydrates (about 24–51 wt.% versus 3–30 wt.% of microalgae [87]) usually made up of sugar monomers or polysaccharides such as glucose, cellulose, and starch. Ash is also a very important fraction in micro- and even more in macroalgae (up to 35 wt.% for *Gracilaria gracilis* [87]) and is mainly composed of alkali and earth alkali metals, halogens, and sulfur with the addition of Fe, Zn Cu, and Mn as trace metals. As a consequence of this composition, micro- and macroalgal biochars show different properties. A study conducted on seven different macroalgae shows that all the biochars produced under slow pyrolysis conditions up to 450 °C are low in carbon content due to the high ash content of the parent material [88]. The presence of high amounts of alkali and earth alkali metals also determined high values of pH and extractable inorganic nutrients including P, K, Ca, and Mg. Scarce information is available on the pH of microalgal biochar but it is reasonable that the same mechanisms (a loss of oxygen containing groups and an increase of ash content) affect their pH as confirmed by the values obtained for lacustrine algae biochar [89]. The typical high content of proteins in microalgae was responsible of the increased N content in the derived biochars. In general, it was found that all the algal biochars are characterized by very low values of BET surface (in the order of few m^2/g). The high content of ash is the main factor responsible for such underdeveloped porosity [90].

2.3.2 Toxicity issues related to the presence of organic and inorganic contaminants in biochar from phytoremediation activities

Biomass from phytoremediation activities and contaminated biomass from previous uses (e.g. construction) presents a potentially attractive feedstock for biochar production due to its low cost and limited, if any, competition from other uses. However, this comes with risks associated with contaminants present in the different parts of the biomass (stem, branches, leaves, etc) [91]. Therefore, before any decision is made regarding the use of a specific type of biomass from a phytoremediation activity it is imperative to have a good understanding of the nature of contaminants present, their form and distribution throughout the biomass, and, importantly, their fate during pyrolysis.

In general, there are two types of contamination that may be present in contaminated biomass (e.g. demolition wood) and phytoremediation activities: organic compounds (e.g. hydrocarbons from fuel spills, paints, wood treatment, or industrial activities) and inorganic contaminants (e.g. heavy metals from mining and industrial activities). The type and nature of contaminants in the harvested biomass depends on the site contamination as well as the type of plant and its ability to take up different contaminants [92]. The fate of these contaminants during the biochar production process varies but, in general, organic contaminants volatilize or decompose during the pyrolysis process and therefore do not negatively affect biochar quality. On the other hand, inorganic contaminants are largely retained in biochar and their concentration increases with an increasing degree of carbonization as more and more volatile matter is driven off [93, 96]. Many of these potentially toxic elements are stabilized within the biochar structure and are thus less likely to affect the environment where biochar is applied, at least in the short to medium term [97].

2.3.3 Biochar from residues of biological and biochemical treatments of biomass

In addition to virgin biomass originating from forestry and agricultural applications, such as wood, bark, straw, husks, etc, there is a range of biomass resulting from various industrial operations utilizing biological and biochemical treatments, such as wastewater treatment (producing sewage sludge), anaerobic digestion of organic residues (producing digestate), and lignocellulosic ethanol production (yielding lignin-rich residues). This is a heterogeneous group of materials with distinct properties and therefore each category will be addressed separately as a potential feedstock for biochar production.

Sewage sludge, the solid by-product of wastewater treatment processes, has been extensively studied as a potential feedstock for production of nutrient-rich biochar [98]. There are different types of sewage sludge depending on the geographic location, the type of treatment process used, and also the point in the process where the sludge originates from, such as primary or secondary treatment [99, 100]. Sewage sludge typically consists of a high proportion of mineral content ranging from fine grit to minerals associated with organic matter. This inorganic fraction of sewage sludge can also contain potentially harmful contaminants, such as heavy metals that may affect its potential for subsequent use [100, 101]. The fate of these contaminants during pyrolysis is affected by the sludge composition, the concentration and type of contaminants, as well as the pyrolysis process conditions. In general, the concentration of inorganic compounds, including contaminants, increases with increasing pyrolysis temperature as a result of enhanced decomposition of organic matter. At higher pyrolysis temperatures, some metals (e.g. Hg, Cd, As, and Se) can be volatilized and their concentration in the resulting biochar is reduced [102]. Furthermore, processing conditions can affect not only the total concentration, but also the leachability of the metals, with certain conditions stabilizing metals in biochar [103]. The application of sewage sludge derived biochar can therefore be safer than that of sewage sludge [104].

Another important aspect of sewage sludge biochar is its often relatively high content of micro- and macro-nutrients, such as nitrogen and phosphorus [105, 107]. The amount and form of nutrients in sludge biochar depends on the source as well as processing conditions. Their availability is also affected by the processing conditions, in particular maximum temperature, and this fact can be used to optimize biochar for different applications [104, 108, 109].

Anaerobic digestate is the solid residue from the anaerobic digestion of organic wastes and is frequently used to produce methane gas for energy applications. While the digestate can be used directly as a soil amendment, there are numerous advantages in using it as a feedstock for biochar production. Digestate typically contains higher concentrations of nutrients [110] and therefore can yield a biochar with higher fertilizer potential. Furthermore, conversion to biochar allows temporal separation of digestate production and its application to soil, as biochar does not decompose and can be stored. The choice of processing parameters, such as temperature, also affects the availability of nutrients, e.g. P [111], and therefore a biochar with controlled release can be produced. Importantly, pyrolysis of digestate (similar to sewage sludge and other materials) can effectively decompose antibiotics and antibiotic resistance genes [112, 113].

Lignocellulosic ethanol production is a fast developing second generation biofuel technology promising sustainable biofuels while avoiding competition for food resources due to utilization of non-food lignocellulosic biomass. The process involves various pre-treatment steps and fermentation, yielding ethanol and a solid lignin-rich residue. This residue is typically burned for process heat or otherwise used as a fuel, e.g. in the form of pellets [114]. Finding applications for this resource is a hot area of research and many applications have been proposed, including chemicals, fuels, and biochar [115]. Due to its polymeric nature, lignin is a very suitable precursor for the formation of solid carbons, providing high yields [80, 116, 117].

The term lignin covers a very wide range of materials whose structure and properties depend on the source of biomass, the extraction, and any further treatment processes [118–121]. The lignin's nature also affects its pyrolysis and the properties of the resulting biochar. Studies investigating the pyrolysis of lignin confirmed that on average it yields more biochar than lignocellulosic biomass and that even biochar produced at relatively low temperatures (approx. 350 C) would meet the guidelines proposed by the International Biochar Initiative and the European Biochar Certificate (molar $H/C_{org} < 0.7$ and $O/C_{org} < 0.4$) [80, 116]. Lignin-derived biochars have been tested and used in a range of applications, including agriculture, catalysis, environmental management, and engineering applications [122, 129], showing promising results.

2.4 Can biomass properties be altered to control biochar properties?

As discussed in the earlier sections of this chapter, biochar properties can be controlled by feedstock selection as well as process conditions. There is, however, another way to expand the range of potential biochar products in the form of feedstock pre-treatment. This area is not as widely studied as the other areas of biochar research; however, a number of important findings have been made and will be discussed in this section. The focus will be on two major categories of pre-treatment; mechanical treatment of biomass and doping of biomass with additives.

2.4.1 Biomass doping for enhanced biochar production

Doping of biomass by different additives has been identified as a promising way to alter biochar properties and to optimize it for specific applications [130]. A wide range of different inorganic additives has been studied with clays, zeolites, alkali, and alkaline earth metals being the most frequently used.

One aim of the use of additives before pyrolysis is to alter the pyrolysis process in a way that preferentially yields the product of interest, in this case biochar. A number of studies have shown that the presence of alkali and alkaline earth metals, as well as phosphorus in biomass feedstock, can result in a significant increase in biochar yield [82, 84, 131–134], in the order of tens of percents. In addition to the increase in biochar yield, studies have shown that these additives also affect the stability of biochar carbon and can make a significant contribution to its carbon sequestration potential [135]. In some cases, multiple benefits, such as increased carbon sequestration and improved sorption capacity can be achieved through the use of additives [136]. Although there is an existing substantial body of research on the use of additives to boost biochar yield, there are still many unknowns in terms of the detailed mechanisms, in particular in relation to the effects on biochar stability where further research is needed. Similarly, research on the use of more complex additives (as opposed to pure compounds) such as various waste streams is urgently needed [137, 138].

Another reason for the use of additives is the desire to tune biochar properties such as sorption capacity, electric conductivity, etc, for a wide range of environmental and engineering applications. Here the additives selected strongly depend on the target application, e.g. targeting the sorption of specific molecules, and can be added either before or after the pyrolysis process. Only the use of additives before pyrolysis as a biomass pre-treatment is discussed here as reviews of other methods can be found elsewhere [130, 139]. The inspiration for the use of mineral enriched biochar for sorption applications comes from the established application of these minerals in such applications [140, 141]. The production of carbon–mineral composites offers synergies and often enhanced performance compared to the individual components. Biochar composites comprising various natural or modified clays, LDHs (layered double hydroxides), have shown promising results in a range of applications, such as the removal of contaminants via sorption [130, 139, 142–145] or catalytic decomposition [145–148]. Depending on the nature of mineral additives and the biochar used as a support, additives can also affect the toxicity of the resulting composite, either lowering it or in some cases increasing it. More research on these aspects is needed [149].

2.4.2 Mechanical pre-treatment of biomass

There are many different options for mechanical pre-treatment of biomass before pyrolysis, including practice size reduction by cutting, milling, or shredding providing particles of different sizes, different particle size distributions, and with different structural properties. Reduction in particle size typically results in lower yields of biochar and higher yields of pyrolysis liquids as a result of higher and faster heat transfer and a lower extent of secondary pyrolysis reactions [150]. In addition to biochar yield, the particle size also affects the stability of the resulting biochar [151, 152] and its performance in various applications [154, 155]. The effect of particle size on biochar stability is not a straightforward one. It has been reported that after an initial increase of biochar stability with an increase in biomass particle size the stability decreases when the particles become too large [153] as a result of incomplete carbonization. The preferred maximum particle size is therefore dependent on the pyrolysis reactor type and configuration that can achieve full carbonization.

Another category of mechanical pre-treatment of biomass before pyrolysis is densification, either in the form of pelleting or briquetting where milled or shredded biomass is transformed by application of pressure and heat that fuse individual particles into higher density pellets or briquettes. Although a number of published studies have investigated the production of biochar from densified biomass [156–161], only very few studies made actual comparisons between the pyrolysis of densified and raw biomass [162], thus leaving a large gap in current knowledge.

2.5 Predictive approaches for biochar properties: current trends and perspectives

It is clear from previous sections that the complexity of biomass chemistry and structure coupled to the nonlinear interactions between the different biomass components as well as the superposition of different elementary processes during thermochemical process give rise to a quite complex and entangled condition. In this case, the elaboration of truly predictive models able to both guide the selection of the proper feedstock and select the most appropriate process conditions to produce a biochar with purposely tailored properties is not yet a reality.

Since the availability of such modeling tools would represent an outstanding advancement in biochar science and technology, many efforts have been made and are currently underway to this aim. An interesting example of these efforts is a recent improvement of the CRECK kinetic mechanism for biomass pyrolysis [163]. The model describes the solid residue as a mixture of pure carbon together with lumped metaplastic compounds representing all the oxygenated and hydrogenated groups bonded to the carbonaceous matrix. The release of these species is modeled, accounting for chemical reactions and transport phenomena, and it describes with reasonable accuracy the change of both the mass loss and final elemental composition of the biochar. However, further developments are still needed to predict the structural and mechanical properties as well as chemical functionalities of the final product relevant to the practical applications and be able to design custom tailored production processes.

Even further from being able to fully predict biochar characteristics are quantitative approaches such as data syntheses and meta-analyses aimed at formulating predictive relationships between biochar production and its properties. They are nevertheless capable of producing interesting insights into the complex relationships between feedstock nature, processing conditions, and final biochar properties. For example, a recent study of Li *et al* [164] tested the potential of this approach to compare the role of temperature and feedstock composition on some of the most relevant properties of biochar. To this aim, they grouped the biomass parent material into four groups, namely, woody, animal, herbaceous waste, and biosolids. From the analysis of the ultimate properties, it seems that the C and N contents are more heavily dependent on the feedstock type than the O and H contents. The wood and herbaceous derived biochars have relatively greater C/N ratios than animal waste and biosolids. O/C ratios became smaller than 0.2 for woody and herbaceous derived biochars produced at temperatures above 600 °C and for animal waste and biosolids biochars obtained above 700 °C. Finally, provided that the pyrolysis temperature is high enough, the H/C ratio in biochar loses the memory of the H/C ratio of the parent material, achieving values similar to those of carbon black and anthracite. The results of this study also reveal that biochar pH was more sensitive to pyrolysis conditions than to feedstock type, even if the feedstock composition and properties cannot be neglected for the final pH value. In general, the biochar made from agricultural waste had the highest pH when produced under the same pyrolysis temperature according to their higher content of alkali and earth alkali metals. The influence of the feedstock, in particular of the ash content and speciation, was more relevant for the CEC than for the pH. Exchangeable cations are the most abundant in biochars from biosolids followed by biochars from agricultural residues. Finally, the morphological properties are known to be strongly feedstock dependent, however, unacceptable predictions can be obtained due to the errors induced by the different heat transfer conditions characterizing the experimental trials used for the meta-analyses.

Recently, global modeling of the production chain based on big data approaches is becoming a valuable alternative to the traditional approaches. The availability of large datasets and the possibility to organize them in databases that can be accessed by means of automated procedures make possible the set-up of data-based monitoring and control cyber-systems capable of predicting in real time the system behavior with sufficient fidelity to allow for the implementation of control strategies aimed at optimizing the final characteristics of the biochar. In this respect, the natural evolution of the monitoring and control of the biochar production aimed at optimizing the product qualities (according to the targeted applications) will, similarly to other production systems, increasingly rely on either data- or model-based digital twins strategies. Significantly, this trend will require the further development of an intense experimental activity aimed at producing the large database of data needed by the validation and/or data consistency verification of the datasets.

Acknowledgements

Paola Giudicianni and Raffaele Ragucci thank Davide Amato for the documentation work and acknowledge the support of the project 'Rizobiorem' Progetti Di Ricerca Di Rilevante Interesse Nazionale-2017 Prot. 2017BHH84R.

References

[1] Sharifzadeh M, Sadeqzadeh M, Guo M, Borhani T N, Konda N M, Garcia M C, Wang L, Hallet J and Shah N 2019 The multi-scale challenges of biomass fast pyrolysis and bio-oil upgrading: review of the state of art and future research directions *Prog. Energy Combust. Sci.* **71** 1–80

[2] Antal M J and Grønli M 2003 The art, science, and technology of charcoal production *Ind. Eng. Chem. Res.* **42** 1619–40

[3] Wang S, Dai G, Yang H and Luo Z 2017 Lignocellulosic biomass pyrolysis mechanism: a state-of-the-art review *Prog. Energy Combust. Sci.* **62** 33–86

[4] Bridgwater A V and Bridge S A 1991 A review of biomass pyrolysis and pyrolysis technologies *Biomass Pyrolysis Liquids Upgrading and Utilization* (Berlin: Springer) pp 11–92

[5] Chang Y M, Tsai W T and Li M H 2015 Chemical characterization of char derived from slow pyrolysis of microalgal residue *J. Anal. Appl. Pyrolysis* **111** 88–93

[6] Demirbas A 2001 Carbonization ranking of selected biomass for charcoal, liquid and gaseous products *Energy Convers. Manage.* **42** 1229–38

[7] Gao N, Li J, Qi B, Li A, Duan Y and Wang Z 2014 Thermal analysis and products distribution of dried sewage sludge pyrolysis *J. Anal. Appl. Pyrolysis* **105** 43–8

[8] Grierson S, Strezov V and Shah P 2011 Properties of oil and char derived from slow pyrolysis of *Tetraselmis chui Bioresour. Technol.* **102** 8232–40

[9] Jin J *et al* 2016 Influence of pyrolysis temperature on properties and environmental safety of heavy metals in biochars derived from municipal sewage sludge *J. Hazard Mater.* **320** 417–26

[10] Jung K W, Jeong T U, Kang H J and Ahn K H 2016 Characteristics of biochar derived from marine macroalgae and fabrication of granular biochar by entrapment in calcium-alginate beads for phosphate removal from aqueous solution *Bioresour. Technol.* **211** 108–16

[11] Karaosmanoglu F, Isigigur-Ergundenler A and Sever A 2000 Biochar from the straw-stalk of rapeseed plant *Energy Fuels* **14** 336–9

[12] Liu W-J, Jiang H and Yu H-Q 2015 Development of biochar-based functional materials: toward a sustainable platform carbon material *Chem. Rev.* **115** 12251–85
[13] Mašek O, Buss W, Roy-Poirier A, Lowe W, Peters C, Brownsort P, Mignard D, Pritchard C and Sohi S 2018 Consistency of biochar properties over time and production scales: a characterisation of standard materials *J. Anal. Appl. Pyrolysis* **132** 200–10
[14] Norouzi O, Jafarian S, Safari F, Tavasoli A and Nejati B 2016 Promotion of hydrogen-rich gas and phenolic-rich bio-oil production from green macroalgae *Cladophora glomerata* via pyrolysis over its bio-char *Bioresour. Technol.* **219** 643–51
[15] Roberts D A, Paul N A, Dworjanyn S A, Bird M I and De Nys R 2015 Biochar from commercially cultivated seaweed for soil amelioration *Sci. Rep.* **5** 9665
[16] Fytili D and Zabaniotou A 2008 Utilization of sewage sludge in EU application of old and new methods—a review *Renew. Sustain. Energy Rev.* **12** 116–40
[17] Chisti Y 2007 Biodiesel from microalgae *Biotechnol. Adv.* **25** 294–360
[18] Lee X J, Ong H C, Gan Y Y, Chen W H and Mahlia T M I 2020 State of art review on conventional and advanced pyrolysis of macroalgae and microalgae for biochar, bio-oil and bio-syngas production *Energy Convers. Manage.* **210** 112707
[19] Di Blasi C 2008 Modeling chemical and physical processes of wood and biomass pyrolysis *Prog. Energy Combust. Sci.* **34** 47–90
[20] Antal M J, Allen S G, Dai X, Shimizu B, Tam M S and Grønli M G 2000 Attainment of the theoretical yield of carbon from biomass *Ind. Eng. Chem. Res.* **39** 4024–31
[21] Antal M J, Croiset E, Dai X F, De Almeida C, Mok W S L, Norberg N, Richard J-R and Al Majthoub M 1996 High yield biomass charcoal *Energy Fuels* 10 652–8
[22] Blackadder W and Rensfelt E 1985 A pressurized thermo balance for pyrolysis and gasification studies of biomass, wood and peat *Fundamentals of Thermochemical Biomass Conversion* (Dordrecht: Springer) pp 747–59
[23] Mahinpey N, Murugan P, Mani T and Raina R 2009 Analysis of bio-oil, biogas, and biochar from pressurized pyrolysis of wheat straw using a tubular reactor *Energy Fuels* **23** 2736–42
[24] Mok W S L and Antal M J 1983 Effects of pressure on biomass pyrolysis. I. Cellulose pyrolysis products *Thermochim. Acta* **68** 155–64
[25] Pindoria R V, Megaritis A, Messenbock R C, Dugwell D R and Kandiyoti R 1998 Comparison of the pyrolysis and gasification of biomass: effect of reacting gas atmosphere and pressure on Eucalyptus wood *Fuel* **77** 1247–51
[26] Kambo H S and Dutta A 2015 A comparative review of biochar and hydrochar in terms of production, physico-chemical properties and applications *Renew. Sustain. Energy Rev.* **45** 359–78
[27] Lehmann J and Joseph S 2009 *Biochar for Environmental Management* (London: Earthscan)
[28] Downie A, Crosky A and Munroe P 2009 Physical properties of biochar *Biochar for Environmental Management Science and Technology* (London: Earthscan) pp 13–32
[29] Shen Y 2015 Chars as carbonaceous adsorbents/catalysts for tar elimination during biomass pyrolysis or gasification *Renew. Sustain. Energy Rev.* **43** 281–95
[30] Qian K, Kumar A, Zhang H, Bellmer D and Huhnke R 2015 Recent advances in utilization of biochar *Renew. Sustain. Energy Rev.* **42** 1055–64
[31] Das O, Sarmah A K and Bhattacharyya D 2015 A novel approach in organic waste utilization through biochar addition in wood/polypropylene composites *Waste Manage. (Oxford)* **38** 132–40

[32] Kercher A K and Nagle D C 2003 Microstructural evolution during charcoal carbonization by x-ray diffraction analysis *Carbon* **41** 15–27
[33] Yargicoglu E N, Sadasivam B Y, Reddy K R and Spokas K 2015 Physical and chemical characterization of waste wood derived biochars *Waste Manage. (Oxford)* **36** 256–68
[34] Sun Y, Gao B, Yao Y, Fang J, Zhang M, Zhou Y, Chen H and Yang L 2014 Effects of feedstock type, production method, and pyrolysis temperature on biochar and hydrochar properties *Chem. Eng. J.* **240** 574–8
[35] Zhang R, Lei K, Bu Q Y, Cao J and Liu D 2018 Effects of alkali and alkaline earth metal species on the combustion characteristics of single particles from pine sawdust and bituminous coal *Bioresour. Technol.* **268** 278–85
[36] Di Blasi C, Galgano A and Branca C 2009 Influences of the chemical state of alkaline compounds and the nature of alkali metal on wood pyrolysis *Ind. Eng. Chem. Res.* **48** 3359–69
[37] McLachlan D S, Cai K and Sauti G 2001 AC and DC conductivity-based microstructural characterization *Int. J. Refract. Met. Hard Mater* **19** 437–45
[38] Rhim Y R, Zhang D, Fairbrother D H, Wepasnick K A, Livi K J, Bodnar R J and Nagle D C 2010 Changes in electrical and microstructural properties of microcrystalline cellulose as function of carbonization temperature *Carbon* **48** 1012–24
[39] Emmerich F G, De Sousa J C, Torriani I L and Luengo C A 1987 Applications of a granular model and percolation theory to the electrical resistivity of heat treated endocarp of babassu nut *Carbon* **25** 417–24
[40] Das O, Sarmah A K and Bhattacharyya D 2016 Nanoindentation assisted analysis of biochar added biocomposites *Composites* B **91** 219–27
[41] Das O, Sarmah A K and Bhattacharyya D 2015 Structure–mechanics property relationship of waste derived biochars *Sci. Total Environ.* **538** 611–20
[42] Han L, Sun K, Jin J, Wei X, Xia X, Wu F, Gao B and Xing B 2014 Role of structure and microporosity in phenanthrene sorption by natural and engineered organic matter *Environ. Sci. Technol.* **48** 11227–34
[43] Devi P and Saroha A K 2015 Effect of pyrolysis temperature on polycyclic aromatic hydrocarbons toxicity and sorption behaviour of biochars prepared by pyrolysis of paper mill effluent treatment plant sludge *Bioresour. Technol.* **192** 312–20
[44] Xie M, Chen W, Xu Z, Zheng S and Zhu D 2014 Adsorption of sulfonamides to demineralized pine wood biochars prepared under different thermochemical conditions *Environ. Pollut.* **186** 187–94
[45] Crabtree R H 2005 Complexes of π-bound ligands *The Organometallic Chemistry of the Transition Metals* (New York: Wiley) pp 125–58
[46] Crabtree R H 2005 Metal–ligand multiple bonds *The Organometallic Chemistry of the Transition Metals* (New York: Wiley) pp 309–41
[47] Tan X, Liu Y, Zeng G, Wang X, Hu X, Gu Y and Yang Z 2015 Application of biochar for the removal of pollutants from aqueous solutions *Chemosphere* **125** 70–85
[48] Ahmad A, Rajapaksha A U, Lim J E, Zhang M, Bolan N, Mohan D, Vithanage M, Lee S S and Ok Y S . 2014 Biochar as a sorbent for contaminant management in soil and water: a review *Chemosphere* **99** 19–33
[49] Gao F, Xue Y, Deng P, Cheng X and Yang K 2015 Removal of aqueous ammonium by biochars derived from agricultural residuals at different pyrolysis temperatures *Chem. Speciation Bioavail.* **27** 92–7

[50] Micháleková-Richveisová B, Frišták V, Pipíška M, Duriška L, Moreno-Jimenez E and Soja G 2017 Iron-impregnated biochars as effective phosphate sorption materials *Environ. Sci. Pollut. Res.* **24** 463–75
[51] Ronsse F, Van Hecke S, Dickinson D and Prins W 2013 Production and characterization of slow pyrolysis biochar: influence of feedstock type and pyrolysis conditions *GCB Bioenergy* **5** 104–15
[52] Lua A C, Yang T and Guo J 2004 Effects of pyrolysis conditions on the properties of activated carbons prepared from pistachio-nut shells *J. Anal. Appl. Pyrolysis* **72** 279–87
[53] Lang T, Jensen A D and Jensen P A 2005 Retention of organic elements during solid fuel pyrolysis with emphasis on the peculiar behaviour of nitrogen *Energy Fuels* **19** 1631–43
[54] Chen H, Lin G, Chen Y, Chen W and Yang H 2016 Biomass pyrolytic polygeneration of tobacco waste: product characteristics and nitrogen transformation *Energy Fuels* **30** 1579–88
[55] Chen T J, Liu R H and Scott N R 2016 Characterization of energy carriers obtained from the pyrolysis of white ash, switchgrass and corn stover—biochar, syngas and bio-oil *Fuel Process. Technol.* **142** 124–34
[56] Luo L, Xu C, Chen Z and Zhang S Z 2015 Properties of biomass-derived biochars: combined effects of operating conditions and biomass types *Bioresour. Technol.* **192** 83–9
[57] Angin D 2013 Effect of pyrolysis temperature and heating rate on biochar obtained from pyrolysis of safflower seed press cake *Bioresour. Technol.* **128** 593–7
[58] Zhao Y J, Feng D D, Zhang Y, Huang Y D and Sun S Z 2016 Effect of pyrolysis temperature on char structure and chemical speciation of alkali and alkaline earth metallic species in biochar *Fuel Process. Technol.* **141** 54–60
[59] Ma Z, Yang Y, Ma Q, Zhou H, Luo X, Liu X and Wang S 2017 Evolution of the chemical composition, functional group, pore structure and crystallographic structure of bio-char from palm kernel shell pyrolysis under different temperatures *J. Anal. Appl. Pyrolysis* **127** 350–9
[60] Harvey O R, Herbert B E, Rhue R D and Kuo L J 2011 Metal interactions at the biochar–water interface: energetics and structure–sorption relationships elucidated by flow adsorption microcalorimetry *Environ. Sci. Technol.* **45** 5550–56
[61] Van Lith S C, Jensen P A, Frandsen F J and Glarborg P 2008 Release to the gas phase of inorganic elements during wood combustion. Part 2: influence of fuel composition *Energy Fuels* **22** 1598–609
[62] Lua A C, Lau F Y and Guo J 2006 Influence of pyrolysis conditions on pore development of oil-palm-shell activated carbons *J. Anal. Appl. Pyrolysis* **76** 96–102
[63] Anicic B, Lin W, Dam-Johansen K and Wu H 2018 Agglomeration mechanism in biomass fluidized bed combustion–reaction between potassium carbonate and silica sand *Fuel Process. Technol.* **173** 182–90
[64] Angin D 2013 Effect of pyrolysis temperature and heating rate on biochar obtained from pyrolysis of safflower seed press cake *Bioresour. Technol.* **128** 593–7
[65] Zhao B, O'Connor D, Zhang J, Peng T, Shen Z, Tsang D C and Hou D 2018 Effect of pyrolysis temperature, heating rate, and residence time on rapeseed stem derived biochar *J. Clean. Prod.* **174** 977–87
[66] Brewer C E, Schmidt-Rohr K, Satrio J A and Brown R C 2009 Characterization of biochar from fast pyrolysis and gasification systems *Environ. Prog. Sustain. Energy* **28** 386–96

[67] Bruun E W, Ambus P, Egsgaard H and Hauggaard-Nielsen H 2012 Effects of slow and fast pyrolysis biochar on soil C and N turnover dynamics *Soil Biol. Biochem.* **46** 73–9

[68] Mohanty P, Nanda S, Pant K K, Naik S, Kozinski J A and Dalai A K 2016 Evaluation of the physiochemical development of biochars obtained from pyrolysis of wheat straw, timothy grass and pinewood: effects of heating rate *J. Anal. Appl. Pyrolysis* **104** 485–93

[69] Cetin E, Gupta R and Moghtaderi B 2005 Effect of pyrolysis pressure and heating rate on radiata pine char structure and apparent gasification reactivity *Fuel* **84** 1328–34

[70] Di Blasi C 2009 Combustion and gasification rates of lignocellulosic chars *Prog. Energy Combust. Sci.* **35** 121–40

[71] Ferreiro A I, Giudicianni P, Grottola C M, Rabaçal M, Costa M and Ragucci R 2017 Unresolved issues on the kinetic modeling of pyrolysis of woody and nonwoody biomass fuels *Energy Fuels* **31** 4035–44

[72] Ferreiro A I, Rabaçal M, Costa M, Giudicianni P, Grottola C M and Ragucci R 2018 Modeling the impact of the presence of KCl on the slow pyrolysis of cellulose *Fuel* **215** 57–65

[73] Rowell R M, Pettersen R, Han J S, Rowell J S and Tshabalala M A 2013 *Handbook of Wood Chemistry and Wood Composites* (Boca Raton, FL: CRC Press) pp 33–72

[74] Scheller H V and Ulvskov P 2010 Hemicelluloses *Annu. Rev. Plant Biol.* **61** 263–89

[75] Xiao B, Sun X F and Sun R 2001 The chemical modification of lignins with succinic anhydride in aqueous systems *Polym. Degrad. Stab.* **71** 223–31

[76] Di Blasi C, Signorelli G and Portoricco G 1999 Product distribution from pyrolysis of wood and agricultural residues *Ind. Eng. Chem. Res.* **6** 2216–24

[77] Debiagi P E A, Pecchi C, Gentile G, Frassoldati A, Cuoci A, Faravelli T and Ranzi E 2015 Extractives extend the applicability of multistep kinetic scheme of biomass pyrolysis *Energy Fuels* **29** 6544–55

[78] Leijenhorst E J, Wolters W, Van De Beld L and Prins W 2016 Inorganic element transfer from biomass to fast pyrolysis oil: review and experiments *Fuel Process. Technol.* **149** 96–111

[79] Giudicianni P, Cardone G and Ragucci R 2013 Cellulose, hemicellulose and lignin slow steam pyrolysis: thermal decomposition of biomass components mixtures *J. Anal. Appl. Pyrolysis* **100** 213–22

[80] Ma Z, Yang Y, Wu Y, Xu J, Peng H, Liu X, Zhang W and Wang S 2019 In-depth comparison of the physicochemical characteristics of bio-char derived from biomass pseudo components: hemicellulose, cellulose, and lignin *J. Anal. Appl. Pyrolysis* **140** 195–204

[81] Gargiulo V, Alfe M, Giudicianni P and Ragucci R 2016 A study on the structural features of the water-insoluble fraction (WIF) isolated from biomass slow steam pyrolysis liquids *J. Anal. Appl. Pyrolysis* **121** 128–37

[82] Gargiulo V, Giudicianni P, Alfè M and Ragucci R 2015 Influence of possible interactions between biomass organic components and alkali metal ions on steam assisted pyrolysis: a case study on *Arundo donax J. Anal. Appl. Pyrolysis* **112** 244–52

[83] Abanades S, Flamant G and Gauthier D 2002 Kinetics of heavy metal vaporization from model wastes in fluidized bed *Environ. Sci. Technol.* **36** 3879–84

[84] Giudicianni P, Cardone G, Cavaliere A and Ragucci R 2014 Effect of feedstock demineralization on physico-chemical characteristics of *Arundo donax* derived biochar *Chem. Eng. Trans.* **37** 85–90

[85] Zhao L, Cao X, Zheng W, Wang Q and Yang F 2015 Endogenous minerals have influences on surface electrochemistry and ion exchange properties of biochar *Chemosphere* **136** 133–9

[86] Ragucci R, Giudicianni P and Cavaliere A 2013 Cellulose slow pyrolysis products in a pressurized steam flow reactor *Fuel* **107** 122–30
[87] Lee X J, Ong H C, Gan Y Y, Chen W H and Mahlia T M I 2020 State of art review on conventional and advanced pyrolysis of macroalgae and microalgae for biochar, bio-oil and bio-syngas production *Energy Convers. Manage.* **210** 112707
[88] Bird M I, Wurster C M, de Paula Silva P H, Bass A M and De Nys R 2011 Algal biochar–production and properties *Bioresour. Technol.* **102** 1886–91
[89] Chiodo V, Zafarana G, Maisano S, Freni S and Urbani F 2016 Pyrolysis of different biomass: direct comparison among Posidonia Oceanica, Lacustrine Alga and White-Pine *Fuel* **164** 220–27
[90] Yu K L, Lau B F, Show P L, Ong H C, Ling T C, Chen W H, Ng E P and Chang J S 2017 Recent developments on algal biochar production and characterization *Bioresour. Technol.* **246** 2–11
[91] Tripathi V, Edrisi S A and Abhilash P C 2016 Towards the coupling of phytoremediation with bioenergy production *Renew. Sustain. Energy Rev.* **57** 1386–89
[92] Dastyar W, Raheem A, He J and Zhao M 2019 Biofuel production using thermochemical conversion of heavy metal-contaminated biomass (HMCB) harvested from phytoextraction process *Chem. Eng. J.* **358** 759–85
[93] Buss W, Graham M C, Shepherd J G and Mašek O 2016 Suitability of marginal biomass-derived biochars for soil amendment *Sci. Total Environ.* **547** 314–22
[94] He J, Strezov V, Kumar R, Weldekidan H, Jahan S, Dastjerdi B H, Zhou X and Kan T 2019 Pyrolysis of heavy metal contaminated *Avicennia marina* biomass from phytoremediation: characterisation of biomass and pyrolysis products *J. Clean. Prod.* **234** 1235–45
[95] Grottola C M, Giudicianni P, Michel J-B and Ragucci R 2018 Torrefaction of woody waste for use as biofuel *Energy Fuels* **32** 10266–71
[96] Grottola C M, Giudicianni P, Pindozzi S, Stanzione F, Faugno S, Fagnano M, Fiorentino N and Ragucci R 2019 Steam assisted slow pyrolysis of contaminated biomasses: effect of plant parts and process temperature on heavy metals fate *Waste Manage. (Oxford)* **85** 232–41
[97] Břendová K, Tlustoš P and Száková J 2015 Can biochar from contaminated biomass be applied into soil for remediation purposes? *Water Air Soil Pollut.* **226** 193
[98] Singh S, Kumar V, Dhanjal D S, Datta S, Bhatia D, Dhiman J, Samuel J, Prasad R and Singh J 2020 A sustainable paradigm of sewage sludge biochar: valorization, opportunities, challenges and future prospects *J. Clean. Prod.* **269** 122259
[99] Rogers H R 1996 Sources, behaviour and fate of organic contaminants during sewage treatment and in sewage sludges *Sci. Total Environ.* **185** 3–26
[100] Sommers L E 1977 Chemical composition of sewage sludges and analysis of their potential use as fertilizers *J. Environ. Qual.* **6** 225–32
[101] Rulkens W 2008 Sewage sludge as a biomass resource for the production of energy: overview and assessment of the various options *Energy Fuels* **22** 9–15
[102] Van Wesenbeeck S, Prins W, Ronsse F and Antal M J 2014 Sewage sludge carbonization for biochar applications. Fate of heavy metals *Energy Fuels* **28** 5318–26
[103] He Y D, Zhai Y B, Li C T, Yang F, Chen L, Fan X P, Peng W F and Fu Z M 2010 The fate of Cu, Zn, Pb and Cd during the pyrolysis of sewage sludge at different temperatures *Environ. Technol.* **31** 567–74

[104] Yuan H, Lu T, Huang H, Zhao D, Kobayashi N and Chen Y 2015 Influence of pyrolysis temperature on physical and chemical properties of biochar made from sewage sludge *J. Anal. Appl. Pyrolysis* **112** 284–89
[105] Faria W M, Figueiredo C C D, Coser T R, Vale A T and Schneider B G 2018 Is sewage sludge biochar capable of replacing inorganic fertilizers for corn production? Evidence from a two-year field experiment *Arch. Agron. Soil Sci.* **64** 505–19
[106] Frišták V, Pipíška M and Soja G 2018 Pyrolysis treatment of sewage sludge: a promising way to produce phosphorus fertilizer *J. Clean. Prod.* **172** 1772–78
[107] Yuan H, Lu T, Wang Y, Chen Y and Lei T 2016 Sewage sludge biochar: nutrient composition and its effect on the leaching of soil nutrients *Geoderma* **267** 17–23
[108] Hossain M K, Strezov V, Chan K Y, Ziolkowski A and Nelson P F 2011 Influence of pyrolysis temperature on production and nutrient properties of wastewater sludge biochar *J. Environ. Manage.* **92** 223–8
[109] Song X D, Xue X Y, Chen D Z, He P J and Dai X H 2014 Application of biochar from sewage sludge to plant cultivation: influence of pyrolysis temperature and biochar-to-soil ratio on yield and heavy metal accumulation *Chemosphere* **109** 213–20
[110] Möller K and Müller T 2012 Effects of anaerobic digestion on digestate nutrient availability and crop growth: a review *Eng. Life Sci.* **12** 242–57
[111] Bruun S, Harmer S L, Bekiaris G, Christel W, Zuin L, Hu Y, Jensen L S and Lombi E 2017 The effect of different pyrolysis temperatures on the speciation and availability in soil of P in biochar produced from the solid fraction of manure *Chemosphere* **169** 377–86
[112] Tian R, Li C, Xie S, You F, Cao Z, Xu Z, Yu J and Wang Y 2019 Preparation of biochar via pyrolysis at laboratory and pilot scales to remove antibiotics and immobilize heavy metals in livestock feces *J. Soils Sediments* **19** 2891–902
[113] Kimbell L K, Kappell A D and McNamara P J 2018 Effect of pyrolysis on the removal of antibiotic resistance genes and class I integrons from municipal wastewater biosolids *Water Research and Technology* **4** 1807–18
[114] Sannigrahi P, Pu Y and Ragauskas A 2010 Cellulosic biorefineries—unleashing lignin opportunities *Curr. Opin. Env. Sustain.* **2** 383–93
[115] Garlapati V K, Chandel A K, Kumar S J, Sharma S, Sevda S, Ingle A P and Pant D 2020 Circular economy aspects of lignin: towards a lignocellulose biorefinery *Renew. Sustain. Energy Rev.* **130** 109977
[116] Ghysels S, Ronsse F, Dickinson D and Prins W 2019 Production and characterization of slow pyrolysis biochar from lignin-rich digested stillage from lignocellulosic ethanol production *Biomass Bioenergy* **122** 349–60
[117] Li J, Li Y, Wu Y and Zheng M 2014 A comparison of biochars from lignin, cellulose and wood as the sorbent to an aromatic pollutant *J. Hazard. Mater.* **280** 450–57
[118] Savy D and Piccolo A 2014 Physical–chemical characteristics of lignins separated from biomasses for second-generation ethanol *Biomass Bioenergy* **62** 58–67
[119] Huang Y, Liu H, Yuan H, Zhuang X, Yuan S, Yin X and Wu C 2018 Association of chemical structure and thermal degradation of lignins from crop straw and softwood *J. Anal. Appl. Pyrolysis* **134** 25–34
[120] Chang G, Huang Y, Xie J, Yang H, Liu H, Yin X and Wu C 2016 The lignin pyrolysis composition and pyrolysis products of palm kernel shell, wheat straw, and pine sawdust *Energy Convers. Manage.* **124** 587–97

[121] Cao X, Zhong L, Peng X, Sun S, Li S, Liu S and Sun R 2014 Comparative study of the pyrolysis of lignocellulose and its major components: characterization and overall distribution of their biochars and volatiles *Bioresour. Technol.* **155** 21–7

[122] Supanchaiyamat N, Jetsrisuparb K, Knijnenburg J T, Tsang D C and Hunt A J 2019 Lignin materials for adsorption: current trend, perspectives and opportunities *Bioresour. Technol.* **272** 570–81

[123] Xiong X, Iris K M, Cao L, Tsang D C, Zhang S and Ok Y S 2017 A review of biochar-based catalysts for chemical synthesis, biofuel production, and pollution control *Bioresour. Technol.* **246** 254–70

[124] Zhang H, Jia D, Yang Z, Yu F, Su Y, Wang D and Shen Q 2017 Alkaline lignin derived porous carbon as an efficient scaffold for lithium–selenium battery cathode *Carbon* **122** 547–55

[125] Xu L, Yao Q, Zhang Y and Fu Y 2017 Integrated production of aromatic amines and N-doped carbon from lignin via *ex situ* catalytic fast pyrolysis in the presence of ammonia over zeolites *Sustain. Chem. Eng.* **5** 2960–69

[126] Lawrinenko M, Wang Z, Horton R, Mendivelso-Perez D, Smith E A, Webster T E, Laird D A and van Leeuwen J H 2017 Macroporous carbon supported zerovalent iron for remediation of trichloroethylene *Sustain. Chem. Eng.* **5** 1586–93

[127] Jiang C, Bo J, Xiao X, Zhang S, Wang Z, Yan G, Wu Y, Wong C-P and He H 2020 Converting waste lignin into nano-biochar as a renewable substitute of carbon black for reinforcing styrene–butadiene rubber *Waste Manage. (Oxford)* **102** 732–42

[128] Yoon K, Cho D W, Tsang Y F, Tsang D C, Kwon E E and Song H 2019 Synthesis of functionalised biochar using red mud, lignin, and carbon dioxide as raw materials *Chem. Eng. J.* **361** 1597–604

[129] Cho D W, Yoon K, Ahn Y, Sun Y, Tsang D C, Hou D, Ok Y S and Song H 2019 Fabrication and environmental applications of multifunctional mixed metal–biochar composites (MMBC) from red mud and lignin wastes *J. Hazard. Mater.* **374** 412–19

[130] Wang L, Ok Y S, Tsang D C, Alessi D S, Rinklebe J, Wang H, Mašek O, Hou R, O'Connor D and Hou D 2020 New trends in biochar pyrolysis and modification strategies: feedstock, pyrolysis conditions, sustainability concerns and implications for soil amendment *Soil Use Manage.* **6** 358–86

[131] Ren N, Tang Y and Li M 2018 Mineral additive enhanced carbon retention and stabilization in sewage sludge-derived biochar *Process Safety Env. Protect.* **115** 70–8

[132] Zhao L, Zheng W, Mašek O, Chen X, Gu B, Sharma B K and Cao X 2017 Roles of phosphoric acid in biochar formation: synchronously improving carbon retention and sorption capacity *J. Environ. Qual.* **46** 393–401

[133] Fuentes M E, Nowakowski D J, Kubacki M L, Cove J M, Bridgeman T G and Jones J M 2008 Survey of influence of biomass mineral matter in thermochemical conversion of short rotation willow coppice *J. Energy Inst.* **81** 234–41

[134] Wang Z, Wang F, Cao J and Wang J 2010 Pyrolysis of pine wood in a slowly heating fixed-bed reactor: potassium carbonate versus calcium hydroxide as a catalyst *Fuel Process. Technol.* **91** 942–50

[135] Mašek O, Buss W, Brownsort P, Rovere M, Tagliaferro A, Zhao L, Cao X and Xu G 2019 Potassium doping increases biochar carbon sequestration potential by 45%, facilitating decoupling of carbon sequestration from soil improvement *Sci. Rep.* **9** 5514

[136] Zhao L, Zheng W, Mašek O, Chen X, Gu B, Sharma B K and Cao X 2017 Roles of phosphoric acid in biochar formation: synchronously improving carbon retention and sorption capacity *J. Environ. Qual.* **46** 393–401

[137] Buss W, Jansson S, Wurzer C and Mašek O 2019 Synergies between BECCS and biochar—maximizing carbon sequestration potential by recycling wood ash *Sustain. Chem. Eng.* **7** 4204–09

[138] Buss W, Jansson S and Mašek O 2019 Unexplored potential of novel biochar-ash composites for use as organo-mineral fertilizers *J. Clean. Prod.* **208** 960–67

[139] Han H, Rafiq M K, Zhou T, Xu R, Mašek O and Li X 2019 A critical review of clay-based composites with enhanced adsorption performance for metal and organic pollutants *J. Hazard. Mater.* **369** 780–96

[140] Cornejo J, Celis R, Pavlovic I and Ulibarri M A 2008 Interactions of pesticides with clays and layered double hydroxides: a review *Clay Miner.* **43** 155–75

[141] Awad A M, Shaikh S M, Jalab R, Gulied M H, Nasser M S, Benamor A and Adham S 2019 Adsorption of organic pollutants by natural and modified clays: a comprehensive review *Sep. Purif. Technol.* **228** 115719

[142] Viglašová E, Galamboš M, Danková Z, Krivosudský L, Lengauer C L, Hood-Nowotny R, Soja G, Rompel A, Matik M and Briančin J 2018 Production, characterization and adsorption studies of bamboo-based biochar/montmorillonite composite for nitrate removal *Waste Manage. (Oxford)* **79** 385–94

[143] Chen L, Chen X L, Zhou C H, Yang H M, Ji S F, Tong D S, Zhong Z, Yu W H and Chu M Q 2017 Environmental-friendly montmorillonite-biochar composites: facile production and tunable adsorption-release of ammonium and phosphate *J. Clean. Prod.* **156** 648–59

[144] Fu C, Zhang H, Xia M, Lei W and Wang F 2020 The single/co-adsorption characteristics and microscopic adsorption mechanism of biochar-montmorillonite composite adsorbent for pharmaceutical emerging organic contaminant atenolol and lead ions *Ecotoxicol. Environ. Safety* **187** 109763

[145] Gao R, Wang Q, Liu Y, Zhu J, Deng Y, Fu Q and Hu H 2018 Co-pyrolysis biochar derived from rape straw and phosphate rock: carbon retention, aromaticity, and Pb removal capacity *Energy Fuels* **33** 413–19

[146] Wang W, Zhang J, Chen T, Sun J, Ma X, Wang Y, Wang J and Xie Z 2020 Preparation of TiO_2-modified biochar and its characteristics of photo-catalysis degradation for enrofloxacin *Sci. Rep.* **10** 6588

[147] Mian M M and Liu G 2018 Recent progress in biochar-supported photocatalysts: synthesis, role of biochar, and applications *RSC Adv.* **8** 14237–48

[148] Lisowski P, Colmenares J C, Mašek O, Lisowski W, Lisovytskiy D, Kamińska A and Łomot D 2017 Dual functionality of TiO_2/biochar hybrid materials: photocatalytic phenol degradation in the liquid phase and selective oxidation of methanol in the gas phase *Sustain. Chem. Eng.* **5** 6274–87

[149] Mumme J, Getz J, Prasad M, Lüder U, Kern J, Mašek O and Buss W 2018 Toxicity screening of biochar-mineral composites using germination tests *Chemosphere* **207** 91–100

[150] Demirbas A 2004 Effects of temperature and particle size on bio-char yield from pyrolysis of agricultural residues *J. Anal. Appl. Pyrolysis* **72** 243–48

[151] Manyà J J, Ortigosa M A, Laguarta S and Manso J A 2014 Experimental study on the effect of pyrolysis pressure, peak temperature, and particle size on the potential stability of vine shoots-derived biochar *Fuel* **133** 163–72

[152] Abbas Q, Liu G, Yousaf B, Ali M U, Ullah H, Munir M A M and Liu R 2018 Contrasting effects of operating conditions and biomass particle size on bulk characteristics and surface chemistry of rice husk derived-biochars *J. Anal. Appl. Pyrolysis* **134** 281–92

[153] Leng L and Huang H 2018 An overview of the effect of pyrolysis process parameters on biochar stability *Bioresour. Technol.* **270** 627–42

[154] Han Y, Cao X, Ouyang X, Sohi S P and Chen J 2016 Adsorption kinetics of magnetic biochar derived from peanut hull on removal of Cr (VI) from aqueous solution: effects of production conditions and particle size *Chemosphere* **145** 336–41

[155] Liang C, Gascó G, Fu S, Méndez A and Paz-Ferreiro J 2016 Biochar from pruning residues as a soil amendment: effects of pyrolysis temperature and particle size *Soil Tillage Res.* **164** 3–10

[156] Pokharel P and Chang S X 2019 Manure pellet, woodchip and their biochars differently affect wheat yield and carbon dioxide emission from bulk and rhizosphere soils *Sci. Total Environ.* **659** 463–72

[157] Salema A A and Ani F N 2012 Pyrolysis of oil palm empty fruit bunch biomass pellets using multimode microwave irradiation *Bioresour. Technol.* **125** 102–7

[158] Zambon I, Colosimo F, Monarca D, Cecchini M, Gallucci F, Proto A R, Lord R and Colantoni A 2016 An innovative agro-forestry supply chain for residual biomass: physicochemical characterisation of biochar from olive and hazelnut pellets *Energies* **9** 526

[159] Santos L B, Striebeck M V, Crespi M S, Ribeiro C A and De Julio M 2015 Characterization of biochar of pine pellet *J. Therm. Anal. Calorim.* **122** 21–32

[160] Colantoni A, Evic N, Lord R, Retschitzegger S, Proto A R, Gallucci F and Monarca D 2016 Characterization of biochars produced from pyrolysis of pelletized agricultural residues *Renew. Sustain. Energy Rev.* **64** 187–94

[161] Sorgonà A, Longo L, Proto A R, Cavalletti P, Cecchini M, Salvati L, Gallucci F and Colantoni A 2016 Characterization of biochar and syngas obtained from pellets of grape vine and sun flower husk using a pyrolysis system *Procedia Social Behav. Sci.* **223** 871–78

[162] Crombie K and Mašek O 2014 Investigating the potential for a self-sustaining slow pyrolysis system under varying operating conditions *Bioresour. Technol.* **162** 148–56

[163] Debiagi P E A, Gentile G, Cuoci A, Frassoldati A, Ranzi E and Faravelli T 2018 A predictive model of biochar formation and characterization *J. Anal. Appl. Pyrolysis* **134** 326–35

[164] Li S, Harris S, Anandhi A and Chen G 2019 Predicting biochar properties and functions based on feedstock and pyrolysis temperature: a review and data syntheses *J. Clean. Prod.* **215** 890–902

[165] Diamantaras K I and Kung S Y 1996 *Principal Component Neural Networks: Theory and Applications* (New York: Wiley)

[166] Hyvarinen A, Karhunen J and Oja E 2001 *Independent Component Analysis* (New York: Wiley)

[167] Jain A and Chang E Y 2004 Adaptive sampling for sensor networks *ACM International Conference Proceeding Series* pp 10–6

[168] Madden S, Franklin M J, Hellerstein J M and Hong W 2002 TAG: a tiny aggregation service for ad-hoc sensor networks *Proceedings of the 5th ACM Symposium on Operating System Design and Implementation (OSDI)* vol 36 (New York: ACM) pp 131–46

[169] Madden S R, Franklin M J, Hellerstein J M and Hong W 2005 TinyDB: an acquisitional query processing system for sensor networks *ACM Trans. Database Syst. (TODS)* **30** 122–73

[170] Tulone D and Madden S 2006 PAQ: time series forecasting for approximate query answering in sensor networks *Proc. 3rd European Workshop on Wireless Sensor Networks* (Berlin: Springer) pp 21–37

Chapter 3

Large scale biochar production and activation

Edoardo Miliotti and David Chiaramonti

Slow pyrolysis or carbonization, converting biomass material into char by heating (400 °C–600 °C) in the absence of oxygen or in a limited oxygen atmosphere, is an ancient technology. Together with the main product, non-condensable gases, water, and condensable vapors (tars) are also produced. The carbonization process can operate in a very wide scale and in a variety of systems: from simple and rudimental batch earth mounds to large continuous, modern, and environmentally friendly industrial systems. Carbonization reactors can be classified in many ways, one of which is the way in which the reactor is heated, i.e. allothermal and autothermal reactors. In the former reactor typology, the heat for the pyrolysis reactions is provided by an external source, often by the combustion of pyrogas, while in the latter case heat is supplied by partial oxidation of the feedstock, due to controlled and limited injection of an oxidizing agent.

Hydrothermal carbonization is instead a process which is carried out in hot compressed water (180 °C–250 °C, 10–40 bar) and therefore it is particularly indicated for treating very wet materials, such as micro/macroalgae, sewage sludge, digestate, and manure, which cannot be easily converted with conventional technologies, unless energy-consuming pre-treatment steps are taken such as mechanical dewatering and drying [1]. HTC is a relatively new technology and the main reactor configuration consists in continuous plug-flow reactors or multi-batch systems with continuously stirred tanks.

Due to the continuous increase of the global population and environmental pollution, the need for clean air and clean water is becoming a relevant issue and governments across the globe are offering subsidies and issuing new stringent environmental regulations and directives for water as well as air purification. In addition to this, emerging countries, such as the Asia Pacific region, are experiencing rapid industrialization. These are the main reasons behind the growth of the global activated carbon market (over 3 M$ in 2015) [2]. Activated carbons (ACs) are produced from lignocellulosic biomass (coconut shell, wood, etc) and from coal,

doi:10.1088/978-0-7503-2660-5ch3

lignite, and peat [2]. In order to decrease the dependence on fossil sources, the production of ACs from renewable and/or residual sources is an optimal opportunity from the perspective of sustainable bioeconomy. ACs are amorphous organic materials with high adsorption potential and are produced industrially via two different processes: physical and chemical activation. In physical activation, an activating agent (steam, carbon dioxide, air, or a combination thereof) at elevated temperatures gasifies part of the carbon structure, creating a porous structure. In chemical activation, the feedstock is impregnated with a reagent (KOH, $ZnCl_2$, H_3PO_4, K_2CO_3, etc) and then it is heated in an inert atmosphere. A combination of physical and chemical methods can be adopted for producing ACs with finely tuned porosity [3]. In general, physical activation is carried out on a pre-carbonized material (char, double-step activation), while in the chemical process carbonization occurs during activation.

3.1 Introduction

The main thermochemical processes adopted for charcoal/biochar production are pyrolysis, gasification, and hydrothermal processing. However, related char yields and characteristics can vary widely between these technologies, in combination with the feedstocks.

Pyrolysis is a thermochemical process in which biomass is heated in the absence of oxygen, or in a limited oxygen atmosphere. In addition to char, other co-products are obtained from this process: non-condensable pyrolysis gases, water, and condensable vapors (tars or bio-oil). Depending on the reaction conditions, selectivity towards specific products can be maximized. The process which enhances char yield is called slow pyrolysis or carbonization, and is generally carried out at atmospheric pressure with reaction temperatures between 400 °C and 600 °C, low heating rates, and long solids retention times in the reactor (from hours to days, depending on the reactor technology). Once pyrolysis vapors are condensed, a biphasic liquid is typically obtained, which is composed of an aqueous phase with soluble organics and a water-insoluble tarry oil, often referred to as bio-oil or biocrude. In most industrial slow pyrolysis configurations gases and vapors (i.e. pyrogas) are oxidized to generate heat, which in turn can be employed to sustain the process, to dry the feedstock, and/or to deliver heat to end users. The other main pyrolysis processes are named intermediate and fast pyrolysis [4] (see figure 3.1 and table 3.1), however, these not addressed in this chapter. Slow pyrolysis is a very mature system, while the first examples of industrial scale fast pyrolysis plants have been developed in recent years and are now commercially operated.

In gasification, lignocellulosic biomass is mainly converted into combustible gases by heating at elevated temperatures (even above 900 °C) under a flow of a gasifying agent; in this case, char is a side-product. The produced gas, after cleaning, is typically sent to an internal combustion engine for electricity production or cogeneration. Gasification is a well-defined and commercialized technology available at many different scales, but its economic feasibility depends highly on incentives for renewable electricity production [5].

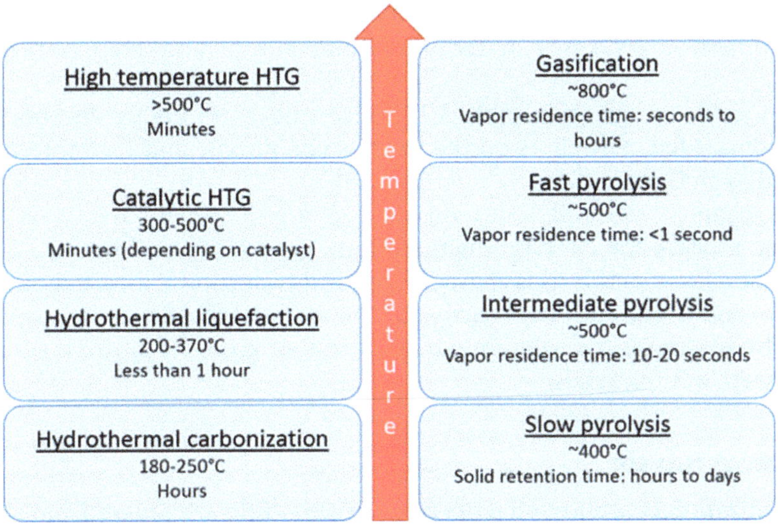

Figure 3.1. Hydrothermal and conventional thermochemical conversion.

Table 3.1. Comparison between typical yields of char-producing processes. Values expressed as mass fraction (dry basis).

Process	Typical yields	Reference
Slow pyrolysis	Solid 35%, liquid 30%, gas 35%	[5]
Intermediate pyrolysis	Solid 20%, liquid 50%, gas 30%	[5]
Fast pyrolysis	Solid 12%, liquid 75%, gas 13%	[5]
Gasification	Solid 10%, liquid 5%, gas 85%	[5]
HTC	Solid 60%, liquid 35%, gas 5%	[1]
HTL	Solid 5%, liquid 80%[a], gas 15%	[1, 6–8]
Catalytic HTG	Gas 66%	[1, 9–11]
High-temperature HTG	Gas > 90%	[1, 9, 10, 12]

[a] Liquid: biocrude 30%, aqueous organics 50%.

Hydrothermal processing is a relatively new technology, which in particular favors the thermochemical conversion of wet materials, such as sewage sludge, algae, and agricultural or food industry residue, as this process is carried out in hot compressed water. The maximum yield of solid product, often called hydrochar, is obtained at low temperatures (180 °C–250 °C) and long residence times (hours) and the process is referred to as hydrothermal carbonization (HTC). Similar to pyrolysis and gasification, hydrothermal processing also includes hydrothermal liquefaction (HTL) and hydrothermal gasification (HTG). The main barrier to hydrothermal processing commercialization is the requirement for high pressure, which is needed to keep water in the liquid state at the reaction temperature. In HTC, the pressure generally ranges between 10 and 40 bar, while it can exceed 350 bar in HTG or supercritical HTL [1].

A summary, comparing the operating conditions of conventional thermochemical and hydrothermal processes in terms of temperature and time is presented in figure 3.1.

Typical yields are given in table 3.1. These values are highly dependent on the type of biomass fed, as well as on the process conditions, and therefore have to be considered as indicative data.

In this chapter, we will focus on slow pyrolysis which, today, is the most convenient process for char production from lignocellulosic biomass. Compared to gasification, slow pyrolysis is characterized by lower capital investment, it is easier to operate, and it does not necessarily rely on external incentives. A section of this chapter also addresses hydrothermal carbonization, given its intrinsic advantage to easily convert wet materials.

3.2 Slow pyrolysis

The carbonization process can be operated at very different scales: from simple and rudimental batch earth mounds to large continuous industrial systems. There are several options to classify carbonization reactors, such as the mode of operation, construction material, feedstock allowable size, portability, biomass loading mode, and heating method. The latter is one of the most important classification criteria, as it determines the energy self-sustainability of the process. Two major classes are defined: allothermal and autothermal reactors (figure 3.2). In the former reactor typology, pyrolysis is performed in the full absence of oxygen, and the heat for the process is provided by an external source, normally by the combustion of pyrogases (non-condensable gases and condensable vapors). The resulting heat can be exchanged to the reactor by direct injection of the flue gas or by a heat transfer surface. In the autothermal case, heat is supplied by partial oxidation of the pyrogases and the char generated by the biomass pyrolysis process itself, due to the controlled injection of an oxidizing agent (typically air). In general, all commercial processes are operated at ambient pressure due to economic and operating issues, but it was proven that the char yield is increased when elevated pressures are applied, due to increased polymerization of tarry vapors [13].

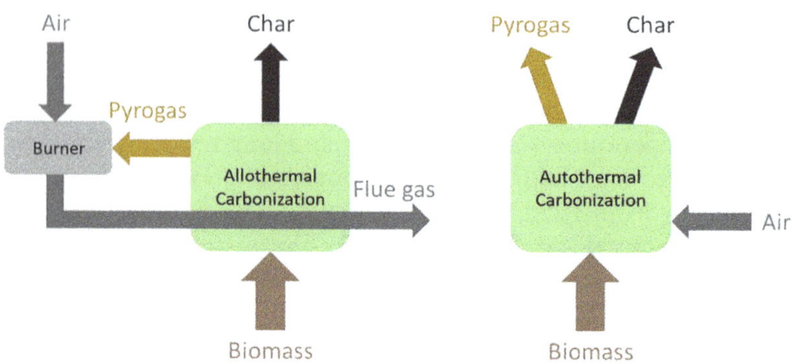

Figure 3.2. Allothermal versus autothermal slow pyrolysis.

Another possible classification, based on reactor technology, was proposed by Emrich [14] and thoroughly reviewed by Garcia-Nunez *et al* [15] and Garcia-Perez *et al* [16]. It can be summarized as follows:

- *Kilns*: traditional systems adopted for char production, without pyrogas recovery.
- *Retorts*: reactors able to convert pile-wood or wood logs above 18 cm in diameter and longer than 30 cm.
- *Converters*: reactors processing small-sized feedstock, such as chipped or pelletized wood.

3.2.1 Kilns

Kilns are batch autothermal reactors that can be subclassified by their construction material.

Earth kilns are the oldest and most rudimental carbonizers and are still used in developing countries due to their extremely low cost and ease of operation. Typically, wood logs are placed in an earth pit or assembled to form a mound and then covered with soil, which ensure partial biomass oxidation by preventing air flow. Thus, these kilns can be realized in the same location as biomass harvesting, reducing transport costs. However, they require a high amount of labor, in particular in the preparation of mound kilns, and are characterized by very long production times, even up to several weeks. During the pyrolysis process, vapors are released into the atmosphere and liquid condensate can easily contaminate the soil, leading to serious environmental impacts. The control of the carbonization process is simple and does not require sophisticated technology as it is carried out by observation of the color of the vapors.

Technological advancement is essentially related to improvement in the construction materials, that is *brick, concrete, or metal*, leading to a long lifespan, higher char quality, and yield. Various types of such kilns have been developed since the nineteenth century, with different shape, capacity, and innovations, such as adopting external heating chambers or pyrogas afterburners.

3.2.2 Retorts

Most industrial carbonization units are characterized by an allothermal retort configuration, in which hot vapors and gases are burnt for producing the heat needed for pyrolysis. In some configurations, pyrolysis vapors are condensed in order to collect a variety of chemicals such as methanol and acetic acid, while non-condensable gases are used for process heating.

The simplest form of retort carbonizer has an allothermal, indirect heating configuration, typically constituted by a closed metal container (often even a barrel) equipped with a pipeline conveying the produced gases and vapors towards the combustion chamber, which can be placed beneath the reactor or developed around it, so to maximize the heat transfer. Wood logs are placed in the container, which is then sealed and heated by additional fuel for starting the process. First, water vapors are produced from biomass drying, then pyrolysis actually starts and pyrogas is generated, spontaneously moving towards the combustion chamber; the process is therefore self-sustainable. As the reactor is sealed, no oxidation occurs and, consequently, the char yield is generally higher with respect to autothermal systems. Several variations have been developed during the decades, including semi-stationary and mobile configurations, semi-batch and continuous operation, with direct heating or partial oxidation, horizontal or vertical reactors, with manual or mechanical feedstock loading and so on.

One of the retorts with the highest throughput (6 kt year^{-1}) of char was the so-called *Lambiotte reactor*, which was first developed in the 1940s. It consists in a vertical continuous fixed bed retort, in which small wood logs are fed from the top and are forced to pass through different reactor zones by gravity; the biomass is subjected to drying, conversion into char, and then cooling. The first version of the Lambiotte retort adopted SIFIC technology, where pyrogas combustion was carried out in an external chamber, but an updated version was developed, employing internal combustion and gas recirculation, the so-called CISR Lambiotte. This plant is characterized by a high level of automation, leading to high char yield and quality. Its main drawback is that its operability is heavily affected by the biomass moisture content.

Another retort operating on the same principle is that called the *Lurgi process*: the reactor is equipped with an air-locking valve in order to prevent partial combustion of the biomass bed and pyrogas combustion occurs in an external chamber. A plant with a capacity of 26 kt year^{-1} was installed in Bunbury, Western Australia [16].

A very simple, yet smart, retort carbonizer is the *Carbo twin retort*. It is a semi-batch system in which two metal tank reactors are placed in an insulated oven fed by the produced pyrogas and are alternately loaded and unloaded by means of an upper crane on a monorail, ensuring continuous operation. This system is highly modular and requires minimal labor.

A similar concept, but with continuous operation, is the *wagon retort*: wood logs are loaded in wagon cars placed on a track which enters a drying chamber, a furnace chamber heated by pyrogas combustion where the carbonization occurs, and then cooling chambers. Cooling is a critical step due to the risk of unintended char combustion and requires long times.

3.2.3 Converters

Differently to kilns and retorts, converters are those reactors that are able to carbonize small-sized biomass, such as wood chips and pellets, which would hinder the process by impeding gas penetration, and thus heat transfer, due to their high bulk density. Thus char is often obtained as a powder and further treatment, such as pelletization with binding agents, may be required to meet market standards. Most of these converters are continuous units, characterized by inner moving parts which convey biomass through the reactor.

A well-known example is represented by the *rotary kiln*. Mainly used in waste incineration and in the cement industry as dryer, it was also adopted for biomass thermochemical conversion [17]. It consists in a cylindrical shell which is inclined and rotated to allow biomass to flow down the entire length of the kiln. The heat for pyrolysis can be supplied by heating the shell from the outside or from the inside by means of hot gas flow, which is generally counter-current with respect to the feedstock. The main advantages offered by rotary kilns are the biomass flexibility, scalability, and maturity of the technology. The main drawback is represented by the complexity in maintaining an inert atmosphere inside the reactor due to rotating seals [18]. A rotary kiln pyrolyzer was installed and has been operated since 2009 in Dürnrohr, Austria, for the production of char from agricultural residues for co-firing with biomass in a 3 MW thermal power plant [19].

In *auger converters*, biomass is typically fed to the reactor by a hopper and then is gradually transported throughout its length by means of an inner screw. The reactor can be indirectly heated by pyrogas combustion or internally heated by flue gas injection. Materials such as sand and steel/ceramic spheres can be adopted in order to improve the heat transfer, in particular when bio-oil is a key co-product (intermediate pyrolysis). A commercial example is the Pyreg process (1 t d^{-1} of char). It is a continuous plant where the reactor is externally heated by hot gases from pyrogas combustion and its average char yield is around 27% w/w [20].

One of the oldest converters is the *multiple hearth furnace* or Herreshoff furnace, which can process a diverse range of biomass even with finely grounded particle size, such as sawdust. It has a vertical layout, consisting of a cylindrical insulated steel shell in which several horizontal plates are placed. Biomass is fed from the top and is moved through the hearth by an axial vertical rotating shaft with radial arms. Holes in each plate ensure the feedstock's descending path. The radial arms of the shaft enhance the biomass contact with hot gases, which are typically produced by an external burner. Due to the high capital costs, larger capacities are required for economic viability.

Other configurations include the *paddle pyrolysis reactors* and *moving agitated beds*.

Another interesting carbonization concept is represented by the *flash carbonization reactor*, first developed by Antal *et al* [21]. This was a batch system characterized by a very fast reaction time, about 30 min, making this technology unique among the other carbonization processes. A metal canister full of biomass is inserted in a pressure vessel which is then sealed. This tank is pressurized at 10 atm by an air compressor and a fire is ignited at its bottom using to electric heaters. Thus, a flame front is developed, moving vertically towards the unconverted biomass, reaching about 400 °C.

Char production by slow pyrolysis represents a major opportunity for the diversification of income of agro-forestry firms, which, in particular in Southern Europe, are generally small and not acquainted with the management of bioenergy systems. A significative barrier is indeed due to the small financial capacities of these companies, which cannot afford high capital investments.

Carbonization is a valid alternative to decentralized bioenergy production, because the former does not rely on public incentives, is characterized by a lower capex and a lower complexity of operation, and allows char to be produced using local biomass. This latter element deserves great attention to pinpoint the problems

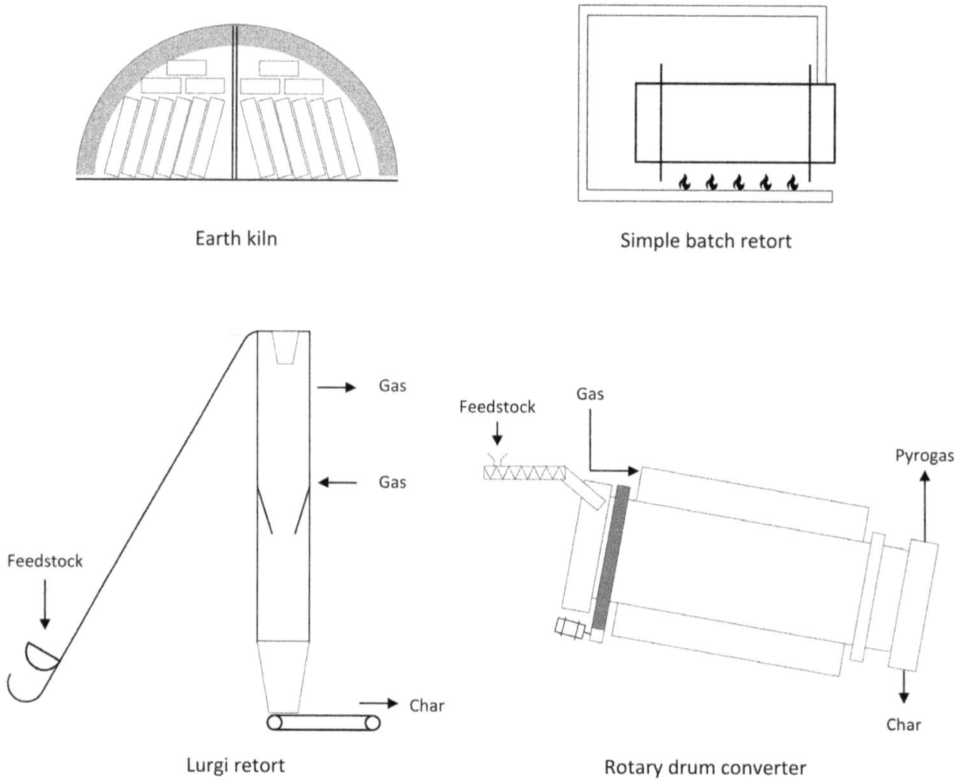

Figure 3.3. Schematic representation of different kinds of carbonizers. Adapted from [15].

related to the business of char importation from developing countries. As an example, the charcoal market is not regulated by the same restrictions as the wood pellet market, and this contributes to ensuring low char production costs. In some developing countries char production is associated with illegal wood harvesting, poor social conditions, and safety and environmental risks, causing uncontrolled deforestation and soil depletion. Furthermore, resources are inefficiently exploited and technological development of the process is hindered.

Figure 3.3 depicts schemes for some of the above-mentioned carbonizers.

3.3 Hydrothermal carbonization

At first sight, hydrothermal carbonization (HTC) is a relatively simple process: biomass is mixed with water and pressurized in order to avoid evaporation during the subsequent heating. The reaction temperature is maintained for the desired residence time, after which the system is cooled and depressurized. The char, called hydrochar, is thus separated from the aqueous phase by mechanical dewatering, e.g. using a filter press, and eventually dried. The aqueous phase can be rich in organic and inorganic content and its management is often critical: depending on the HTC feedstock, anaerobic digestion is one of the options to exploit the dissolved organics, while precious inorganics such as phosphates can be recovered with dedicated processes. During the process a low amount of a CO_2-rich gas phase is produced. Several designs have been patented, involving batch, semi-batch, and continuous operation, and thorough reviews have been published on this subject [1, 10, 22]. Although the reaction mechanisms occurring during hydrothermal carbonization are still not completely understood, they can be considered as hydrolysis, dehydration, decarboxylation, polymerization, and aromatization [23]. In this context, water plays a key role as it simultaneously acts as a heat transfer medium, reactant, catalyst, and solvent. The reason why hydrothermal conversion processes occur at lower temperatures compared to dry thermochemical processes is that the presence of hot compressed water outside and in particular inside the biomass structure leads to an enhancement in heat and mass transfer, increasing temperature homogeneity.

Most hydrothermal carbonization tests have been conducted batch-wise in lab-scale autoclaves. However, various pilot and demonstration-scale plants have been built recently and are operated by different firms. The main reactor configuration consists of continuous plug-flow reactors, multi-batch systems with/without continuously stirred tanks, or a combination thereof. Many companies possess their own patented HTC technology and offer commercial or pre-commercial solutions including feasibility studies, design, engineering, construction, testing, and after-sales services. Most of these companies are European, in particular from Germany, but commercial interest in HTC is also growing in Asia (China, Japan, and Indonesia) [24–31].

3.4 Activated carbon production

Activated carbons (ACs) are typically produced from lignocellulosic biomass (coconut shell, wood, etc) and from coal, lignite, and peat [2]. In order to decrease the dependence on fossil sources, the production of ACs from renewable and/or residual sources is an optimal opportunity from the perspective of sustainable bioeconomy. ACs are amorphous organic materials with high adsorption potential, industrially produced via two different processes: physical and chemical activation. In physical activation, an activating agent (steam, carbon dioxide, air, or a combination thereof) gasifies part of the carbon structure at elevated temperatures creating porosities. In chemical activation, the feedstock is impregnated with a reagent (KOH, $ZnCl_2$, H_3PO_4, K_2CO_3, etc) and then it is heated in an inert atmosphere. A combination of physical and chemical methods can be adopted for producing ACs with finely tuned porosity [3]. In general, physical activation is carried out on a pre-carbonized material (char, double-step activation), while, in the chemical process carbonization occurs during activation [3]. Compared to char, which commonly has a specific surface area (measured by N_2 adsorption at 77 K, the BET method) around 150 $m^2\ g^{-1}$, ACs can reach 3000 $m^2\ g^{-1}$.

Most activated carbons are produced via physical activation of carbonized materials between 800 °C and 1100 °C and the industrially preferred gasifying agent is superheated steam, followed by carbon dioxide.

Activation with these gases occurs according to the following endothermic reactions:

$$C + H_2O = CO + H_2 \tag{3.1}$$

$$C + CO_2 = 2CO. \tag{3.2}$$

Using steam in place of CO_2 has the advantage of removing char trapped products, such as tar, acids, and ketones, developing not only micro- but also mesoporosity [32]. A wider pore size distribution is indeed required as it increases the capability of the activated carbon to adsorb molecules of different sizes. In addition, mesopores (2–50 nm) are also needed to facilitate access to the micropores (⩽2 nm) [3]. Furthermore, steam activation is industrially preferred as off-gases from activation can be easily burned to produce the required steam in a water boiler [33].

Chemical activation is generally performed for producing powdered AC with a well-defined porosity and is commonly carried out between 450 °C and 800 °C on uncarbonized materials. The first step of the process involves the impregnation of the lignocellulosic precursor with the chemical activating agent, which interacts with the carbon skeleton during activation, developing the microporosity. Compared to physical activation, chemical activation leads to a higher yield (27%–47% w/w compared to 6%–10% w/w) and a greater porosity, but soon after the activation step the AC must be washed to remove the chemicals, generating economic and environmental issues.

As activated carbons with low ash content are preferred, lignocellulosic biomass is the most common material adopted as an AC precursor. Wood powder (e.g. sawdust) is often adopted in the chemical activation processes as impregnation of the activating chemical is favored by the small particle size of the precursor. In contrast, due to their hardness and volatile content, coconut shells and fruit stones are preferred for the production of granular ACs via physical activation.

Commonly, activated carbons are industrially produced using three kinds of continuous reactors: rotary kilns, multiple hearth furnaces, and fluidized bed reactors [3].

3.4.1 Rotary kilns

The rotary kilns used for carbon activation are essentially similar to those used for biomass carbonization, except for the fact that now a gasifying agent is introduced in a co-current or counter-current mode. These are the most popular due to their ability to produce a more controlled and microporous structure thanks to enhanced temperature control and longer residence time. They are often equipped with inner flights, which allows better heat exchange and reagent mixing due to their showering action. Often, direct-fired kilns are used by burning the activation off-gases, pyrolysis gases (from the previous carbonization step), or fuel gases (e.g. natural gas) in the presence of steam and excess combustion air.

3.4.2 Multiple hearth furnaces

Multiple hearth furnaces have already been described in the carbonization section. When they are adopted for activation, steam or another gasifying agent is injected at different levels, ensuring good temperature control. However, they are characterized by shorter residence times, leading to more mesoporous carbons. The regeneration of spent AC is commonly carried out in this kind of reactor.

3.4.3 Fluidized beds

Fluidized beds are more often used for fast pyrolysis rather than slow pyrolysis and can also be adopted for char activation. In the latter case the fluidizing gas is the gasification agent. They are characterized by excellent heat and mass transfer, leading to short residence times. Commonly this kind of reactor has a large capacity and is operated continuously, as it is difficult to operate at partial load and has a long start-up time. Moreover, in order to obtain a suitable fluidization, it must be fed with materials with reduced particle size. For this reason fluidized beds are adopted for producing powdered activated carbons. Because of the attrition occurring during fluidization, fluidized beds are commonly used with abrasion-resistant precursors [34].

3.5 Conclusions

Despite the production of biochar through slow pyrolysis being a mature process, there is still great room for innovation, in particular in terms of reactor configuration, control of process conditions, and selection of feedstocks to match product specification, to drive and optimize the process towards the targeted characteristics for the biochar at the outlet of the system. In this context, activated carbons are a well-known type of product and process.

Moving from small-scale to industrial scale units favors the improvement of the process and environmental performance. However, a main advantage of slow pyrolysis technologies is their ability to adjust to the scale needed, i.e. from the farm scale to industrial scale. This flexibility is a major advantage of the process.

In recent years, the interest in hydrothermal processes, and in particular HTC, has been growing, as these processes can treat very wet dedicated or residual materials, as algae or sludges. Pilot and demo HTC plants are moving towards the full commercialization of this technology.

References

[1] Peterson A A, Vogel F, Lachance R P, Fröling M, Antal M J Jr and Tester J W 2008 Thermochemical biofuel production in hydrothermal media: a review of sub- and supercritical water technologies *Energy Environ. Sci.* **1** 32–65

[2] Grand View Research 2018 Activated carbon market size, share and trends analysis report by product (powdered activated carbon, granular activated carbon), by application, by end-use, and segment forecasts, 2018–2024 *Market Analysis Report* https://grandviewresearch.com/industry-analysis/activated-carbon-market

[3] Marsh H and Rodríguez-Reinoso F 2006 *Activated Carbon* (Amsterdam: Elsevier)
[4] Bridgwater A V 2012 Review of fast pyrolysis of biomass and product upgrading *Biomass Bioenergy* **38** 68–94
[5] Bridgwater T 2007 *Biomass Pyrolysis*, Task 34 Overview, IEA Bioenergy: T34:2007:01
[6] Knorr D, Lukas J and Schoen P 2013 *Production of Advanced Biofuels via Liquefaction – Hydrothermal Liquefaction Reactor Design*, NREL technical report, Subcontract Report NREL/SR-5100-60462, Atlanta, Georgia
[7] Goudriaan F, Naber J and van der Berg E *Conversion of Biomass Residues to Transportation Fuels with the HTU ® Process* http://www.adktroutguide.com/files/HTU_Process.pdf (Accessed: 10 March 2017)
[8] Miliotti E 2015 Analysis, study and design of a biomass hydrothermal liquefaction plant *Master Thesis* University of Florence
[9] Matsumura Y, Minowa T, Potic B, Kersten S R A, Prins W, van Swaaij W P M, van de Beld B, Elliott D C, Neuenschwander G G and Kruse A 2005 Biomass gasification in near- and super-critical water: status and prospects *Biomass Bioenergy* **29** 269–92
[10] Pavlovič I, Knez Ž and Škerget M 2013 Hydrothermal reactions of agricultural and food processing wastes in sub- and supercritical water: a review of fundamentals, mechanisms, and state of research *J. Agric. Food Chem.* **61** 8003–25
[11] Elliott D, Santosa D, Valkenburg C, Neuenschwander G, Jones S, Hart T, Rotness L J, Tjokro Rahardjo S and Zacher A 2009 Catalytic hydrothermal gasification of lignin-rich biorefinery residues and algae final report *PNL Technical Report* http://pnl.gov/main/publications/external/technical_reports/PNNL-18944.pdf
[12] Boukis N, Galla U, D'Jesus P, Müller H and Dinjus E 2005 Gasification of wet biomass in supercritical water. Results of pilot plant experiments *14th Eur. Biomass Conf., Paris, France* pp 964–67
[13] Antal M J and Grønli M 2003 The art, science, and technology of charcoal production *Ind. Eng. Chem. Res.* **42** 1619–40
[14] Emrich W 1985 *Handbook of Charcoal Making* (Heidelberg: Springer)
[15] Garcia-Nunez J A, Pelaez-Samaniego M R, Garcia-Perez M E, Fonts I, Abrego J, Westerhof R J M and Garcia-Perez M 2017 Historical developments of pyrolysis reactors: a review *Energy Fuels* **31** 5751–75
[16] Garcia-Perez M, Lewis T and Kruger C 2011 *Methods for Producing Biochar and Advanced Biofuels in Washington State. Part 1: Literature Review of Pyrolysis Reactors. First Project Report* (Pullman, WA: Washington State University)
[17] Zajec L 2009 Slow pyrolysis in a rotary kiln reactor: optimization and experiment *Master Thesis* School of Renewable Energy Science in affiliation with University of Iceland and the University of Akureyri
[18] Boateng A A 2016 *Rotary Kilns* (Boston, MA: Butterworth-Heinemann)
[19] Kern S, Halwachs M, Proll T, Kampichler G and Hofbauer H 2010 Rotary kiln pyrolysis—first results of a 3 MW pilot plant *18th Eur. Biomass Conf. Exhib., Lyon, France* pp 3–7
[20] Ronsse F 2013 Commercial biochar production and its certification *Biochar-Interreg4B.Eu* 22 http://biochar-interreg4b.eu/images/file/02_Presentation-Frederik-Ronsse.pdf%5Cnpapers2://publication/uuid/0A8C3957-9ACA-4807-A0B7-52DA14D80A89
[21] Antal M J, Mochidzuki K and Paredes L S 2003 Flash carbonization of biomass *Ind. Eng. Chem. Res.* **42** 3690–99

[22] Elliott D C 2011 Hydrothermal processing *Thermochemical Processing of Biomass Conversion into Fuels, Chemicals and Power* ed R C Brown (Chichester: Wiley) pp 200–31
[23] Funke A and Ziegler F 2010 Hydrothermal carbonization of biomass: a summary and discussion of chemical mechanisms for process engineering *Biofuels, Bioprod. Biorefining* **4** 160–77
[24] Child M 2014 Industrial-scale hydrothermal carbonization of waste sludge materials for fuel production *Master Thesis* Lappeenranta University of Technology
[25] Ingelia n.d. *Ingelia* http://ingelia.com/?lang=en (Accessed: 8 February 2017)
[26] SunCoal Industries GmbH n.d. Making energy from organic waste *Suncoal Industries* http://suncoal.de/en (Accessed: 9 February 2017)
[27] TerraNova Energy GmbH n.d. Renewable fuel systems for hydrothermal carbonization *TerraNova Energy* http://terranova-energy.com/en/ (Accessed: 6 February 2017)
[28] Pels J R and Bergman P C A 2006 TORWASH Proof of Principle – Phase 1, ECN technical report, ECN-E-06-021
[29] Yoshikawa K 2017 Commercial scale demonstration of high quality solid fuel and fertilizer production from biomass and wastes employing HTC *1st Int. Syposium Hydrothermal Carbonization* (London: Queen Mary University)
[30] CarboREM 2018 *CarboREM* http://carborem.com/it/
[31] C-Green Technology AB n.d. The smart way to recycle wet waste into biocoal and biogas *C-Green* https://c-green.se/ (Accessed: 24 June 2020)
[32] Sajjadi B, Chen W Y and Egiebor N O 2019 A comprehensive review on physical activation of biochar for energy and environmental applications *Rev. Chem. Eng.* **35** 735–76
[33] Arena N, Lee J and Clift R 2016 Life cycle assessment of activated carbon production from coconut shells *J. Clean. Prod.* **125** 68–77
[34] McDougall G J 1991 The physical nature and manufacture of activated carbon *J. South African Inst. Min. Metall.* **91** 109–20

Chapter 4

Microwave heating-assisted pyrolysis of biomass for biochar production

Sherif Farag and Jamal Chaouki

The volumetric energy conversion mechanism of microwaves has become a modern alternative to the superficial heat transfer of conventional heating in the thermal and catalytic conversion of biomass and waste. The selective, uniform, fast, controllable, and efficient heating ability of microwaves are the main merits behind its rapid growth during the last few decades. It has been evidenced that microwave heating noticeably enhances product yield, composition, and structure compared to conventional heating methods.

This chapter discusses the characteristics of biochar derived from microwave pyrolysis against conventional heating. The impact of the process parameters—including temperature, heating rate, residence time, and the presence of a catalyst/microwave absorber—are discussed. The up-to-date developed microwave processes to produce biochar at the lab, pilot, or industrial scale are demonstrated. Last but not least, recommendations to overcome the common challenges related to scaling-up microwave pyrolysis processes are elucidated.

4.1 Microwave fundamentals

Electromagnetic radiation behaves like waves moving at the speed of light and photons carrying radiated energy. It is comprised of both alternating electric and magnetic fields that are orthogonal to each. The electromagnetic spectrum covers a wide range of implementations, e.g. radio waves, microwaves, infrared, and visible light. Each application holds a specific frequency to avoid overlapping each other. For example, microwaves occur from 0.3 GHz to 300 GHz. The rapid development of microwave technology throughout the twentieth century has established it in a significant number of applications, including communications, navigation, radar detection, power transmission, and microwave heating (MH).

doi:10.1088/978-0-7503-2660-5ch4

The dominant mechanism of MH—which relies on direct volumetric energy conversion within the irradiated material—is fundamentally different from the superficial heat transfer of classical heating (CH) [1, 2]. As a result, it can avoid most of the issues and limitations associated with CH—such as the temperature gradient inside and outside the exposed material, energy consumption, and product selectivity. Since MH depends on the dielectric and magnetic properties of the materials, it can heat selectively where the payload is composed of multiphases/multicomponents. This aspect can lead to initiating reactions that cannot be initiated when CH is applied, achieving the existing reactions under conditions far from those of the CH-based processing. Thus, materials with a novel microstructure could be produced. In addition, there is a noticeable reduction in the amount of heat energy needed to achieve a particular aim. This has positive effects on the operating costs, the potential for thermal hazards, and other aspects.

4.2 The main parameters to describe microwave heating

The crucial parameters that describe microwave propagation and conversion inside an irradiated material are:

- *For non-magnetic materials*: Complex permittivity (ε), which measures the ability of the material to absorb and store electric potential energy.
- *For magnetic materials*: Complex permeability (μ), which represents the response of a material to an alternating magnetic field.

Complex permittivity and complex permeability are expressed by [3–6]

$$\varepsilon = \varepsilon' - j\varepsilon'' \tag{4.1}$$

$$\mu = \mu' - j\mu'', \tag{4.2}$$

where ε' is called a dielectric constant and refers to the ability of a dielectric material to pass microwaves through it. ε'' is a dielectric loss factor and indicates the ability of the material to dissipate microwave energy.

Loss tangent ($\tan\delta$) describes the ability of the heated material to convert the absorbed microwave energy into heat energy and is expressed by the ratio between ε'' and ε'. For most of the conventional materials, ε', ε'', and $\tan\delta$ are well documented in the scientific literature. Typical values of the selected materials are tabulated in [7–11]. It should be noted that the dielectric properties of a material are affected by the microwave frequency and temperature of the material [12, 13]. Therefore the effects of these two factors must be taken into consideration during the design of a microwave-based system.

4.3 Microwave-assisted pyrolysis of biomass and waste

In pyrolysis applications, MH can avert most of the issues and limitations associated with conventional pyrolysis. In CH pyrolysis, heat is transferred superficially from a heating source to the outer surface of the target material and subsequently into the core. Heat capacity, thermal conductivity, and the dimension of the workload, among other factors, affect the heating rate of the material. On top of that the decomposition of the chemical bonds follows the same direction as that of the heat transfer. The creation of a temperature gradient—a surface hotter than the core—leads to an initial layer of char on the outer surface that moves toward the core, as shown in figure 4.1(a). Such a char layer reduces the heat transfer within the processed material, which eventually leads to further degradations that negatively affect the quantity and/or quality of the final product.

In MH pyrolysis, in contrast, microwaves penetrate the entire material at almost the same time—of course, the penetration limits must be considered. Therefore, a uniform temperature distribution within the workload should be obtained. Due to the heat transfer at the outer surface of the heated material, however, a core hotter than the surface is always found, as shown in figure 4.1(b). As a result, a char layer is created in the core and grows to the surface direction. This means that the creation of the char and vapor products follow the same path as the heat transfer. In such a case, the product quality is enhanced compared to the CH processes. Farag *et al* have investigated the impact of this aspect on the yield and composition of the liquid product [9]. Herein, the impacts on the yield, quality, and end-application of biochar are debated.

4.4 The effects of microwaves on biochar properties

Biochar is a carbon-rich solid—i.e. a strong microwave absorber, high tanδ—with an aromatic surface and can be produced from various thermochemical degradation approaches such as the pyrolysis of agricultural and forestry waste. Owing to its high surface area, varying surface adsorption sites, as well as high stability, biochar has been widely applied for gas adsorption, wastewater treatment, and agricultural sector, as discussed in the following sections [14].

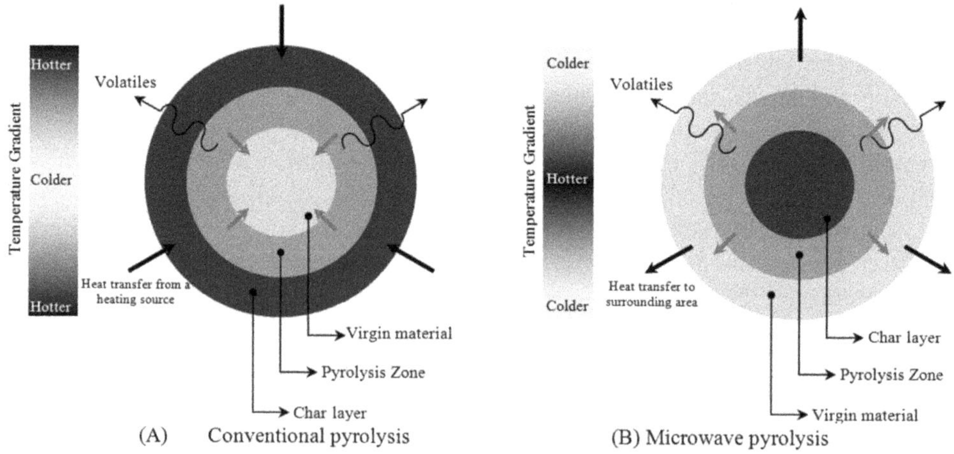

Figure 4.1. Schematic diagram of (a) conventional pyrolysis and (b) microwave pyrolysis. Reproduced from [10].

In microwave-assisted biochar production, the feedstock interacts with the electromagnetic radiation. Then the heat is generated within the solid material itself, thereby ensuring a lower loss of the energy compared to CH. In addition, the heating caused by microwave irradiation offers several benefits, such as fast heating rate, heating selectivity, environmental-friendliness, and more homogeneous products [15].

The unique advantages of MH make the physicochemical properties of biochar obtained by such an energy conversion mechanism much better than those of the superficial heating mechanism. Over the last decades, greats efforts have been documented in the scientific literature.

Ho *et al* [16] have observed that the yields of biochar obtained from torrefaction with either CH or MH of coffee grounds and microalgal residues were similar, i.e. CH: 55.4–98.2 wt.% and 56.4–92.4 wt.% for coffee grounds and microalgal residue, respectively, and MH: 48.8–94.3 wt.% and 57.3–89.7 wt.% for coffee grounds and microalgal residue, respectively. It can be concluded that no significant variation was observed between the biochar yields between these two heating mechanisms. However, a shorter processing time was obtained for MH and thus MH-assisted torrefaction is a better route to produce biochar than CH-based torrefaction in order to lower the energy consumption. Szewczuk-Karpisz *et al* [17] studied the influence of the heating mechanism on the adsorption ability of hay-derived biochar. The results showed that the biochar samples prepared by either MH or CH were found to demonstrate medium porosity. Specifically, a larger micropore of 187 $m^2\ g^{-1}$ was achieved for MH-prepared biochar than that prepared using CH (149 $m^2\ g^{-1}$), resulting in a higher Cu (II) ion adsorption capability. The SEM analysis in figure 4.2 indicates that evident differences in the surface morphology in terms of the shape and size of the particle and the number and arrangement of cracks, crevices, and spherical holes were detected in the biochar samples obtained by two different heating mechanisms.

Szewczuk-Karpisz *et al* have also found that the oxygen content in the biochar obtained using MH was higher than that obtained using CH, which implies that microwave irradiation is not as effective as CH in removing oxygen-containing functional groups from the surface. This was evidenced by the more significant number of oxygen-containing functional groups present on the surface of

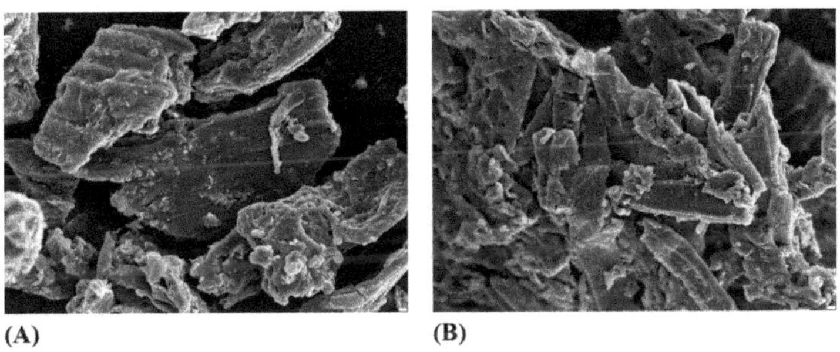

Figure 4.2. SEM images of the hay-derived biochar prepared using (A) CH and (B) MH. Reproduced from [17].

MH-prepared biochar. Another difference between MH- and CH-produced biochar samples is related to the carbon and ash content, i.e. biochar obtained using MH had a carbon and mineral content of 79.1% and 32.3%, respectively, and biochar obtained using CH had a carbon and mineral content of 84.2% and 26.1%, respectively.

Further comparison between the MH and CH mechanisms on the yield and physiochemical properties of biochar is summarized in table 4.1.

As can be seen in table 4.1, MH-prepared biochar has been found to be of better quality than that obtained from conventional pyrolysis since the biochar prepared using CH tends to have cracks and fissures on the surface caused by the large difference in temperature gradient and thus are easily broken. A higher surface area and pore volume are also commonly identified in the biochar obtained from pyrolysis using MH when compared to those obtained from pyrolysis using CH [23]. In addition, the micropores present in the surface of MH-prepared biochar are highly uniform and clean, and therefore it can be concluded that the use of MH to substitute with CH as the heating mechanism offers an alternative solution to produce porous materials that can be widely used in wastewater treatment, the agricultural sector, and gas adsorption, as discussed in detail below.

4.5 Applications of biochar from microwave pyrolysis

4.5.1 Wastewater treatment

One of the most common applications of biochar is to remove organic pollutants from wastewater. With the use of microwave-assisted pyrolysis, the physiochemical properties of biochar can be altered. Thus, a specially engineered char product with a high removal efficiency of pollutants can be produced. Yek *et al* [24] produced orange peel waste-derived biochar by *in situ* microwave-assisted pyrolysis and

Table 4.1. The comparison between MH and CH in terms of the yield and physiochemical properties of biochar.

Feedstock	MH	CH	Ref.
Cylindrical wood blocks	BET surface area of 450 $m^2\ g^{-1}$	BET surface area of 184 $m^2\ g^{-1}$	[18]
Pine sawdust	Biochar yield of 13–26 wt.%	Biochar yield of 17–33 wt.%	[19]
Willow chips	Biochar yield of 27.3 wt.% BET surface area of 3.87 $m^2\ g^{-1}$ Pore volume of 2.07 $cm^3\ g^{-1}$	Biochar yield of 39.8–98.1 wt.% BET surface area of 0.17–1.14 $m^2\ g^{-1}$ Pore volume of 0.58–1.44 $cm^3\ g^{-1}$	[20]
Switchgrass	BET surface area of 76.3 $m^2\ g^{-1}$	BET surface area of 0.33 $m^2\ g^{-1}$	[21]
Straw pellets	Biochar yield of 33.7 wt.% BET surface area of 1.14 $m^2\ g^{-1}$ Pore volume of 0.37 $cm^3\ g^{-1}$	Biochar yield of 46.4–93.9 wt.% BET surface area of 0.51–0.93 $m^2\ g^{-1}$ Pore volume of 0.17–1.57 $cm^3\ g^{-1}$	[20]
Sewage sludge	Biochar yield of 10–13 wt.%	Biochar yield of 10.4 wt.%	[22]

activation by CO_2 or steam, which was also compared to the CH method, and the results indicate that a higher heating rate and reaction temperature and lower residence time were achieved from pyrolysis using microwave heating. Moreover, the biochar produced by microwave pyrolysis contained a higher content of fixed carbon and BET surface area compared to that produced by pyrolysis with conventional heating. When comparing the activation methods using CO_2 or steam, it was observed that the biochar obtained from activation by steam resulted in a considerably higher adsorption efficiency (136 mg g^{-1}) than that using biochar activated by CO_2 (91 mg g^{-1}). Therefore, a higher removal rate of Congo Red dye (93%) was achieved by steam activation-produced biochar compared to that prepared using CO_2 activation (89%). Furthermore, the underlying mechanism for the MH to facilitate the activation of biochar was discussed. As suggested by the existing literature, the basic reaction pathways occurring in the pore formation induced by the physical activation include gasification, Boudouard reaction, and the water–gas shift reaction, as shown in the following equations:

$$C + CO_2 \leftrightarrow 2CO, \quad \Delta H = +159 \text{ kJ/mol} \tag{4.3}$$

$$C + H_2O \rightarrow CO + H_2, \quad \Delta H = +117 \text{ kJ/mol} \tag{4.4}$$

$$CO + H_2O \leftrightarrow CO_2 + H_2, \quad \Delta H = -41 \text{ kJ/mol}. \tag{4.5}$$

As can be seen, an extra supply of heat is required to ensure the proceeding of the reactions, and thus MH that offers a fast heating mechanism and the capacity to lower the activation energy needed for the Boudouard reaction is more advantageous than CH using an electric furnace [25].

Shukla *et al* [26] investigated the effectiveness of using rice husk biochar obtained from microwave-assisted pyrolysis at a microwave power of 900 W, a temperature of 600 °C, and a N_2 flow rate of 0.2 l min^{-1} for wastewater treatment. It was found that the BET surface area of rice husk-derived biochar was 190 m^2 g^{-1} and mesopores were identified as the predominant type of pore in this solid product. In the adsorption kinetic study, the adsorption efficiencies for the phosphates and nitrates were 55%–65% and 65%–75%, respectively, at an adsorption equilibrium of 6 h, which implies the strong ability of microwave pyrolysis-prepared biochar to remove pollutants from wastewater. In another study, Lam *et al* [27] compared the influence of the heating mechanism of MH and CH in pyrolysis treatment to produce biochar from waste palm shell. The authors reported that the use of MH produced a higher yield (45 wt.%) of biochar with a high BET surface area (679.22 m^2 g^{-1}) compared to that when using CH for the pyrolysis. This highly microporous biochar obtained by microwave pyrolysis was then utilized to treat landfill leachate, and the treatment efficiency was compared to that obtained using commercially available activated carbon. The highest adsorption efficiency of 595 mg g^{-1} was achieved by using microwave pyrolysis-produced biochar after 24 h of processing time, along with a chemical oxygen demand (COD) reduction rate of 65%. The results were found to be similar to those obtained using commercial activated carbon (an adsorption efficiency of 663 mg g^{-1} and a COD reduction rate of 70%). Even though a high removal efficiency of organic pollutants can be achieved using microwave pyrolysis-produced biochar, the high production cost must be considered prior to its industrial-scale application [14].

It is known that biochar needs to be activated before using it as an absorbent for removing organic pollutants. The most common activation methods can be categorized into physical and chemical approaches. Recently, several studies have evaluated the influence of chemical activation by microwave irradiation on the properties of biochar and its ability to remove pollutants. A previous study conducted by Sun *et al* [28], examined the removal rate of 2,4-dichlorophenoxy acetic acid by oak or apple tree derived biochar under microwave heating, an oil bath, and room temperature. The authors reported that the removal efficiency of 2,4-dichlorophenoxy acetic acid was significantly higher using microwave irradiation, irrespective of the type of biochar used. Specifically, the removal efficiency obtained from MH at 1 min was found to be 2.3-fold higher than at room temperature. When compared to the oil bath at 90 °C for 1–30 min, a better adsorption performance of 2,4-dichlorophenoxy acetic acid was also achieved using microwave irradiation. The high removal efficiency of 2,4-dichlorophenoxy acetic acid is mainly attributed to the generation of hot spots and radicals (including hydroxyl, anionic superoxide, and singlet oxygen radicals) caused by the microwave treatment. Although the use of microwave treatment demonstrates a beneficial impact on the removal of pollutants, the gas emission could be a technical challenge to microwave irradiation, in particular when the organic pollutants are either volatile or semi-volatile.

4.5.2 Agricultural sector

Apart from use as an absorbent, microwave pyrolysis-prepared biochar can be applied for the agricultural sector. Mahari *et al* [29] carried out microwave-assisted pyrolysis experiments of palm kernel shell at 750 W for 35 min to produce porous and high surface area (270 cm^2 g^{-1}) of biochar, and a higher yield of biochar (28 wt.%) was observed than performing pyrolysis using CH (23 wt.%). The resulting biochar was then applied as the additive to grow mushrooms, and a higher moisture content of mushroom substrate (99%) was found when adding biochar compared to without the addition of biochar (96%). This increase in the moisture retention ability of biochar could be due to the high porosity and surface area of biochar. Additionally, the application of microwave pyrolysis-produced biochar as the neutralizing agent was assessed. The results showed that the pH level of the mushroom substrate was maintained at 7 at the initial growth stage and a slight decrease in the pH level from 7 to 6.8 was found at the half-colonization stage, which was possibly caused by the mycelium-induced degradation of organics. Following this, an increase in the pH value from 6.8 to 7–7.2 was observed when the final colonization stage was reached. Overall, it can be concluded that the biochar produced by microwave pyrolysis is effective in stabilizing the pH level and demonstrates a beneficial impact on the growth of mushrooms by retaining the moisture content and acting as a neutralizing agent. Su *et al* [30] produced palm kernel shell-derived biochar through microwave pyrolysis at 700 W for 30 min, followed by physical activation using steam for 30 min. The measured average pore diameter and BET surface area of the biochar obtained by microwave pyrolysis were detected to be 1.803 nm and 419 $m^2\ g^{-1}$. Subsequently, this biochar was applied as the biological carrier to grow nitrifying bacteria, in which the water quality was enhanced through the removal of 67% of ammonia and 68% of total suspended solids. In addition, in the study of plant cultivation it was found that a high growth rate of lettuce of 0.0562% d^{-1} was achieved with the use of microwave pyrolysis-produced biochar, resulting from an increase in the nitrogen uptake up to 29.7 mg l^{-1} and a pH value up to 6.8, which is accompanied by a 100% survival rate for catfish and an acceptable biological oxygen demand (BOD) level of 3.94 mg l^{-1}. According to these observances, the authors stated that the biochar obtained by microwave-assisted pyrolysis could be used as an exciting approach to reduce the concentration of ammonia and simultaneously produce nitrate for growing plants.

4.5.3 Gas adsorption

Huang *et al* [31] investigated the CO_2 adsorption ability of biochar obtained from microwave pyrolysis of rice straw and compared with that obtained from pyrolysis using conventional heating. At optimal operational conditions (a microwave power of 200 W and temperature of 300 °C) for preparing biochar, an amount of 80 mg g^{-1} of CO_2 was adsorbed at 20 °C. In comparison, the biochar obtained from pyrolysis by CH at 550 °C only absorbed 50 mg g^{-1} of CO_2. Most importantly, the authors claimed that the pyrolysis by MH meets the zero CO_2 emission without any post-processing when considering the CO_2 absorption by biochar. Nevertheless, it is worthwhile to mention that the CO_2 adsorption capacity of microwave pyrolysis-derived biochar is lower than commercial activated carbon; however, it can be significantly improved after activation and impregnation. In addition, Huang *et al* [32] performed co-torrefaction experiments on sewage sludge and leucaena wood using MH to prepare biochar for CO_2 adsorption purposes, and it was observed that the high carbon containing biochar resulted in a high CO_2 adsorption ability. The amount of absorbed CO_2 from pure leucaena wood derived biochar (53 mg g^{-1}) was found to be four times higher than that obtained from pure sewage sludge produced biochar.

4.6 Conclusions

In conclusion, the nature of microwave-assisted pyrolysis (fast heating rate, short processing time, and high heat transfer efficiency) ensures it is a promising technology to produce valuable engineered biochar from various types of biomass. To further improve the quality of biochar in terms of surface morphology and chemical composition, the subsequent activation stage by the physical (steam and CO_2) or chemical approach (e.g. acid and alkali) must be carried out. During this stage modification by microwave irradiation is also reported to help improve the adsorption efficiency of organic pollutants for biochar. After an appropriate activation step, engineered biochar obtained by microwave pyrolysis has been used for wastewater treatment, the agricultural sector, and gas adsorption.

However, no study so far has investigated the effectiveness of using microwave pyrolysis-produced biochar for energy storage and this application has been previously evaluated for biochar obtained from conventional heating-assisted pyrolysis.

References

[1] Motasemi F and Afzal M T 2013 A review on the microwave-assisted pyrolysis technique *Renew. Sustain. Energy Rev.* **28** 317–30

[2] Farag S, Kouisni L and Chaouki J 2014 Lumped approach in kinetic modeling of microwave pyrolysis of kraft lignin *Energy Fuels* **28** 1406–17

[3] Farag S and Chaouki J 2015 A modified microwave thermo-gravimetric-analyzer for kinetic purposes *Appl. Therm. Eng.* **75** 65–72

[4] Farag S 2013 Production of chemicals by microwave thermal treatment of lignin *PhD Thesis* École Polytechnique de Montréal, University of Montreal

[5] Farag S and Chaouki J 2017 Recover of inorganic chemicals of the pulp and paper making processes using microwaves and related techniques *US Patent* US20190119852A1

[6] Farag S *et al* 2012 Temperature profile prediction within selected materials heated by microwaves at 2.45 GHz *Appl. Therm. Eng.* **36** 360–69

[7] Chaouki J *et al* 2020 The development of industrial (thermal) processes in the context of sustainability: the case for microwave heating *Can. J. Chem. Eng.* **98** 832–47

[8] Doucet J *et al* 2014 Distributed microwave pyrolysis of domestic waste *Waste Biomass Valoriz.* **5** 1–10

[9] Farag S *et al* 2016 Impact of the heating mechanism on the yield and composition of bio-oil from pyrolysis of kraft lignin *Biomass Bioenergy* **95** 344–53

[10] Farag S and Chaouki J 2015 Microwave heating assisted biorefinery of biomass *Innovative Solutions in Fluid–Particle Systems and Renewable Energy Management* ed T Katia (Hershey, PA: IGI Global) pp 131–66

[11] Mai A 2019 Microwaves-assisted demetallization and desulfurization of heavy petroleum oil using a developed demetallization agent *PhD Thesis* Polytechnique Montreal, University of Montreal

[12] Motasemi F *et al* 2014 Microwave dielectric characterization of switchgrass for bioenergy and biofuel *Fuel* **124** 151–7

[13] Salema A A *et al* 2013 Dielectric properties and microwave heating of oil palm biomass and biochar *Ind. Crops Prod.* **50** 366–74

[14] Foong S Y *et al* 2020 Valorization of biomass waste to engineered activated biochar by microwave pyrolysis: progress, challenges, and future directions *Chem. Eng. J.* **389** 124401

[15] Gil M, Pasieczna-Patkowska S and Nowicki P 2019 Application of microwave heating in the preparation of functionalized activated carbons *Adsorption* **25** 327–36

[16] Ho S-H *et al* 2018 Characterization of biomass waste torrefaction under conventional and microwave heating *Bioresour. Technol.* **264** 7–16

[17] Szewczuk-Karpisz K *et al* 2020 Hay-based activated biochars obtained using two different heating methods as effective low-cost sorbents: solid surface characteristics, adsorptive properties and aggregation in the mixed Cu(II)/PAM system *Chemosphere* **250** 126312

[18] Miura M *et al* 2004 Rapid pyrolysis of wood block by microwave heating *J. Anal. Appl. Pyrolysis* **71** 187–99

[19] Wang X-H *et al* 2009 Properties of gas and char from microwave pyrolysis of pine sawdust *BioResources* **4** 2009
[20] Mašek O *et al* 2013 Microwave and slow pyrolysis biochar—comparison of physical and functional properties *J. Anal. Appl. Pyrolysis* **100** 41–8
[21] Mohamed B A *et al* 2016 Microwave-assisted catalytic pyrolysis of switchgrass for improving bio-oil and biochar properties *Bioresour. Technol.* **201** 121–32
[22] Domínguez A *et al* 2005 Investigations into the characteristics of oils produced from microwave pyrolysis of sewage sludge *Fuel Process. Technol.* **86** 1007–20
[23] Li J *et al* 2016 Biochar from microwave pyrolysis of biomass: a review *Biomass Bioenergy* **94** 228–44
[24] Yek P N Y *et al* 2020 Engineered biochar via microwave CO_2 and steam pyrolysis to treat carcinogenic Congo red dye *J. Hazard. Mater.* **395** 122636
[25] Hunt J *et al* 2013 Microwave-specific enhancement of the carbon–carbon dioxide (Boudouard) reaction *J. Phys. Chem.* C **117** 26871–80
[26] Shukla N, Sahoo D and Remya N 2019 Biochar from microwave pyrolysis of rice husk for tertiary wastewater treatment and soil nourishment *J. Clean. Prod.* **235** 1073–79
[27] Lam S S *et al* 2020 Engineering pyrolysis biochar via single-step microwave steam activation for hazardous landfill leachate treatment *J. Hazard. Mater.* **390** 121649
[28] Sun Y *et al* 2020 Tailored design of graphitic biochar for high-efficiency and chemical-free microwave-assisted removal of refractory organic contaminants *Chem. Eng. J.* **398** 125505
[29] Wan Mahari W A *et al* 2020 Applying microwave vacuum pyrolysis to design moisture retention and pH neutralizing palm kernel shell biochar for mushroom production *Bioresour. Technol.* **312** 123572
[30] Su M H *et al* 2020 Simultaneous removal of toxic ammonia and lettuce cultivation in aquaponic system using microwave pyrolysis biochar *J. Hazard. Mater.* **396** 122610
[31] Huang Y-F *et al* 2015 Microwave pyrolysis of rice straw to produce biochar as an adsorbent for CO_2 capture *Energy* **84** 75–82
[32] Huang Y-F, Chiueh P-T and Lo S-L 2019 CO_2 adsorption on biochar from co-torrefaction of sewage sludge and leucaena wood using microwave heating *Energy Procedia* **158** 4435–40

Chapter 5

Biochar characterization methods

Ondřej Mašek, Anna Bogush, Anjali Jayakumar, Christian Wurzer and Clare Peters

Biochar comprises a vast heterogeneous set of materials derived from a diverse range of feedstock. In addition, it has numerous applications in the context of carbon storage, soil amendment and remediation, building materials, environmental management, etc. For this reason, it is very important to have a good understanding of the different properties of biochar relevant for their intended application. At the start, the distinction between physical and chemical characterization, and functional characterization of biochar will be introduced, followed by discussion of a range of key characterization techniques.

There is a huge number of analytical techniques that can be used on solid carbon materials such as biochar. Some of these have been in use in other fields, such as fuels or activated carbon characterization, for a very long time, while others are more recent and specific to biochar, e.g. carbon stability. This chapter will focus on reviewing some of the most commonly used characterization techniques, highlighting their main areas of application, as well as discussing their potential challenges and limitations. In addition, selected advanced characterization methods and methods relevant to emerging biochar applications will also be introduced.

While the aim of the chapter is not to provide an exhaustive introduction and discussion of all possible ways to characterize biochar in all its aspects, it will provide readers with useful pointers to navigate this complex area.

5.1 Introduction

Biochar comprises a vast heterogeneous set of materials derived from a diverse range of feedstock. In addition, it has numerous applications in the context of carbon storage, soil amendment and remediation, engineering and building materials, environmental management, etc. For this reason, it is very important to have a good understanding of the different properties of biochar relevant for their intended applications.

doi:10.1088/978-0-7503-2660-5ch5

There is a huge number of analytical techniques that can be used for characterization of solid carbon materials such as biochar. Some of these have been adopted from use in other fields, such as fuel (coal, charcoal, biomass) or activated carbon characterization, and have been known for a very long time, while others are more recent and specific to biochar, e.g. carbon stability. This chapter focuses on reviewing some of the most commonly used characterization techniques, highlighting their main areas of application, as well as discussing their potential challenges and limitations. In addition, selected advanced characterization methods and methods relevant to emerging biochar applications are also introduced. While the aim of the chapter is not to provide an exhaustive introduction and discussion of all possible ways to characterize biochar in all its aspects, it should provide readers with useful pointers to navigate this complex area.

Before moving on to the individual characterization techniques, it is useful to discuss the general principles of sampling of solid materials and their preparation for analysis.

5.1.1 Sampling

Before any characterization or application of biochar can be conducted, sampling of the produced lot and division into representative subsamples is required. Representative sampling of biochar is particularly important as sampling errors due to the constitutional and distributional heterogeneity of biochar might be significantly larger than analytical errors [1]. As biochar feedstocks are generally heterogeneous, the resulting biochar with differences can be traced back even to different plant parts [2] or feedstock particle sizes [3]. Guidelines for correct sampling procedures have been proposed by Bucheli *et al* [4] or can be found in the EBC [5] and IBI [6] guidelines.

5.1.2 General sample preparation for analysis

In general, the biochar used in laboratory settings has to be pre-treated to obtain the sufficient sample homogeneity required for most characterization methods and applications, often using only milligrams of biochar (e.g. proximate and ultimate analysis, FTIR, NMR). However, simple treatment techniques such as grinding, sieving [7], or washing [8] might already affect biochar properties significantly. Special care is also required in adsorption experiments or soil applications, where parameters such as the particle size of the used biochar can have substantial impacts on a variety of properties, i.e. surface area [9], PAH content [4], CEC, ash content, or pH [7]. Further, the final application performance of the material has been reported to be significantly affected by simple pre-treatment methods such as sieving or grinding [7, 10]. As biochar characterization should be conducted on a similar material as used in the target application, the same preparation protocol should be followed. Additional preparation steps required for specific analysis methods (i.e. milling to a fine powder—ultimate analysis, proximate analysis) should be done on subsamples, maintaining homogeneity of the bulk sample. It has to be highlighted that in the case of biochar characterization, certain analysis requirements might influence the actual parameter of interest. Sigmund *et al* [11] observed the influence of degassing conditions on biochars' BET surface area measurements, Liao and Thomas [7] examined the influence of post-processing parameters such as sieving and grinding on plant performance and soil water retention, and Enders and Lehmann [12] analysed the influence of biochar digestion methods on total elemental analysis using ICP. Most prominently, ball milling of biochar is used as a sample preparation technique and as a stand-alone biochar modification technique to increase the specific surface area [13] and adsorption performance [14]. Therefore, special attention has to be taken to carefully plan the eventual washing, sieving, or grinding steps upfront (including eventual sample mass losses) for both the characterization and application of biochar to obtain reliable results.

5.2 Biochar compositional analysis

5.2.1 Elemental (CHNSO) analysis

Ultimate (or elemental analysis) refers to characterization in terms of mass concentrations of carbon, hydrogen, nitrogen, sulfur, and oxygen. For typical analysis using an elemental analyser, samples are weighed into tin capsules and combusted for analysis of CHNS or pyrolysed for analysis of O. For CHNS, the weighed sample in the tin capsule is placed in the auto-sampler drum, where it is dearated to remove any atmospheric nitrogen. The sample is introduced into a vertical quartz tube heated at 950 °C with a constant flow of helium (carrier gas). The helium stream is enriched with high purity oxygen to achieve a strong oxidizing environment which guarantees complete combustion/oxidation of C, H, and S. To achieve complete oxidation the combustion gas mixture is driven through an oxidation catalyst (Cr_2O_3) zone, followed by copper, which reduces nitrogen oxides. The resultant gas mixture is swept into a chromatographic column, from which any eluted pure combustion gas passes through a detector which generates an electrical output proportional to the amount of eluted gas.

Potential issues with the ultimate analysis of biochar:

(a) The potential high proportions of very stable carbon in biochar can be challenging to fully combust to give accurate, reproducible results. It is therefore recommended to use small quantities (1–2 mg of biochar), to add vanadium pentoxide within the tin capsules with biochar, and to use a longer injection time of oxygen, as this will aid combustion.

(b) Moisture uptake—even with overnight drying of biochar at 105 °C, placing biochar into a desiccator within the oven, then taking the biochar samples out of the desiccator one at a time to weigh into tin capsules, the rapid moisture uptake and resulting change in mass can produce significant errors in the overall results and vary between different biochars.

(c) Many researchers do not measure oxygen, they calculate it by difference. It is important to ensure that the calculation is carried out on an ash-free (af) basis, i.e. O% = 100 – C% – N% – H% – S% – Ash%.

For total organic carbon measurements, the samples are acidified prior to analysis to remove inorganic carbon.

5.2.2 ICP-OES/MS

Inductively coupled plasma-optical emission spectroscopy (ICP-OES) and inductively coupled plasma mass spectrometry (ICP-MS) are multi-elemental analysis techniques. Both these techniques use an ICP source to dissociate the sample into atoms or ions and excite them. Analysed elements can be identified by their characteristic emission lines by ICP-OES and measuring a mass of atoms by ICP-MS. ICP-OES is mainly used for major and minor elemental analysis due to its moderate detection limits (0.2–100 ppb). On the other hand, ICP-MS has low detection limits (0.0005–1.0 ppb) and is used for trace element analysis.

The elemental composition of biochar highly depends on the raw material and the pyrolysis condition (e.g. temperature) [15, 16]. ICP-OES/MS analysis of biochar requires sample preparation such as incineration/combustion with subsequent total acid digestion of biochar ash. The series of standard methods (e.g. DIN 22022–1&2, DIN 22022–7, DIN 51729, DIN EN ISO 11885, DIN EN ISO 17294–2, DIN EN 1483, USEPA3050B) are applied for determination of major (e.g. P, Mg, Ca, K, Na, Fe, Si, S) and trace (e.g. Pb, Cd, Cu, Ni, Hg, Zn, Cr, B, Mn) element concentrations in biochar [5, 17]. Also, [15] indicated that quality assurance and quality control measurement of biochar should include the generation of a set of representative biochar reference materials such as standard biochar materials [18] and biochar reference methods.

5.2.3 X-ray fluorescence

X-ray fluorescence (XRF) is a nondestructive analytical technique for the qualitative and quantitative analysis of the elemental composition of different materials including biochar. The sample is excited by a primary x-ray source and the characteristic fluorescent (secondary) x-ray emission, which is unique for each present element, is measured. This analytical technique has a detection limit in parts per million (ppm) and can be considered as complementary analysis to ICP-OES/MS. This method can be used for the qualitative and quantitative analysis of the elemental composition of biochar and biochar ash [17, 19]. Also, the XRF method is very robust, fast, and easy to use. It needs very simple sample preparation without complicated and time consuming digestion procedures. The XRF data are reported as weight percents of the common elemental oxide. Synchrotron XRF elemental mapping is a very powerful method and can provide information about the occurrence and distribution of nutrients (e.g. P, K, Mg, etc) and potential pollutants (e.g. Zn, Pb, etc) in biochar [20].

5.2.4 XAS (XANES and EXAFS)

X-ray absorption spectroscopy (XAS) is a powerful technique that provides local geometric and/or electronic structural information of analysed samples through the analysis of their x-ray absorption spectra of selected atoms [21, 22]. The XAS analysis is performed at synchrotron radiation facilities that have intense and tunable x-ray beams. Extended x-ray absorption fine structure (EXAFS) and x-ray absorption near edge structure (XANES) are two types of XAS. The XANES spectra may contain a fine structure that shows the electronic and geometrical environment of the absorbing atom and, therefore, can provide information on the element oxidation states and on-site symmetry. With EXAFS, the local atomic environment to the element, such as the interatomic distances and coordination numbers of shells surrounding the absorbing atom, can be obtained. Also, both these techniques can give important information about element speciation for crystalline and amorphous materials, even if the investigated element is present at very low concentrations. The XAS data analysis is usually performed using Athena and Artemis software, for example, using principal component analysis (PCA) and linear combination fitting (LCF) for species identification and quantification [23]. These techniques become very important in biochar-related research and have already been used for the investigation of element speciation (e.g. nutrients, potential pollutants, etc) in biochar and understanding the processes of removing pollutants from soil and water using biochar [20, 24–27].

5.2.5 XPS

X-ray photoelectron spectroscopy (XPS) is a quantitative spectroscopic technique that analyses the surface chemistry of a sample [28]. A sample surface with a depth of 1–10 nm is irradiated/excited with a beam of x-rays causing photoelectron emission. Then, XPS spectra are obtained/recorded by counting emitted photoelectrons over a range of electron kinetic energies. This technique can quantitatively measure the intensity of a photoelectron peak, the elemental composition and identity, and elemental chemical states of elements present on the sample surface. XPS can detect almost all elements excluding hydrogen and helium. The detection limit is in the parts per thousand (ppt) range, but it is possible to analyse the element concentration in ppm if longer collection times are applied. This method is very important for biochar-related research, for example, for determination of the bonding energies of carbon on the biochar surface and the anion exchange capacity of biochar, investigation of the surface chemistry of fresh and aged biochars, understanding the transformation of surface functional groups resulting from application of different pyrolysis conditions, etc [29–36].

5.3 Structural characterization of biochar

The structural characteristics of biochar refer to the properties of the physical structure of biochar, such as porosity, surface area, pore size distribution, crystalline or amorphous structures, etc. Biochar macrostructure is largely derived from the macrostructure of the starting material, while the microstructure and various structural properties are strongly affected by the processing conditions used in biochar production and any subsequent post-processing [37]. This section highlights some of the key structural characterization techniques and their applications for biochar characterization.

5.3.1 X-ray μ-tomography

X-ray microtomography or x-ray micro-computed tomography (micro-CT) is a 3D nondestructive microscopy method. This technique uses x-rays to create cross-sections of an investigated sample and provides very fine internal 3D structure imaging data that help to better understand a material's structure, properties, and performance. It is an easy-to-use method and does not need sample preparation. The pixel size of micro-CT is in the micrometer range. The reconstruction of the slice images for 3D visualization, analysis, and advanced materials characterization are generally performed using appropriate software such as AVIZO (https://www.thermofisher.com) and/or DATOS (Wunstorf, Germany). Also, the free Image J software (National Institute of Health, Bethesda, MD, USA, http://rsb.info.nih.gov/ij/) is useful for image cropping, normalization, edge enhancement, and denoising [38]. Micro-CT can also be used to characterize biochar as a function of pyrolysis temperature, helping to understand the evolution of biochar properties (e.g. pore structure, morphological features, the presence of metals, etc) during its production [39]. The micro-CT technique is also useful in understanding the effects of biochar application on soil structure and water retention [38]. Several researchers have investigated the porosity, pore size distribution, and pore connectivity using micro-CT [40–44].

5.3.2 Electron microscopy (SEM/EDX)

Scanning electron microscopy (SEM) is an imaging technique that allows observations of biochar surface morphology. In SEM an electron beam is accelerated through high voltage (1–30 keV) and focused on the specimen under a low vacuum, which is then scanned across the sample area. The emitted signals are used to produce the final image, allowing imaging at lateral resolutions of 1–10 nm and up to 500,000 magnifications [45]. For SEM analysis of biochar, additional sample preparation is often required using techniques such as gold sputtering to increase the biochars' conductance [46]. SEM images of biochar allow a visual assessment of the macroscopic pore structures and surface morphological features, such as attached mineral compounds, sintered carbon structures, and a general assessment of the shape of biochar particles [47, 48]. While SEM images can assist in the assessment of biochar morphology, the obtained information is limited to qualitative assessment of individual particles and is often only used as a confirmation method for additional in-depth characterization of biochar [49]. A more comprehensive utilization of SEM can be achieved in combination with energy dispersive x-ray spectroscopy (EDX). EDX allows a quantitative measurement of the surface elemental composition of specific areas selected in the SEM image. Mapping of the surface composition data can yield important information for a variety of applications such as the ageing processes of biochar in soils [50] or localizing adsorbed metal species after adsorption [51]. Emerging applications include the characterization of mineral biochar composites, e.g. magnetic biochars, to assess the homogeneity of the distribution of mineral particles on the carbon structure [52, 53].

5.3.3 Surface area

The surface area of biochar is an important material characteristic as it provides a measure of potential interaction sites with the surrounding matrix. Biochars' generally high surface area stems from pores spanning over different size ranges from micro- (<2 nm), meso- (2–50 nm), and macropores (>50 nm), up to pore diameters of several tens of micrometers. Although a wide variety of experimental methods exist, no single technique can measure the whole range of pore sizes [54]. As micro- and mesopores generally contribute the largest share to biochars' surface area, nanometric physiosorption techniques using N_2, CO_2, Ar, Kr, or H_2O as adsorbates are the most common methods due to their coverage of the important micro- to macropore range [55]. Different adsorbates present advantages and disadvantages in terms of the applicable pore size range and the reliability of the assessment. The most common method is the use of N_2 at its boiling point of −196.15 °C. Here, the sample has to be degassed in a vacuum at elevated temperatures (105 °C–300 °C) before repetitive dosing of N_2 is used to determine the adsorption and desorption isotherm on which the calculation of characterization parameters, i.e. surface area, pore volume, pore size distribution—is based. For biochars, the measurement using N_2 poses several experimental difficulties. Degassing at elevated temperatures is necessary to remove volatiles trapped in biochar pore structure, and while identification of the correct degassing temperature is crucial, it is not yet standardized [56]. The adsorption of N_2 at −196.15 °C also poses several difficulties as the amorphous pore structure of biochar can lead to prolonged adsorption times of up to several days, making accurate measurements time consuming and costly [57]. Once the experimental isotherm is established, there are a number of ways to calculate the surface area of biochars. The most commonly used method is the one based on the theory of Brunauer–Emmet–Teller (BET). However, despite its common use, the applicability of BET for porous materials such as biochar is questionable and often challenged [58]. In particular for high temperature biochars displaying relevant microporosity, additional requirements have to be fulfilled to allow an assessment of the surface area which are often ignored [59]. Significant advancements in physiosorption analysis led to the development of alternative computational models based on the density functional theory (DFT). These novel approaches can offer a more reliable surface area assessment and pore specificity, enabling a differentiation of surface area and pore volumes according to pore sizes from one model. Advanced models developed for carbon materials with heterogeneous surfaces such as 2D NLDFT [60] or QSDFT [61] can therefore provide significantly more information from the same experimental isotherm than the BET theory. As these models are already included in modern instrument software, their use does not present an increased workload and will supersede the current BET method in the near future.

5.3.4 Raman spectroscopy

Raman spectroscopy is a type of vibrational spectroscopy, which draws information on the vibrational states of a molecule based on an inelastic scattering process, which brings about a change in the polarizability of a molecule [62]. Here, when a sample is irradiated with monochromatic light (lasers), scattered beams with an energy less than one vibrational unit of energy (unique to each material) are detected [62]. Raman spectroscopy, with high sensitivity, high spatial resolution, and ease of sample preparation is largely used for qualitatively (structure analysis, identification of different organic, inorganic phases) and quantitatively characterizing carbon allotropes [63, 64]. Biochar samples show two major bands in their Raman spectra, commonly called the G band (1580–1600 cm^{-1} from vibrations of in-plane graphitic sp^2 carbon) and D band (1350–1370 cm^{-1} from vibrations of in-plane sp^2-bonded carbon with defects in their structure), and in some cases, additional peaks are detected in highly disordered biochar [63, 65, 66]. The intensities, peak-widths and deconvolution of these Raman bands give useful information on biochar graphitization degrees (ordered and disordered carbon rings), crystallinity, nano-crystallite sizes, structural defects, evolution of the biochar structure with pyrolysis conditions, and determining long-term structural changes in biochar [67–69]. The requirement of high-power lasers, sample degradation, and issues of fluorescence make the instrument expensive and less widely used. However, advanced techniques such as surface enhanced Raman spectroscopy (SERS), stimulated Raman spectroscopy (SRS), tip-enhanced Raman spectroscopy (TERS), etc are now available to address many of these limitations, made possible by constantly evolving technology related to Raman instrumentation, sample preparation methodologies and software, making it even more powerful as a characterization tool [62].

5.3.5 X-ray diffraction (XRD)

Diffraction of x-rays by the unique atomic and molecular planes of a material (x-rays penetrate inside a material because of their short wavelengths) can give useful information on its crystalline or amorphous (absence of a crystal structure) nature [66, 70]. Biochar samples have both amorphous (carbon rings) and crystalline phases (minerals commonly present in biochar such as quartz, calcite, etc) [68]. The distance between graphitic planes of crystallites and the in-plane graphitic structures are represented by the indices (0 0 2) for $2\theta = 22.5°$ and (1 0 0) for $2\theta = 43°$, respectively [66, 70]. Width and intensity of these peaks are indicative of crystallite sizes and degree of crystallization; for example, peaks at (1 0 0) indicate the size of the aromatic layers. X-ray diffraction is particularly useful when characterizing biochar-based composites; this allows the determination of various crystal structures and gives information on how the amorphous and crystalline phases evolve with changing experimental conditions such as temperature, feedstock, dopants, etc. XRD, however, is more challenging for characterizing the structure of amorphous materials and also materials with a small crystalline fraction in a largely amorphous matrix. The current detection limits are close to 0.2%– 5% crystallinity by mass [71]. New x-ray sources, innovative optical devices to better collimate and focus the beams, low-noise detectors which are radiation-resistant, position-sensitive and fast, are some of the new advancements to enhance the accuracy, sensitivity, and efficiency of XRD equipment [72].

5.4 Biochar stability

Biochar stability determination is important to assess biochar's potential for long-term carbon sequestration, in particular within soil. Depending on the feedstock and processing conditions, biochar of different stability can be produced. Incubation studies assessing the nature of how biochar prevails in soil are time consuming, requiring at least months if not years. Therefore, a range of proxies has been proposed, based on easy to measure biochar characteristics, such as elemental ratios (section 5.4.1) from ultimate analysis or fixed carbon from proximate analysis (section 5.4.2); oxidation recalcitrance (sections 5.4.2 and 5.4.3). The range of potential techniques is much higher and well beyond the scope of this chapter. Further information can be found elsewhere [73].

5.4.1 Elemental ratios (O/C and H/C)

The original van Krevelen diagram [74] compared elemental ratios of H/C with O/C for the study of coal in terms of structure and reaction processes. More recently, it has been applied to biochar as a means of assessing the evolution of biochar composition during pyrolysis where H/C is indicative of aromaticity (content of carbon with aromatic ring structure) and consequently resistance to microbial and chemical degradation, and O/C indicating extent of deoxygenation. A good correlation of elemental ratios and stability observed by other methods has been reported [75, 76]. One issue with this technique is the potential high content of inorganic C within biochar with high ash contents. Use of C_{org} rather than total C can avoid the issue [6], however C_{org} is not routinely measured and the issues surrounding the measurement of total C (see section 5.2.1) also apply to measurement of C_{org}.

5.4.2 TGA-based methods (proximate analysis and R50 index)

Fixed carbon has been proposed as a measure of biochar stability. The content of fixed carbon has traditionally been determined as part of proximate analysis of coal and char using a muffle furnace, but nowadays it is more commonly determined using a thermogravimetric analyser (TGA). Proximate analysis characterizes solid samples in terms of moisture, volatile matter, fixed carbon, and ash. A number of different methods exist, e.g. ASTM D1762-84, D3173-11, D3174-12, and D7582-15. Although the different methods differ in temperature steps used, in general the methods involves weighing of small samples of finely ground material (13–20mg for biomass and 7–13mg for biochar and activated biochar) into crucibles. Then, in the first stage, the samples are heated under N_2 atmosphere up to 104 °C–110 °C, and held there for a period of time, to drive off any moisture. The samples are then further heated to between 550 °C and 950 °C and held there under N_2, to drive off volatiles. Subsequently, the remaining sample is combusted in air at temperatures between 500 °C and 900 °C to remove the fixed carbon, leaving only ash. The TGA evaluation program allows masses attributed to moisture, volatiles, fixed C, and ash to be determined. Fixed carbon should then be calculated as a mass percentage on a dry basis (db) according to

$$FC(\%, db) = (mass_{FC}/(mass_{FC} + mass_{Volatiles} + mass_{Ash})) \times 100\%. \tag{5.1}$$

Thus, on a dry basis, the sum of percentages of FC, volatiles and ash equals 100%. Some researchers present FC on a dry, ash-free basis, in which case the sum of percentages of FC and volatiles equals 100%.

The recalcitrance index, R_{50}, is a relative measure of the oxidation recalcitrance of biochar compared to graphite determined using a TGA to assess the stability of carbon in biochar [77]. According to the method proposed, small amounts of biochar (7–15mg) are weighed into 70 μL small crucibles and heated under air at 10 °C min^{-1} to a temperature of 1000 °C, held there for 5 min, then cooled back to room temperature. After correction for moisture and ash contents the temperature (T_{50}) corresponding to 50% mass loss due to oxidation/volatilization is calculated. For biochar x, the recalcitrance index is calculated using

$$R_{50,\,x} = T_{50,\,x}/T_{50,\,graphite}. \tag{5.2}$$

The T_{50} is determined for graphite using the same procedure as for the biochar, with a typical value of approximately 880 °C.

5.4.3 Edinburgh stability tool

The Edinburgh stability tool [78] is an accelerated aging method used as a proxy for the environmental ageing of biochar of approximately 100 years and can be performed in under 1 week. In this method dried and milled biochar equivalent to 0.1 g C (calculated based on known C content) is weighed into a test-tube before 7 ml of 5% hydrogen peroxide (H_2O_2) is added in two steps to the test-tube at room temperature. Some biochar will be more reactive to the H_2O_2 than other biochar. After ensuring no bubbling out of the test-tube at room temperature, the test-tube is placed in an oven or heating block at 80 °C for 48 h. The hydrophobic nature of some biochar can be an issue with not all biochar remaining in suspension thus the test-tubes are regularly agitated to ensure all biochar has remained in the H_2O_2 long enough to fully react. Any remaining H_2O_2 is then evaporated by drying overnight at 105 °C. Following cooling in a desiccator, the residual mass is determined. The 'aged' biochar is subsequently analysed for total C content. The biochar carbon stability (Æ) is expressed as the proportion of initial carbon remaining after the oxidative treatment,

$$Æ(\%) = ((Br \times BrC)/(Bt \times BtC)) \times 100,$$

where Br is the residual mass of biochar following oxidation and BrC is its %C content, Bt the initial mass of biochar prior to treatment and BtC the %C content of the biochar.

This method allows comparison of stability of different biochar in a simple way that can be carried out in most laboratories without need for specialized equipment. Its predictions are comparable to those obtained by other methods [75, 79].

5.4.4 Nuclear magnetic resonance (NMR) spectroscopy

A strong relationship between biochar C stability and biochar aromaticity (the proportion of aromatic carbon to nonaromatic carbon) and biochar degree of aromatic condensation were found in many studies. Therefore, solid state NMR offers a potential way to assess the stability of biochar. There are two main ways NMR has been used for assessing biochar stability: (1) measurement of aromaticity using dipolar dephasing and (2) determination of degree of aromatic condensation using ^{13}C-labeled probe molecules, such as benzene, sorbed to biochar. Biochar stability predictions obtained by NMR have shown linear correlation with other techniques, such as the R_{50} and elemental ratios discussed above in this section.

5.5 Other key biochar characteristics

This section describes several other important characterization methods that do not fit into any of the above categories but that provide important information on biochar properties and allow its matching to target applications.

5.5.1 Electrical and electrochemical properties

The electrochemical properties of biochar arise from the presence of functional groups such as quinone/hydroquinone redox pair, several inorganic mineral phases such as Fe, Mn (found in several oxidation sates) and persistent free radicals formed during the high temperature pyrolysis [80–82]. The electrical conductivity of biochar is due to π-electron delocalization over many aromatic rings, the graphite-like sheets and electron shuttling via several redox functional groups. Electroanalytical techniques study the chemical response of a system from an electrical stimulation with techniques such as cyclic voltammetry, linear sweep voltammetry, chronoamperometry, etc, in a three/two/four-electrode set-up for quantitative and qualitative characterization of the electrochemical properties of biochar, while two/four-probe techniques are commonly used in measuring electrical conductivity [80, 82–84]. These techniques are easy to use, fast, and give information (directly or extractable) on electron donating capacity (EDC), electron accepting capacity (EAC), electrical conductivities, energy-storage capacities (from faradaic and non-faradaic processes), kinetics and mechanisms of electron transfer involving biochar. The major limitations of electroanalytical tools remain the fact that the information from these experiments are very sensitive to the experimental conditions and the electrochemical cell set-up such as temperature, pH, mediators used, electrode preparation, concentration and types of electrolytes/mediators, low solubility of biochar (many redox-active sites are deep inside the biochar structure, not accessible to electrolytes due to several mass-transfer and diffusion limitations), etc. These issues can largely be addressed by using techniques such as mediated electrochemical analysis, where chemical mediators of known concentration interact with biochar in redox reactions, helping in more precisely quantifying the biochar redox properties [85]. Electron paramagnetic resonance spectroscopy (EPFR) for the study of persistent free radicals (PFRs) in biochar, which can potentially generate reactive oxygen species

(ROS) considered crucial for biodegradation, has been gaining attention in recent times [86–88]. There are also upcoming techniques such as spectro-electrochemistry which combine *in situ* spectroscopic techniques such as FTIR, Raman spectroscopy, etc, with electroanalytics, to better monitor and understand molecular level changes that take place during these complex redox reactions mediated by biochar [89]. These properties and analytical techniques are crucial for understanding and using biochar in applications related to energy and environment, particularly in dealing with the biogeochemical cycles associated with water, soil, sediments, etc [85, 90–92].

5.5.2 pH

The pH of biochar is one of its most often characterized features, due to its importance for a number of biochar applications, whether agricultural, environmental, or engineering. Although measurement of pH is a relatively simple laboratory technique, using pH electrodes, characterization of biochar pH presents specific challenges. First, due to its solid nature, biochar needs to be suspended in deionized water before its pH can be measured. The most commonly used biochar to water ratios (weight:volume) are 1:5, 1:10, and 1:20. Due to the high water absorption ability of some biochar, it is recommended to use the two higher ratios, i.e. 1:10 or 1:20, as at 1:5 ratio biochar can form a slurry with little solution for immersing the pH electrode. The International Biochar Initiative guidelines recommend 1:20 (w:v) ratio [6].

While there are variations in the biochar pH measurement methods, they all have in common the need to prepare suspension of finely ground biochar in water (or solution of $CaCl_2$), followed by a period of shaking (between 1 h and 24 h), period of approx. 30 min allowing the suspension to stand, followed by pH measurement.

In addition to the common measurement of bulk biochar pH in suspension, it is also now possible to measure the pH of biochar *in situ*, e.g. in soil, using planar optode technology. This approach utilizes special pH sensitive foils that are in direct contact with the material to be characterized. The pH is then read using a special camera and associated software. This approach enables measurement of pH with high spatial (several tens of micrometers) and temporal resolution, not only of biochar alone, but more importantly biochar in the surrounding environment and its effect on pH change in its surroundings [93].

5.5.3 Surface functional groups (FTIR)

Pyrogenic nature of biochar makes it a storehouse of aliphatic and aromatic structures with several oxygen-containing functional groups such as ketones, quinone, carboxylic groups, etc [94]. Fourier transform infrared spectroscopy (FTIR), which analyses the vibrational excitation of these bonds induced by IR irradiation, is increasingly being used to study the surface bonding of biochar. In FTIR, a part of the incident radiation, which matches with the unique vibrational energy state of a molecule, is absorbed by the molecule, the absence of which is detected in the radiation escaping the sample after interaction. Unlike techniques such as x-ray photoelectron spectroscopy, nuclear magnetic resonance, etc, FTIR, despite its lower specificity, is commonly used nowadays for preliminary biochar characterization for information on both its organic and inorganic phases, because of the ease involved in the sample preparation, simple instrumentation, quick results, and relatively low-cost [95]. Diffuse reflectance (DR), attenuated total reflectance (ATR), and photoacoustic (PAS) detectors are commonly used in combination with FTIR to enable testing a wide range of liquid and solid samples of varying thickness and refractory indices (including dark, opaque biochar samples), often helping in reducing the scattering effects and diffraction processes [81, 94, 95]. FTIR is particularly useful in detecting changes in bonding and molecular structures in biochar samples by measuring the blue/red shift of peaks recorded in FTIR spectra and this can to a certain extent explain the dependence of surface chemistry to pyrolysis conditions and various chemical/physical treatment steps [70, 96, 97]. The commonly recorded peaks in biochar-based samples are around 4000–2700 cm^{-1} (OH (including adsorbed moisture), CH groups), 1800–1000 cm^{-1} (aliphatic, aromatic, carboxylates, carbonates, ash and clay constituents), and peaks below 1000 cm^{-1} (C–H groups, other organic and inorganic phases) [97, 98]. Titration methods such as Boehm titration are also used in studying acidic groups in biochar, even though these techniques are more chemical-intensive, involve complex sample preparation, and are less accurate because of a lack of standardized titration methods for heterogeneous samples such as biochar [99].

5.5.4 Magnetic properties

Interest is increasing in magnetic biochar due to its use as an adsorbent which can subsequently be removed from cleaned wastewater by its attraction to magnets. The simplest method to quantify magnetization is to measure the magnetic susceptibility, which is defined as the ease with which a material can be magnetized. A sample is placed in a small magnetic field, typically less than 1 mT and the induced magnetization determined. The volume susceptibility is defined as $\kappa = M/H$ where M is the magnetization induced by an applied field, H, with SI values of κ being dimensionless. To allow direct comparison between samples the mass specific magnetic susceptibility should be calculated and is defined as $\chi = \kappa/\rho$, where ρ is the density, giving χ units of $m^3\ kg^{-1}$. All materials, to some extent, display and contribute to magnetic susceptibility values: ferrimagnetic minerals have the highest magnetic susceptibility, e.g. $600 \times 10^{-6}\ m^3\ kg^{-1}$ for magnetite; antiferromagnetic minerals are lower, e.g. $0.6 \times 10^{-6}\ m^3\ kg^{-1}$ for haematite and $0.7 \times 10^{-6}\ m^3\ kg^{-1}$ for goethite; paramagnetic minerals span a range of magnetic susceptibilities, e.g. approx. $0.05 \times 10^{-6}\ m^3\ kg^{-1}$ for montmorillonite (clay) and $0.9 \times 10^{-6}\ m^3\ kg^{-1}$ for nontronite (Fe-rich clay); diamagnetic minerals display smaller, negative susceptibilities, e.g. quartz ($-0.006 \times 10^{-6}\ m^3\ kg^{-1}$) and water ($-0.009 \times 10^{-6}\ m^3\ kg^{-1}$) [100].

5.6 Conclusions

This chapter provided a brief overview of many, but by no means all, options for characterization of biochar's physical and chemical properties as well as carbon sequestration potential. Most of the techniques are ubiquitous, widely used in other fields of research, and with well-established analytical protocols. Despite this, when these techniques are applied to biochar special precautions and sometimes adjustments to the procedures are needed to accommodate the specifics of biochar, such as variable hydrophobicity, hydroscopic nature, large sorption capacity, low density, electrostatic nature, etc. Several examples of related issues were provided in relevant sections of this chapter. In addition, there are some aspects of biochar characterization that are unique, such as assessment of long-term stability. These techniques, some of them adopted from other fields and repurposed for assessment of stability, while others newly developed for this purpose, are still in the development and validation phase and no clear consensus has yet been reached as to which one or set of techniques yields results most representative of real stability of biochar in the environment. While some quantification of biochar stability has been attempted, most techniques focus on relative stability assessment, and then compare different biochar against each other. Here further work is needed on calibration that would allow reliable estimation of biochar stability in different applications.

References

[1] Singh B, Arbestain M C and Lehmann J 2017 *Biochar – A Guide to Analytical Methods* (Boca Raton, FL: CRC Press)

[2] Popescu C-M, Hill C a S, Anthony R, Ormondroyd G and Curling S 2015 Equilibrium and dynamic vapour water sorption properties of biochar derived from apple wood *Polym. Degrad. Stab.* **111** 263–8

[3] Hong Z *et al* 2020 Effects of temperature and particle size on the compositions, energy conversions and structural characteristics of pyrolysis products from different crop residues *Energy* **190** 116413

[4] Bucheli T D *et al* 2014 On the heterogeneity of biochar and consequences for its representative sampling *J. Anal. Appl. Pyrolysis* **107** 25–30

[5] European Biochar Foundation (EBC) 2016 Guidelines for a Sustainable Production of Biochar

[6] International Biochar Initiative (IBI) 2015 Standardized product definition and product testing guidelines for biochar that is used in soil *Int. Biochar. Initia.* 23

[7] Liao W and Thomas S 2019 Biochar particle size and post-pyrolysis mechanical processing affect soil pH, water retention capacity, and plant performance *Soil Systems* **3** 14

[8] Thomas Klasson K, Uchimiya M and Lima I M 2014 Uncovering surface area and micropores in almond shell biochars by rainwater wash *Chemosphere* **111** 129–34

[9] Fazeli Sangani M, Abrishamkesh S and Owens G 2020 Physicochemical characteristics of biochars can be beneficially manipulated using post-pyrolyzed particle size modification *Bioresour. Technol.* **306** 123157

[10] Kang S, Jung J, Choe J K, Ok Y S and Choi Y 2018 Effect of biochar particle size on hydrophobic organic compound sorption kinetics: applicability of using representative size *Sci. Total Environ.* **619–20** 410–18

[11] Sigmund G, Hüffer T, Hofmann T and Kah M 2017 Biochar total surface area and total pore volume determined by N_2 and CO_2 physisorption are strongly influenced by degassing temperature *Sci. Total Environ.* **580** 770–75

[12] Enders A and Lehmann J 2012 Comparison of wet-digestion and dry-ashing methods for total elemental analysis of biochar *Commun. Soil Sci. Plant Anal.* **43** 1042–52

[13] Peterson S C, Jackson M A, Kim S and Palmquist D E 2012 Increasing biochar surface area: optimization of ball milling parameters *Powder Technol.* **228** 115–20

[14] Naghdi M *et al* 2017 A green method for production of nanobiochar by ball milling—optimization and characterization *J. Clean. Prod.* **164** 1394–405

[15] Bachmann H J *et al* 2016 Toward the standardization of biochar analysis: the COST action TD1107 interlaboratory comparison *J. Agric. Food Chem.* **64** 513–27

[16] Bucheli T D *et al* 2014 On the heterogeneity of biochar and consequences for its representative sampling *J. Anal. Appl. Pyrolysis* **107** 25–30

[17] Zhang J, Lü F, Zhang H, Shao L, Chen D and He P 2015 Multiscale visualization of the structural and characteristic changes of sewage sludge biochar oriented towards potential agronomic and environmental implication *Sci. Rep.* **5** 9406

[18] Mašek O, Buss W and Sohi S 2018 Standard biochar materials *Environ. Sci. Technol.* **52** 9543–44

[19] Waqas M, Aburiazaiza A S, Miandad R, Rehan M, Barakat M A and Nizami A S 2018 Development of biochar as fuel and catalyst in energy recovery technologies *J. Clean. Prod.* **188** 477–88

[20] Buss W, Bogush A, Ignatyev K and Mašek O 2020 Unlocking the Fertilizer Potential of Waste-Derived Biochar *ACS Sustain. Chem. Eng.* **8** 12295–303

[21] Koningsberger D C and Prins R (ed) 1987 *X-Ray Absorption: Principles, Applications, Techniques of EXAFS, SEXAFS and XANES* (New York: Wiley)

[22] Newville M 2014 Fundamentals of XAFS *Rev. Mineral Geochem.* **78** 33–74

[23] Ravel B and Newville M 2005 ATHENA, ARTEMIS, HEPHAESTUS : data analysis for X-ray absorption spectroscopy using IFEFFIT *J. Synchrotron Radiat.* **12** 537–941

[24] Wagner A, Kaupenjohann M, Hu Y, Kruse J and Leinweber P 2015 Biochar-induced formation of Zn-P-phases in former sewage field soils studied by P K-edge XANES spectroscopy *J. Plant Nutr. Soil Sci.* **178** 582–85

[25] Qian T, Wang Y, Fan T, Fang G and Zhou D 2016 A new insight into the immobilization mechanism of Zn on biochar: the role of anions dissolved from ash *Sci. Rep.* **6** 33630

[26] Rechberger M V *et al* 2019 Enhanced Cu and Cd sorption after soil aging of woodchip-derived biochar: what were the driving factors? *Chemosphere* **216** 463–71

[27] Liu P, Ptacek C J, Blowes D W, Finfrock Y Z and Liu Y 2020 Characterization of chromium species and distribution during Cr(VI) removal by biochar using confocal micro-x-ray fluorescence redox mapping and x-ray absorption spectroscopy *Environ. Int.* **134** 105216

[28] Briggs D and Seah M P 1990 *Practical Surface Analysis, Auger and X-ray Photoelectron Spectroscopy* 2nd ed. (New York: Wiley)

[29] Cheng C-H, Lehmann J, Thies J E, Burton S D and Engelhard M H 2006 Oxidation of black carbon by biotic and abiotic processes *Org. Geochem.* **37** 1477–88

[30] Nguyen B T, Lehmann J, Kinyangi J, Smernik R, Riha S J and Engelhard M H 2008 Long-term black carbon dynamics in cultivated soil *Biogeochemistry* **89** 295–308

[31] Lin Y, Munroe P, Joseph S, Kimber S and Van Zwieten L 2012 Nanoscale organo-mineral reactions of biochars in ferrosol: an investigation using microscopy *Plant Soil* **357** 369–80

[32] Singh B, Fang Y, Cowie B C C and Thomsen L 2014 NEXAFS and XPS characterisation of carbon functional groups of fresh and aged biochars *Org. Geochem.* **77** 1–10

[33] Lawrinenko M and Laird D A 2015 Anion exchange capacity of biochar *Green Chem.* **17** 4628–36
[34] Lawrinenko M, Laird D A, Johnson R L and Jing D 2016 Accelerated aging of biochars: Impact on anion exchange capacity *Carbon NY* **103** 217–27
[35] Yuan H-Y, Ding L-J, Zama E F, Liu P-P, Hozzein W N and Zhu Y-G 2018 Biochar modulates methanogenesis through electron syntrophy of microorganisms with ethanol as a substrate *Environ. Sci. Technol.* **52** 12198–207
[36] Hu Q, Zhu Y, Hu B, Lu S and Sheng G 2018 Mechanistic insights into sequestration of U (VI) toward magnetic biochar: batch, XPS and EXAFS techniques *J. Environ. Sci.* **70** 217–25
[37] Wang L *et al* 2020 New trends in biochar pyrolysis and modification strategies: feedstock, pyrolysis conditions, sustainability concerns and implications for soil amendment *Soil Use Manag* sum.12592
[38] Yu X and Lu S 2020 Double effects of biochar in affecting the macropore system of paddy soils identified by high-resolution x-ray tomography *Sci. Total Environ.* **720** 137690
[39] Jones K, Ramakrishnan G, Uchimiya M and Orlov A 2015 New applications of x-ray tomography in pyrolysis of biomass: biochar imaging *Energy Fuels* **29** 1628–34
[40] Conte P and Nestle N 2015 Water dynamics in different biochar fractions *Magn. Reson. Chem.* **53** 726–34
[41] Jeffery S *et al* 2015 Biochar application does not improve the soil hydrological function of a sandy soil *Geoderma* **251–252** 47–54
[42] Schnee L S, Knauth S, Hapca S, Otten W and Eickhorst T 2016 Analysis of physical pore space characteristics of two pyrolytic biochars and potential as microhabitat *Plant Soil* **408** 357–68
[43] Hyväluoma J, Kulju S, Hannula M, Wikberg H, Källi A and Rasa K 2018 Quantitative characterization of pore structure of several biochars with 3D imaging *Environ. Sci. Pollut. Res.* **25** 25648–58
[44] Campos P, Miller A Z, Knicker H, Costa-Pereira M F, Merino A and De la Rosa J M 2020 Chemical, physical and morphological properties of biochars produced from agricultural residues: implications for their use as soil amendment *Waste Manag.* **105** 256–67
[45] Inkson B J 2016 *Scanning Electron Microscopy (SEM) and Transmission Electron Microscopy (TEM) for Materials Characterization* (Amsterdam: Elsevier)
[46] Das D, Bordoloi U, Muigai H H and Kalita P 2020 A novel form stable PCM based bio composite material for solar thermal energy storage applications *J. Energy Storage* **30** 101403
[47] de la Rosa J M, Paneque M, Miller A Z and Knicker H 2020 Relating physical and chemical properties of four different biochars and their application rate to biomass production of *Lolium perenne* on a Calcic Cambisol during a pot experiment of 79 days *Sci. Total Env.* **499** 175–84
[48] Chintala R *et al* 2014 Molecular characterization of biochars and their influence on microbiological properties of soil *J. Hazard. Mater.* **279** 244–56
[49] Chia C H, Gong B, Joseph S D, Marjo C E, Munroe P and Rich A M 2012 Vibrational spectroscopy imaging of mineral-enriched biochar by FTIR *Raman and SEM—EDX* **62** 248–57
[50] Jorio A *et al* 2012 Microscopy and spectroscopy analysis of carbon nanostructures in highly fertile Amazonian anthrosoils *Soil Tillage Res.* **122** 61–6

[51] Wang Z *et al* 2015 Investigating the mechanisms of biochar's removal of lead from solution *Biores. Technol.* **177** 308–17

[52] Han Z *et al* 2015 Magnetite impregnation effects on the sorbent properties of activated carbons and biochars *Water Res.* **70** 394–403

[53] Wei D *et al* 2018 Biochar-based functional materials in the purification of agricultural wastewater: fabrication, application and future research needs *Chemosphere* **197** 165–80

[54] Brewer C E *et al* 2014 New approaches to measuring biochar density and porosity *Biomass Bioenergy* **66** 176–85

[55] Thommes M and Cychosz K A 2014 Physical adsorption characterization of nanoporous materials: progress and challenges *Adsorption* **20** 233–50

[56] Sigmund G, Hüffer T, Hofmann T and Kah M 2017 Biochar total surface area and total pore volume determined by N_2 and CO_2 physisorption are strongly influenced by degassing temperature *Sci. Total Environ.* **580** 770–5

[57] Braida W J, Pignatello J J, Lu Y, Ravikovitch P I, Neimark A V and Xing B 2003 Sorption hysteresis of benzene in charcoal particles *Environ. Sci. Technol.* **37** 409–17

[58] Sing K S W 2004 Characterization of porous materials: past, present and future *Colloids Surf.* A **241** 3–7

[59] Thommes M *et al* 2015 Physisorption of gases, with special reference to the evaluation of surface area and pore size distribution (IUPAC Technical Report) *Pure Appl. Chem.* **87** 1051–69

[60] Jagiello J, Kenvin J, Celzard A and Fierro V 2019 Enhanced resolution of ultra micropore size determination of biochars and activated carbons by dual gas analysis using N_2 and CO_2 with 2D-NLDFT adsorption models *Carbon NY* **144** 206–15

[61] Dieguez-Alonso A, Funke A, Anca-Couce A, Girolamo Rombolà A, Ojeda G, Bachmann J and Behrendt F 2018 Towards biochar and hydrochar engineering—influence of process conditions on surface physical and chemical properties, thermal stability, nutrient availability, toxicity and wettability *Energies* **11** 496

[62] Smith E and Dent G 2019 *Modern Raman Spectroscopy: A Practical Approach* (New York: Wiley)

[63] Xu J *et al* 2020 Raman spectroscopy of biochar from the pyrolysis of three typical Chinese biomasses: a novel method for rapidly evaluating the biochar property *Energy* **202** 117644

[64] Azargohar R, Nanda S, Dalai A K and Kozinski J A 2019 Physico-chemistry of biochars produced through steam gasification and hydro-thermal gasification of canola hull and canola meal pellets *Biomass Bioenergy* **120** 458–70

[65] Sun Y *et al* 2020 Tailored design of graphitic biochar for high-efficiency and chemical-free microwave-assisted removal of refractory organic contaminants *Chem. Eng. J.* **398** 125505

[66] Belhachemi M, Khiari B, Jeguirim M and Sepúlveda-Escribano A 2019 Characterization of biomass-derived chars *Char and Carbon Materials Derived from Biomass* (Amsterdam: Elsevier) pp 69–108

[67] Guizani C, Haddad K, Limousy L and Jeguirim M 2017 New insights on the structural evolution of biomass char upon pyrolysis as revealed by the Raman spectroscopy and elemental analysis *Carbon NY* **119** 519–21

[68] Tsaneva V N, Kwapinski W, Teng X and Glowacki B A 2014 Assessment of the structural evolution of carbons from microwave plasma natural gas reforming and biomass pyrolysis using Raman spectroscopy *Carbon NY* **80** 617–28

[69] de Sousa D V, Guimarães L M, Félix J F, Ker J C, Schaefer C E R G and Rodet M J 2020 Dynamic of the structural alteration of biochar in ancient anthrosol over a long timescale by Raman spectroscopy *PLoS One* **15** e0229447

[70] Rodriguez J A, Lustosa Filho J F, Melo L C A, de Assis I R and de Oliveira T S 2020 Influence of pyrolysis temperature and feedstock on the properties of biochars produced from agricultural and industrial wastes *J. Anal. Appl. Pyrolysis* **149** 104839

[71] Newman J A, Schmitt P D, Toth S J, Deng F, Zhang S and Simpson G J 2015 Parts per million powder x-ray diffraction *Anal. Chem.* **87** 10950–55

[72] Fetisov G V 2020 X-ray diffraction methods for structural diagnostics of materials: progress and achievements *Phys.-Usp.* **63** 2–32

[73] Budai A *et al* 2013 Biochar carbon stability test method: an assessment of methods to determine biochar carbon stability *International Biochar Initiative*

[74] Krevelen D W 1950 Graphical-statistical method for the study of structure and reaction processes of coal *Fuel* **29** 269–84

[75] Crombie K, Mašek O, Sohi S P, Brownsort P and Cross A 2013 The effect of pyrolysis conditions on biochar stability as determined by three methods *GCB Bioenergy* **5** 122–31

[76] Spokas K A 2010 Review of the stability of biochar in soils: predictability of O:C molar ratios *Carbon Manage.* **1** 289–303

[77] Harvey O R, Kuo L-J, Zimmerman A R, Louchouarn P, Amonette J E and Herbert B E 2012 An index-based approach to assessing recalcitrance and soil carbon sequestration potential of engineered black carbons (biochars) *Environ. Sci. Technol.* **46** 1415–21

[78] Cross A and Sohi S P 2013 A method for screening the relative long-term stability of biochar *GCB Bioenergy* **5** 215–20

[79] Leng L, Huang H, Li H, Li J and Zhou W 2019 Biochar stability assessment methods: a review *Sci. Total Environ.* **647** 210–22

[80] Chacón F J, Cayuela M L, Roig A and Sánchez-Monedero M A 2017 Understanding, measuring and tuning the electrochemical properties of biochar for environmental applications *Rev. Environ. Sci. Biotechnology* **16** 695–715

[81] Sun T *et al* 2017 Rapid electron transfer by the carbon matrix in natural pyrogenic carbon *Nat. Commun.* **8** 14873

[82] Sun T *et al* 2018 Simultaneous quantification of electron transfer by carbon matrices and functional groups in pyrogenic carbon *Environ. Sci. Technol.* **52** 8538–47

[83] Zhao C, Shi Y, Xie J, Lei F and Zhang L 2019 Direct measurement of electrical conductivity of porous biochar monolith for supercapacitors *Mater. Res. Express* **6** 095526

[84] Gabhi R, Basile L, Kirk D W, Giorcelli M, Tagliaferro A and Jia C Q 2020 Electrical conductivity of wood biochar monoliths and its dependence on pyrolysis temperature *Biochar* **2** 369–78

[85] Klüpfel L, Keiluweit M, Kleber M and Sander M 2014 Redox properties of plant biomass-derived black carbon (biochar) *Env. Sci. Technol* **48** 5601–11

[86] Liao S, Pan B, Li H, Zhang D and Xing B 2014 Detecting free radicals in biochars and determining their ability to inhibit the germination and growth of corn, wheat and rice seedlings *Env. Sci. Technol.* **48** 8581–87

[87] Pignatello J J, Mitch W A and Xu W 2017 Activity and reactivity of pyrogenic carbonaceous matter toward organic compounds *Environ. Sci. Technol.* **51** 8893–908

[88] Ruan X *et al* 2019 Formation, characteristics, and applications of environmentally persistent free radicals in biochars: a review *Bioresour. Technol.* **281** 457–68

[89] Lee K J, Elgrishi N, Kandemir B and Dempsey J L 2017 Electrochemical and spectroscopic methods for evaluating molecular electrocatalysts *Nat. Rev. Chem.* **1** 0039
[90] Yuan Y *et al* 2017 Applications of biochar in redox-mediated reactions *Bioresour. Technol.* **246** 271–81
[91] Wang C, Liu Y, Gao X, Chen H, Xu X and Zhu L 2018 Role of biochar in the granulation of anaerobic sludge and improvement of electron transfer characteristics *Bioresour. Technol.* **268** 28–35
[92] Liu W-J, Jiang H and Yu H-Q 2019 Emerging applications of biochar-based materials for energy storage and conversion *Energy Environ. Sci.* **12** 1751–79
[93] Buss W, Shepherd J G, Heal K V and Mašek O 2018 Spatial and temporal microscale pH change at the soil–biochar interface *Geoderma* **331** 50–2
[94] Johnston C T 2017 Biochar analysis by Fourier-transform infra-red spectroscopy *Biochar: A Guide to Analytical Methods* p 199
[95] Bekiaris G, Peltre C, Jensen L S and Bruun S 2016 Using FTIR-photoacoustic spectroscopy for phosphorus speciation analysis of biochars *Spectrochim. Acta* A **168** 29–36
[96] Zolfi Bavariani M, Ronaghi A and Ghasemi R 2019 Influence of pyrolysis temperatures on FTIR analysis, nutrient bioavailability, and agricultural use of poultry manure biochars *Commun. Soil Sci. Plant Anal.* **50** 402–11
[97] Yang G-X and Jiang H 2014 Amino modification of biochar for enhanced adsorption of copper ions from synthetic wastewater *Water Res.* **48** 396–405
[98] Jia X *et al* 2020 The antimony sorption and transport mechanisms in removal experiment by Mn-coated biochar *Sci. Total Environ.* **724** 138158
[99] Graber E R, Tsechansky L, Fidel R B, Thompson M L and Laird D A 2017 Determining acidic groups at biochar surfaces via the Boehm titration *Biochar: A Guide to Analytical Methods* vol 85 (Boca Raton, FL: CRC Press) pp 85–94
[100] Thompson R and Oldfield F 1986 *Environmental Magnetism* (Dordrecht: Springer)

IOP Publishing

Biochar

Emerging applications

Alberto Tagliaferro, Carlo Rosso and Mauro Giorcelli

Chapter 6

Cellulose nanocrystals as natural feedstocks for advanced carbon materials

Mattia Bartoli, Michael Chae and David C Bressler

Cellulose nanocrystals are one of the emerging biomaterials that have gained increased attention year by year. Due their astonishing performance, cellulose nanocrystals have been used as they are in plenty of applications. However, they have recently found a new use as template for advance carbonaceous materials. Their high shape tuneability together with the possibility of realizing submicrometric needle-like carbon structures have focused the attention of researchers on the carbonization of cellulose nanocrystals.

In the following chapter, we briefly overview the recent achievements in the use of cellulose nanocrystals for the production of advanced carbon materials for mechanical and electrochemical applications.

6.1 Cellulose nanocrystals: production and properties

The demand for chemicals and fuels continues to grow every year, which has led to the development of technologies that can provide more sustainable approaches when compared to current petrochemical-based products [1–3]. Biomass represents the perfect source to simultaneously satisfy the increasing demand for energy and raw materials [4–6] while reducing the environmental impact of industrial production [7]. Of the various biomass sources available, cellulose has received an increasing amount of attention over the past two decades [8]. Cellulose is a fibrous, highly crystalline, water-insoluble polymer, which is found in the protective cell walls of plants, particularly in the stalks, stems, trunks, and all woody portions of plant tissues [9]. Cellulose is the most abundant wood component (up to 40%–50% of total weight) and it is formed by long linear chains of D-glucopyranose units linked by β-(1,4) glycosidic bonds (as shown in figure 6.1).

Cellulose has an average molecular weight between 1.800 and 2.500 kDa [10]. These values are reduced for cellulose processed by the pulp and paper industry,

doi:10.1088/978-0-7503-2660-5ch6

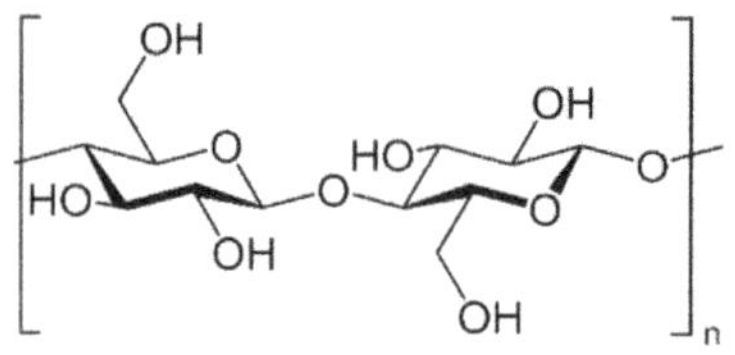

Figure 6.1. The basic structure of cellulose.

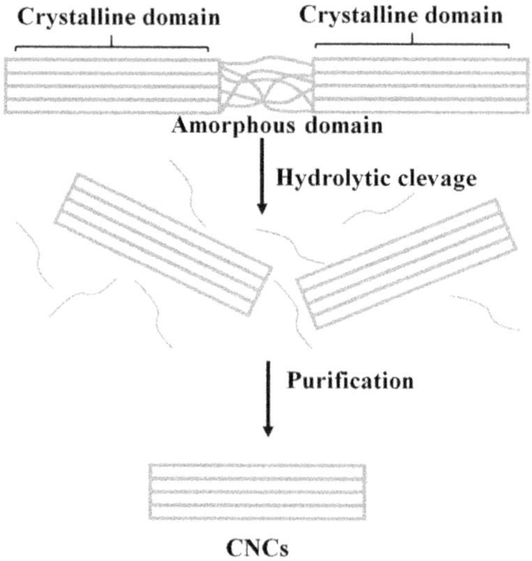

Figure 6.2. Schematic of CNCs production.

reaching an average molecular weight between 180 and 540 kDa [11]. Cellulose is one of the principal raw materials used in the textiles industry for the production of cotton [12]. Cellulose has been employed for the production of polymers such as rayon [13] and activated carbon [14], but its main use remains in the production of paper [15]. Furthermore, neat and modified cellulose have been used in many applications ranging from fillers for polymers [16, 17] or drugs [18], the stationary chiral phase for liquid chromatography [19], or materials for building insulation [20]. The latter applications highlight the growing interest in the application of cellulose and its derivatives in materials science applications [21]. However, traditional cellulose-based fibers do not fulfill all of the required application properties as issues with their durability and uniformity have prevented their extensive use.

A further improvement in this field has been achieved through the production of nanostructured bioderived organic materials such as cellulose nanocrystals (CNCs) [22]. CNCs have piqued a lot of interest from both academic and industrial players [23] due to their astonishing properties such as having a higher storage modulus than glass fibers. CNCs are nano-sized structures isolated after hydrolytic cleavage of highly crystalline domains of neat cellulose, as schematized in figure 6.2.

CNCs are one of the most interesting products produced through the hydrolytic degradation of cellulose [24, 25] and can be generated using a wide range of feedstocks [26] including seaweeds [27, 28], wood pulp [29], bamboo [30], pistachio shells [31], rice straw [32], sugarcane bagasse [33], and several other waste streams [34–38]. The most widely applied method for CNC production is based on acid hydrolysis. It is believed that the initiation of the hydrolysis pathway starts with the hydrolytic cleavage of glycosidic bonds in the amorphous domain generating individual crystallites. The most common CNC production method used employs a controlled hydrolysis mediated by sulfuric acid, and this process has been shown to be improved by the addition of organic acids [39] or by systematically tuning the process conditions for maximizing the CNC yield [40, 41].

Incorporation of several additional strategies in the CNC production process have also been explored with promising results. For example, the acid treatment has been combined with ball milling for the production CNCs with good yields along with the minimization of wet waste streams [42, 43]. Melikoğlu *et al* [44] have reported an alkaline treatment step for the extraction of cellulose that was subsequently used to generate CNCs with crystallinity of up to 78%. In addition, Beyene *et al* [45] used a hydrothermal treatment to increase the crystallinity of wood pulp that was subsequently used to generate CNCs. It this case, an approximately four-fold increase in CNC yield was observed after a hydrothermal pretreatment at 200 °C. Finally, a continuous process was developed by Leite *et al* [46] based on a combination of gelatine and CNCs for the mass production of bioderived films.

Beyond chemical and thermal treatments, great efforts have been devoted to the implementation of enzymatic treatments to generate CNCs, often in conjunction with acid hydrolysis [47]. The most employed enzymes are cellulases [48, 49], including endoglucanases [25, 50]. Enzyme treatments were reported to facilitate the generation of CNCs of good quality with low enzymatic loading. Compared to acid mediated hydrolysis, the use of enzymes for the production of CNCs may prove to be cheaper and easier due to the absence of harsh chemicals and the reduced energy requirements for mechanical mixing and heating [51]. Furthermore, enzymes can selectively degrade the amorphous domains of cellulose, leaving the crystalline domains intact and preserving the hydroxyl surface functionalities.

Alternatively, several processing methods have been explored to replace the standard acid hydrolysis approach. Novo *et al* [52] developed a more environmentally friendly procedure that employed subcritical water, and high temperature and pressure. Although the exclusion of the acidic reagent improved the sustainability of the process, the CNC yield achieved was lower than those reported for standard acidic hydrolysis. Furthermore, ionic liquids were used by Mao *et al* [53] for the production of CNCs mediated by the use of 1-butyl-3-methylimidazolium hydrogen sulfate. In this report, the authors claimed a process yield of up to 84%, as well as an improved dispersibility compared to CNCs obtained with the traditional sulfuric acid method. Saito and Isogai [54] proposed a non-acidic method based on the use of an oxidant agent for the isolation of CNCs. In this approach, 2,2,6,6-tetramethylpiperidine-1-oxyl radicals were used to introduce carboxylic and carbonyl functionalities onto cellulose promoting the disassembling of cellulose fibers while facilitating the isolation of highly crystalline fragments.

While several different processes have been examined for the production of CNCs, the CNC yields obtained have been shown to vary dramatically. However, it is also worth mentioning that the various CNC production platforms may provide benefits other than high overall CNC yield. For example, the incorporation of enzymes in the CNC process has been suggested to facilitate the production of a sugar stream that could be fermented into other value-added products, such as ethanol [24, 25, 55]. Alternatively, hydrothermal treatment has been shown to generate furfural, a valuable platform chemical [45]. Furthermore, the various production methodologies and feedstocks have been shown to produce CNCs with variable shape and size, that could respond differently to various applications (as shown in figure 6.3). For example, bulky CNCs have been shown to assemble themselves into aggregates, driven by the formation of networks of strong hydrogen bonds [56]. As reported by Beck *et al* [57], in a dried solid state, CNCs aggregate as round structures with an average size ranging from 10 μm to 20 μm (as shown in figure 6.4).

Despite their promising properties, CNCs have very poor conductivity and dispersibility in non-polar media [59, 60]. These drawbacks impede their application in polymer composites and lead to a requirement for plasticizers [61, 62] or further chemical modification [63]. Alternatively, thermochemical conversion routes have shown promise for the production of carbonaceous materials. Specifically, CNCs were used as a feedstock for a thermochemical conversion process to generate micro- and nanostructured carbons. In the following section, we review the main route for the production of carbon-based materials from CNCs.

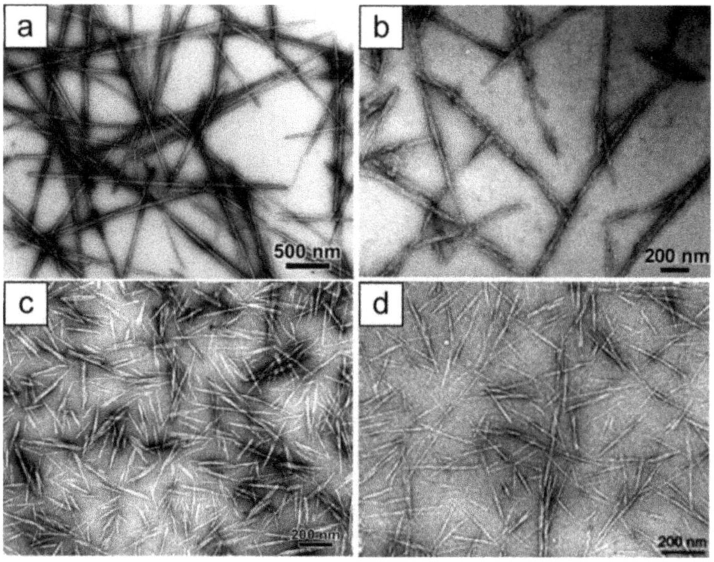

Figure 6.3. Transmission electron microscope micrographs of dried dispersions of CNCs derived from (a) tunicate, (b) bacteria, (c) ramie, and (d) sisal, as reported by Habibi *et al* [58]. Reprinted with permission from [58]. Copyright (2010) American Chemical Society.

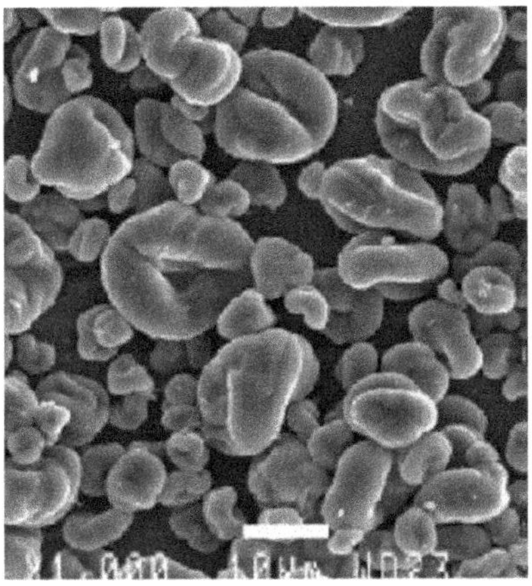

Figure 6.4. Scanning electron microscope micrographs of granules of spray-dried CNCs as reported by Beck *et al* [57]. Reprinted with permission from [57]. Copyright (2012) American Chemical Society.

6.2 Cellulose nanocrystals as the feedstock for new carbonaceous materials

The interest in carbon structures on the micro- and nano-scales has grown year by year since the discovery of allotropic carbon species (i.e. carbon nanotubes, fullerene, carbon fibers, etc) [64]. CNCs are a good feedstock candidate for the production of electroactive nanostructured carbon, as shown by Wu *et al* [65]. In this study, the authors report the use of CNCs/melamine resin precursor to the produce nitrogen-enriched nanorods used for energy storage devices. The resulting hybrid material underwent a pyrolytic conversion promoting the formation of nitrogen-doped carbon nanorods with micro-, meso-, and macropores. The authors achieved a capacitance of up to 329 F g^{-1}. The materials also exhibited a high cycling stability with a high current density of up to 20 A g^{-1}.

Similarly, Wu *et al* [66] prepared CNC-based high-performance supercapacitors using poly(pyrrole)/CNCs composites without carbonization. The authors introduced acidic functionalities on the CNC surface though TEMPO-mediated oxidation, as described above. This treatment promoted the formation of a strong hydrogen bonding network that enhanced the adsorption of pyrrole monomers. This procedure allowed the nano-templating promoted by CNCs that simultaneously controlled the deposition and growth of the polymer layer, enhancing the electrochemical properties of the composite precursor. The hybrid nanostructures showed a high specific capacitance of up to 248 F g^{-1}, comparable to systems based on carbon nanotubes and graphene.

However, the electrochemical applications of CNCs are not limited to the production of supercapacitors. An interesting application is represented by the carbonization of CNCs for the production of an anode used for batteries. The elevated tuneability of carbonized CNCs (c-CNCs) allows the production of both soft and hard carbon for use in both lithium- and sodium-based batteries. This proves that c-CNCs can efficiently promote the intracalation of different ions into their graphitic domains.

Kim *et al* [67] investigated the effect of the morphology and crystallinity of CNC-derived carbon nanomaterials on the electrochemical reactions in lithium-based batteries. The authors reported the effect of the amorphous region of CNCs on the increment of the specific capacity of a cell, together with the overall electrical conductivity of the electrode. The anode comprising c-CNCs showed a high capacity of up to 370 mAh g^{-1} using a low current density of up to 0.1 mA g^{-1}. CNCs exhibited a remarkable improvement in coulombic efficiency and cycle stability (up to 120 cycles). The authors ascribed the properties observed to the substantially decreased local current density and the improved lithium ion storage in the additional c-CNC layers.

Similarly, Xu *et al* [68] also produced nitrogen self-doped porous carbon derived from chitosan/CNC biocomposites for lithium ion battery applications. The authors investigated carbonization temperatures from 800 °C up to 1500 °C, as well as the nitrogen-doping content, and evaluated the effect on electrochemical performance of the porous carbon anode. The authors claimed that high nitrogen-doping content played a crucial role in the electrochemical performance of samples at 800 °C and 1500 °C; c-CNCs were most stable at 1200 °C. Using a treatment temperature of 1200 °C the highest average specific capacity was generated reaching 333 mAh g^{-1} (using a current density of 100 mA g^{-1}), a retention capacity of 251 mAh g^{-1} (using a current density of 2000 mA g^{-1}), together with an excellent cycling stability of 327 mAh g^{-1} after 50 cycles.

Zhang *et al* [69] explored the use of c-CNCs for the production of modified anodes for advanced lithium batteries. The authors modified c-CNCs by tailoring with Fe_3O_4/Fe nanoparticles. The produced mesoporous carbon fibers adhered with metal particles resulting in highly stable anode materials for high-rate lithium batteries, as demonstrated by their excellent cycling performance at higher current and their fast charging capability. The tailored CNC-based anode showed a high reversible capacity of up to 472 mAh g^{-1} after 500 cycles with a rapid charge to 100% in 28 min. After 160 cycles at varying current densities from 1 to 10 A g^{-1}, the CNC-based anode still showed a high discharge capacity of up to 525 mAh g^{-1} and a coulombic efficiency close to 100%. The authors attributed this behavior to the synergistic effects of several factors including the mesoporous structure of the graphitized biocarbon fibers together with the Fe_3O_4.

Beyond lithium batteries, some studies have reported the use of c-CNCs for the production of sodium-based batteries. Kim *et al* [70] produced anodes though the carbonization of CNCs in a wide temperature range from 800 °C to 2500 °C. The authors correlated the structural variations in the CNC-based carbon anodes with the sodiation mechanism by investigating the galvanostatic voltage profiles, which showed that sodium ion adsorption took place in the more disordered carbonaceous structures followed by intercalation into the more ordered internal structures, with an average interlayer spacing of >0.37 nm. CNCs carbonized at 1500 °C were characterized as having the highest reversible specific capacity of up to 311 mAh g^{-1} using a current density of 10 mA g^{-1}. Furthermore, this material showed an outstanding rate capability of up to 273 mAh g^{-1} together with an excellent specific capacity retention of up to 92% even after 400 cycles at 100 mA g^{-1} with an initial coulombic efficiency of 85%.

Zhu *et al* [71] used a lower pyrolytic temperature for the carbonization of CNCs. Using a temperature of 1000 °C, the authors achieved materials with excellent performance, including a high reversible capacity of up to 340 mAh g^{-1} using a current density of 100 mA g^{-1}. Furthermore, the rate capability and cycling stability of the c-CNCs were also excellent. This behavior was due to the formation of nano-sized graphitic carbon from the ordered CNCs at the low carbonization temperature of 1000 °C, as shown by both molecular dynamic simulations and *in situ* transmission electron microscopic measurements.

Regarding c-CNCs, the electrical properties are not the only characteristics being investigated. The scientific community has been fascinated by the great shape tunability of c-CNCs. A comprehensive study on the thermal evolution of CNCs was reported by Eom *et al* [72]. The authors studied the structure evolution mechanism of graphitic domains during carbonization of CNCs in the temperature range from 1000 °C to 2500 °C. Interestingly, the authors noticed that the direct carbonization of CNC in an inert environment led to an irregular morphology due to molecular fusion, whereas oxidative stabilization at 250 °C under air and subsequent carbonization preserved the pristine needle-like structure of the CNCs during carbonization.

The shape tuneability of c-CNCs was explored by Bartoli *et al* [73]. The authors studied the carbonization of CNCs at temperatures lower than 1000 °C, showing the preservation of a twisted ball shape of aggregated CNCs at all pyrolytic temperatures examined. A few carbonized needle-like CNCs were observed using a carbonization temperature of 1000 °C, as shown in figure 6.5. Nonetheless, the authors explored the possibility of tuning the morphology of CNCs through addition of dispersing media, such as water, ethanol, or poly(ethylene glycol), prior to the carbonization. Through this approach, the authors reported the CNC surface modification with nano-carbon hemispheres together with the production of straight carbon needles with a length in the hundreds of nanometers and a thickness of only a few nanometers.

Shopsowitz *et al* [74] described a similar approach using silica as the feedstock for the carbonized nematic material with a fibrous structure. This work was based on a previous study by Cho *et al* [75] that reported a multi-step protocol for the conversion of commercial microcrystalline cellulose into carbonized fibers at the nanoscale. For the first time, the authors described a nematic mesoporous carbon with a high specific surface area, close to 1400 m^2 g^{-1}, that preserved the left-handed helical structure of the CNC precursor.

The polymorphism of c-CNCs is one of the most attractive aspects of these carbon structures with regard to material science applications. Beard and Eichhorn [76] reported the use of CNCs for the manufacturing of a highly porous thermoplastic composite carbon aerogel. Their procedure was based on the carbonization of a poly(ethylene oxide)/CNC precursor followed by a freeze-drying step to process the porous conductive carbon aerogel-like material.

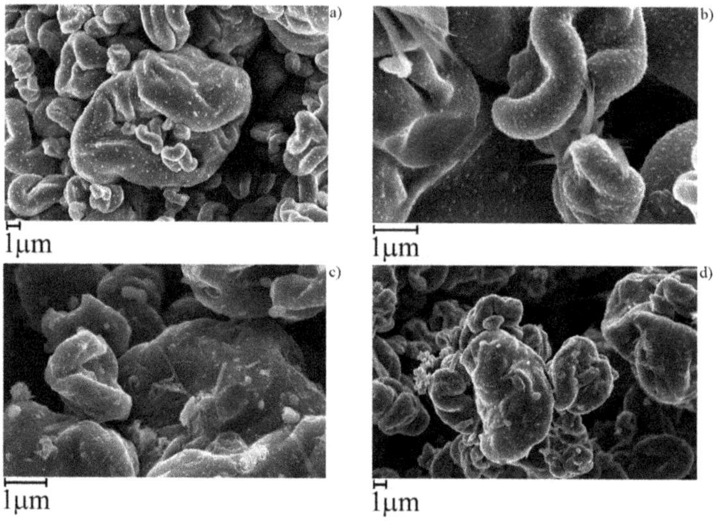

Figure 6.5. Scanning electron microscopic captures of c-CNCs at (a) 400 °C, (b) 600 °C, (c) 800 °C, and (d) 1000 °C as reported by Bartoli *et al* [73]. Reprinted from [73], Copyright (2019), with permission from Elsevier.

Similarly, Chang *et al* [77] produced carbon fibers from a mix of poly(acrylonitrile)/CNCs. This fibrous material showed a remarkable tensile strength in the range of 1.8–2.3 GPa and a tensile modulus of up to 265 GPa, which is comparable to traditional carbon fibers. Bartoli *et al* [78] used carbonized round-like CNCs for the production of epoxy based composites. The reinforced plastics produced showed a remarkable improvement in maximum elongation together with an increase in ultimate tensile strength and Young's modulus.

Composites derived from c-CNCs have also found applications beyond the mere production of reinforced plastic. Kong *et al* [79] produced CNC-based aerogel tailored with Mo_2C@sulfur. After carbonization, this material was used as an electrocatalyst for the production of hydrogen with appreciable outputs. Souz *et al* [80] used CNCs for the production of luminescence carbon nanodots of spherical shape with an average diameter of 4–8 nm. A similar system was developed by Zhang *et al* [81] by combining carbon dots with CNCs for the production of films. These thin materials were able to provide an accurate and sensitive response to surrounding temperature variation in the range from −50 °C to around 110 °C.

With numerous applications ranging from material science to energy storage, the great versatility of c-CNCs has been demonstrated as a great chance to magnify the value of CNCs. For example, c-CNCs showed remarkable performance in energy storage technology applications. Even if energy storage remains as an unresolved challenge for the twenty-first century, the mass production of carbon-based composites is a very easy way to exploit the full potential of c-CNCs, leading to the development of a new and lucrative market for CNCs. These aspects, combined with their biological origin, make c-CNCs a promising alternative for well-established carbon-based materials such as carbon black or carbon fibers.

6.3 Perspectives on the cellulose nanocrystals and related carbon materials

CNCs represent an extremely promising tool for the production of a plethora of different carbonaceous materials. It is relevant to highlight that the great potential of these materials has only been touched upon and further developments are expected in the near future. Subsequent studies will likely demonstrate the credibility of CNCs carbonization for the production of a new generation of carbon microstructures, ranging from fibers to spherules. Furthermore, the relevant fields of application will guarantee a continued focus on CNCs carbonization, which will facilitate the realization of new and improved performance measures.

References

[1] Shafiee S and Topal E 2009 When will fossil fuel reserves be diminished? *Energy Policy* **37** 181–9

[2] Bunker S G 1996 Raw material and the global economy: oversights and distortions in industrial ecology *Soc. Nat. Resour.* **9** 419–29

[3] Utama N A, Fathoni A M, Kristianto M A and McLellan B C 2014 The end of fossil fuel era: supply–demand measures through energy efficiency *Procedia Environ. Sci.* **20** 40–5

[4] McKendry P 2002 Energy production from biomass (part 1): overview of biomass *Bioresour. Technol.* **83** 37–46

[5] Simmons B A 2012 *Plant Biotechnology and Agriculture* ed A Altman and P M Hasegawa (San Diego, CA: Academic) pp 495–505

[6] Cherubini F 2010 The biorefinery concept: using biomass instead of oil for producing energy and chemicals *Energy Convers. Manage.* **51** 1412–21

[7] Rafaschieri A 1999 Life cycle assessment of electricity production from poplar energy crops compared with conventional fossil fuels *Energy Conv. Manage.* **40** 1477–93

[8] Stamm A J 1964 *Wood and Cellulose Science* (New York: Ronald)

[9] O'Sullivan A C 1997 Cellulose: the structure slowly unravels *Cellulose* **4** 173–207

[10] Sjostrom E 2013 *Wood Chemistry: Fundamentals and Applications* (Amsterdam: Elsevier)

[11] Kraemer E O 1938 Molecular weights of celluloses and cellulose derivates *Ind. Eng. Chem.* **30** 1200–3

[12] Myers D and Stolton S 1999 *Organic Cotton: From Field to Final Product* (London: Intermediate Technology Publications) pp 21–35

[13] Wyss G d 1925 The ripening of viscose *Ind. Eng. Chem.* **17** 1043–5

[14] Carroll A and Somerville C 2009 Cellulosic biofuels *Annu. Rev. Plant Biol.* **60** 165–82

[15] Kline J E 1982 *Paper And Paperboard: Manufacturing And Converting Fundamentals* (San Francisco, CA: Miller Freeman)

[16] Kunioka M, Inuzuka Y, Ninomiya F and Funabashi M 2006 Biobased contents of biodegradable poly (ε-caprolactone) composites polymerized and directly molded using aluminium triflate from caprolactone with cellulose and inorganic filler *Macromol. Biosci.* **6** 517–23

[17] Dufresne A and Vignon M R 1998 Improvement of starch film performances using cellulose microfibrils *Macromolecules* **31** 2693–6

[18] Zhang W, Zhang X, Liang M and Lu C 2008 Mechanochemical preparation of surface-acetylated cellulose powder to enhance mechanical properties of cellulose-filler-reinforced NR vulcanizates *Compos. Sci. Technol.* **68** 2479–84
[19] Ishikawa A and Shffiata T 1993 Cellulosic chiral stationary phase under reversed-phase condition *J. Liquid Chromatogr. Related Technol.* **16** 859–78
[20] Breum N, Schneider T, Jørgensen O, Rasmussen T V and Eriksen S S 2003 Cellulosic building insulation versus mineral wool, fiberglass or perlite: installer's exposure by inhalation of fibers, dust, endotoxin and fire-retardant additives *Ann. Occup. Hyg.* **47** 653–69
[21] Huber T, Müssig J, Curnow O, Pang S, Bickerton S and Staiger M P 2012 A critical review of all-cellulose composites *J. Mater. Sci.* **47** 1171–86
[22] Moon R J, Martini A, Nairn J, Simonsen J and Youngblood J 2011 Cellulose nanomaterials review: structure, properties and nanocomposites *Chem. Soc. Rev.* **40** 3941–94
[23] Duran N, Paula Lemes A and Seabra A B 2012 Review of cellulose nanocrystals patents: preparation, composites and general applications *Recent Patents Nanotechnol.* **6** 16–28
[24] Tsao G T and Chou T Y 1981 Process for recovering and utilizing cellulose using sulfuric acid *U.S. Patent* 4,266,981
[25] Dai J, Chae M, Beyene D, Danumah C, Tosto F and Bressler D C 2018 Co-production of cellulose nanocrystals and fermentable sugars assisted by endoglucanase treatment of wood pulp *Materials* **11** 1645
[26] Brinchi L, Cotana F, Fortunati E and Kenny J M 2013 Production of nanocrystalline cellulose from lignocellulosic biomass: technology and applications *Carbohydrate Polym.* **94** 154–69
[27] Doh H, Lee M H and Whiteside W S 2020 Physicochemical characteristics of cellulose nanocrystals isolated from seaweed biomass *Food Hydrocolloids* **102** 105542
[28] Kamelnia E, Divsalar A, Darroudi M, Yaghmaei P and Sadri K 2019 Production of new cellulose nanocrystals from *Ferula gummosa* and their use in medical applications via investigation of their biodistribution *Ind. Crops Prod.* **139** 111538
[29] Dong S, Bortner M J and Roman M 2016 Analysis of the sulfuric acid hydrolysis of wood pulp for cellulose nanocrystal production: a central composite design study *Ind. Crops Prod.* **93** 76–87
[30] Rasheed M, Jawaid M, Parveez B, Zuriyati A and Khan A 2020 Morphological, chemical and thermal analysis of cellulose nanocrystals extracted from bamboo fibre *Int. J. Biol. Macromol.* **160** 183–91
[31] Kasiri N and Fathi M 2018 Production of cellulose nanocrystals from pistachio shells and their application for stabilizing Pickering emulsions *Int. J. Biol. Macromol.* **106** 1023–31
[32] Thakur M, Sharma A, Ahlawat V, Bhattacharya M and Goswami S 2020 Process optimization for the production of cellulose nanocrystals from rice straw derived α-cellulose *Mater. Sci. Energy Technol.* **3** 328–34
[33] Oliveira F B d, Bras J, Pimenta M T B, Curvelo A A d S and Belgacem M N 2016 Production of cellulose nanocrystals from sugarcane bagasse fibers and pith *Ind. Crops Prod.* **93** 48–57
[34] El Achaby M, El Miri N, Hannache H, Gmouh S, Ben youcef H and Aboulkas A 2018 Production of cellulose nanocrystals from vine shoots and their use for the development of nanocomposite materials *Int. J. Biol. Macromol.* **117** 592–600
[35] El Achaby M, Kassab Z, Aboulkas A, Gaillard C and Barakat A 2018 Reuse of red algae waste for the production of cellulose nanocrystals and its application in polymer nanocomposites *Int. J. Biol. Macromol.* **106** 681–91

[36] Hafemann E, Battisti R, Marangoni C and Machado R A F 2019 Valorization of royal palm tree agroindustrial waste by isolating cellulose nanocrystals *Carbohydrate Polym.* **218** 188–98
[37] Xiao Y, Liu Y, Wang X, Li M, Lei H and Xu H 2019 Cellulose nanocrystals prepared from wheat bran: characterization and cytotoxicity assessment *Int. J. Biol. Macromol.* **140** 225–33
[38] Yousef S, Hamdy M, Tatariants M, Tuckute S, El-Abden S Z, Kliucininkas L and Baltusnikas A 2019 Sustainable industrial technology for recovery of cellulose from banknote production waste and reprocessing into cellulose nanocrystals *Resour. Conserv. Recycl.* **149** 510–20
[39] Wang H, Xie H, Du H, Wang X, Liu W, Duan Y, Zhang X, Sun L, Zhang X and Si C 2020 Highly efficient preparation of functional and thermostable cellulose nanocrystals via H_2SO_4 intensified acetic acid hydrolysis *Carbohydrate Polym.* **239** 116233
[40] Xing L, Hu C, Zhang W, Guan L and Gu J 2020 Transition of cellulose supramolecular structure during concentrated acid treatment and its implication for cellulose nanocrystal yield *Carbohydrate Polym.* **229** 115539
[41] Park N-M, Choi S, Oh J E and Hwang D Y 2019 Facile extraction of cellulose nanocrystals *Carbohydrate Polym.* **223** 115114
[42] Du L, Wang J, Zhang Y, Qi C, Wolcott M P and Yu Z 2017 A co-production of sugars, lignosulfonates, cellulose, and cellulose nanocrystals from ball-milled woods *Bioresour. Technol.* **238** 254–62
[43] Gao A, Chen H, Tang J, Xie K and Hou A 2020 Efficient extraction of cellulose nanocrystals from waste *Calotropis gigantea* fiber by SO_4^{2-}/TiO_2 nano-solid superacid catalyst combined with ball milling exfoliation *Ind. Crops Prod.* **152** 112524
[44] Melikoğlu A Y, Bilek S E and Cesur S 2019 Optimum alkaline treatment parameters for the extraction of cellulose and production of cellulose nanocrystals from apple pomace *Carbohydrate Polym.* **215** 330–7
[45] Beyene D, Chae M, Vasanthan T and Bressler D C 2020 A biorefinery strategy that introduces hydrothermal treatment prior to acid hydrolysis for co-generation of furfural and cellulose nanocrystals *Front. Chem.* **8** 323
[46] Leite L S F, Ferreira C M, Corrêa A C, Moreira F K V and Mattoso L H C 2020 Scaled-up production of gelatin-cellulose nanocrystal bionanocomposite films by continuous casting *Carbohydrate Polym.* **238** 116198
[47] Karim Z, Afrin S, Husain Q and Danish R 2017 Necessity of enzymatic hydrolysis for production and functionalization of nanocelluloses *Crit. Rev. Biotechnol.* **37** 355–70
[48] Beyene D, Chae M, Dai J, Danumah C, Tosto F, Demesa A G and Bressler D C 2018 Characterization of cellulase-treated fibers and resulting cellulose nanocrystals generated through acid hydrolysis *Materials* **11** 1272
[49] Pereira B and Arantes V 2020 Production of cellulose nanocrystals integrated into a biochemical sugar platform process via enzymatic hydrolysis at high solid loading *Ind. Crops Prod.* **152** 112377
[50] Teixeira R S S, Silva A S A d, Jang J-H, Kim H-W, Ishikawa K, Endo T, Lee S-H and Bon E P S 2015 Combining biomass wet disk milling and endoglucanase/β-glucosidase hydrolysis for the production of cellulose nanocrystals *Carbohydrate Polym.* **128** 75–81
[51] Anderson S R, Esposito D, Gillette W, Zhu J, Baxa U and Mcneil S E 2014 Enzymatic preparation of nanocrystalline and microcrystalline cellulose *Tappi J.* **13** 35–42
[52] Novo L P, Bras J, García A, Belgacem N and Curvelo A A d S 2016 A study of the production of cellulose nanocrystals through subcritical water hydrolysis *Ind. Crops Prod.* **93** 88–95

[53] Mao J, Heck B, Reiter G and Laborie M-P 2015 Cellulose nanocrystals' production in near theoretical yields by 1-butyl-3-methylimidazolium hydrogen sulfate ([Bmim]HSO_4)—mediated hydrolysis *Carbohydrate Polym.* **117** 443–51
[54] Saito T and Isogai A 2006 Introduction of aldehyde groups on surfaces of native cellulose fibers by TEMPO-mediated oxidation *Colloids Surf.* A **289** 219–25
[55] Beyene D, Chae M, Dai J, Danumah C, Tosto F, Demesa A G and Bressler D C 2017 Enzymatically-mediated Co-production of cellulose nanocrystals and fermentable sugars *Catalysts* **7** 322
[56] Lu P and Hsieh Y-L 2010 Preparation and properties of cellulose nanocrystals: rods, spheres, and network *Carbohydrate Polym.* **82** 329–36
[57] Beck S, Bouchard J and Berry R 2012 Dispersibility in water of dried nanocrystalline cellulose *Biomacromolecules* **13** 1486–94
[58] Habibi Y, Lucia L A and Rojas O J 2010 Cellulose nanocrystals: chemistry, self-assembly, and applications *Chem. Rev.* **110** 3479–500
[59] Lu Y, Weng L and Cao X 2005 Biocomposites of plasticized starch reinforced with cellulose crystallites from cottonseed linter *Macromol. Biosci.* **5** 1101–7
[60] Angles M N and Dufresne A 2000 Plasticized starch/tunicin whiskers nanocomposites. 1. Structural analysis *Macromolecules* **33** 8344–53
[61] Dubief D, Samain E and Dufresne A 1999 Polysaccharide microcrystals reinforced amorphous poly (β-hydroxyoctanoate) nanocomposite materials *Macromolecules* **32** 5765–71
[62] Chazeau L, Cavaille J and Perez J 2000 Plasticized PVC reinforced with cellulose whiskers. II. Plastic behavior *J. Polym. Sci.* B **38** 383–92
[63] Peng B L, Dhar N, Liu H and Tam K 2011 Chemistry and applications of nanocrystalline cellulose and its derivatives: a nanotechnology perspective *Can. J. Chem. Eng.* **89** 1191–206
[64] Malik S 1985 Structural and electronic properties of nano-carbon materials such as graphene, nanotubes and fullerenes *Nature* **318** 162–3
[65] Wu X, Shi Z, Tjandra R, Cousins A J, Sy S, Yu A, Berry R M and Tam K C 2015 Nitrogen-enriched porous carbon nanorods templated by cellulose nanocrystals as high performance supercapacitor electrodes *J. Mater. Chem.* A **3** 23768–77
[66] Wu X, Chabot V L, Kim B K, Yu A, Berry R M and Tam K C 2014 Cost-effective and scalable chemical synthesis of conductive cellulose nanocrystals for high-performance supercapacitors *Electrochim. Acta* **138** 139–47
[67] Kim P J, Kim K and Pol V G 2019 A comparative study of cellulose derived structured carbons on the electrochemical behavior of lithium metal-based batteries *Energy Storage Mater.* **19** 179–85
[68] Xu K, Du G, Zhong T, Chen D, Lin X, Zheng Z and Wang S 2020 Green sustainable, facile nitrogen self-doped porous carbon derived from chitosan/cellulose nanocrystal biocomposites as a potential anode material for lithium-ion batteries *J. Taiwan Inst. Chem. Eng.* **109** 79–89
[69] Zhang S, He W, Zhang X, Yang G, Ma J, Yang X and Song X 2015 Fabricating Fe_3O_4/Fe/biocarbon fibers using cellulose nanocrystals for high-rate Li-ion battery anode *Electrochim. Acta* **174** 1175–84
[70] Kim Y E, Yeom S J, Lee J-E, Kang S, Kang H, Lee G-H, Kim M J, Lee S G, Lee H-W and Chae H G 2020 Structure-dependent sodium ion storage mechanism of cellulose nanocrystal-based carbon anodes for highly efficient and stable batteries *J. Power Sources* **468** 228371

[71] Zhu H *et al* 2017 Low temperature carbonization of cellulose nanocrystals for high performance carbon anode of sodium-ion batteries *Nano Energy* **33** 37–44

[72] Eom Y, Son S M, Kim Y E, Lee J-E, Hwang S-H and Chae H G 2019 Structure evolution mechanism of highly ordered graphite during carbonization of cellulose nanocrystals *Carbon* **150** 142–52

[73] Bartoli M, Giorcelli M, Rovere M, Jagdale P, Tagliaferro A, Chae M and Bressler D C 2019 Shape tunability of carbonized cellulose nanocrystals *SN Applied Sciences* **1** 1661–76

[74] Shopsowitz K E, Hamad W Y and MacLachlan M J 2011 Chiral nematic mesoporous carbon derived from nanocrystalline cellulose *Angew. Chem. Int. Ed.* **50** 10991–5

[75] Cho S Y, Yun Y S and Jin H-J 2014 Carbon nanofibers prepared by the carbonization of self-assembled cellulose nanocrystals *Macromol. Res.* **22** 753–6

[76] Beard J D and Eichhorn S J 2018 Highly porous thermoplastic composite and carbon aerogel from cellulose nanocrystals *Mater. Lett.* **221** 248–51

[77] Chang H, Luo J, Liu H C, Zhang S, Park J G, Liang R and Kumar S 2019 Carbon fibers from polyacrylonitrile/cellulose nanocrystal nanocomposite fibers *Carbon* **145** 764–71

[78] Bartoli M, Rosso C, Giorcelli M, Rovere M, Jagdale P, Tagliaferro A, Chae M and Bressler D C 2019 Effect of incorporation of microstructured carbonized cellulose on surface and mechanical properties of epoxy composites *J. Appl. Polym. Sci.* **137** 48896

[79] Kong X, Chen S, Zou Y, Lyu S, She X, Lu Y, Sun J, Zhang H and Yang D 2018 Cellulose nanocrystals (CNC) derived Mo_2C@sulfur-doped carbon aerogels for hydrogen evolution *Int. J. Hydrogen Energy* **43** 13720–6

[80] Souza D R d S, Mesquita J P d, Lago R M, Caminhas L D and Pereira F V 2016 Cellulose nanocrystals: a versatile precursor for the preparation of different carbon structures and luminescent carbon dots *Ind. Crops Prod.* **93** 121–8

[81] Zhang L, Lyu S, Zhang Q, Wu Y, Melcher C, Chmely S C, Chen Z and Wang S 2019 Dual-emitting film with cellulose nanocrystal-assisted carbon dots grafted $SrAl_2O_4$, Eu^{2+}, Dy^{3+} phosphors for temperature sensing *Carbohydrate Polym.* **206** 767–77

IOP Publishing

Biochar

Emerging applications

Alberto Tagliaferro, Carlo Rosso and Mauro Giorcelli

Chapter 7

Biochar-based circular economy

Harn Wei Kua, Ondřej Mašek and Souradeep Gupta

For the past decade or so, increasing attention has been placed on the concept of the circular economy (CE) as a framework to guide industries and countries toward attaining sustainability. The underlying principles of CE includes a reduction in the requirement for virgin resource input, utilization of resources within planetary limits and in a way that promotes environmental well-being and socioeconomic benefits, long-term value creation and retention of products, and waste minimization through loop-closing of products.

The roles of biochar in CE can be related to resource use efficiency, material recycling/upcycling, and cascade uses. For example, biochar has been shown to affect the efficiency of the use of nutrients supplied in the form of fertilizers, and also to improve soil water management, both of which contribute to increased crop yields. There are many potential combinations of biochar applications that can yield viable sequences of cascade uses, but the ideal sequences would, in each step/application in the sequence, yield biochar that is suitable for the following step/application, and the biochar actually has improved properties for subsequent applications.

Creating an industrial symbiotic network, in which biochar is circulated within the network for different uses and functions, is possibly a good way to keep increasing biochar value within the CE. However, the sustainability of this CE will require the creation and implementation of appropriate policies that enable all the relevant entities in the network to benefit from their respective role in circulating biochar to generate values for the network.

7.1 A circular economy based on bio-waste recycling and recovery

For the past decade or so, increasing attention has been put on the concept of the circular economy (CE) as a framework that guides industries and countries toward attaining sustainability. The underlying principles of the CE include a reduction in the requirement for virgin resource input, utilization of resources within planetary limits and in a way that promotes environmental well-being and socioeconomic benefits, long-term value creation and retention of products, and waste minimization through loop-closing of products. The essence of CE is that strategies, such as waste recycling, must lead to, and be supported by, a sustainable economy ecosystem; the economic growth generated by this ecosystem must also be decoupled from the negative consequences on the natural environment and society [1–4].

In the past few years, substantial progress had been made in the design and planning of CE strategies. Notably, the authors of [5] categorized CE strategies into the following types:

- ‘Smarter product use and manufacture’ through refusal of new products unless they are really necessary, rethinking of product designs and material choices, and reuse of products/materials.
- ‘Extend lifespan of products and their parts’ through reuse, repair, refurbishment, remanufacturing, and repurposing of these products.
- ‘Useful application of materials’ through recycling and recovery.

Out of these three categories of CE strategies, the ‘useful application of materials’ can be applied to the CE ecosystem derived from bio-wastes. This kind of ecosystem can also be called a ‘bio-economy’, which includes sectors, processes, goods, services, and general activities that are based on, and affect, biological resources [6]. A bio-CE is built on the underlying concept of ‘zero waste’, in which precious resources are first converted into value-added products (based on the needs of the economy), before the residues generated from the interlinked processes of production and consumption are utilized in a sustainable way—mainly through waste recycling and recovery (that is, ‘the useful application of materials’). In essence, Kumar *et al* [7] proposed that there are three prerequisites for a sustainable bio-CE—a sustainable resource base (for example, avoiding fossil-based resources as much as possible), sustainable production/consumption processes/products (for example, using sustainable energy in the process and using as few materials for packaging of products as possible), and a circular flux of materials.

7.2 Biochar as part of a circular economy

CE is a wide-ranging concept affecting most, if not all, parts of the economy, and it is therefore not surprising that biochar technology can find its role in many aspects of CE. The roles of biochar in CE can be related to resource use efficiency, material recycling/upcycling, and cascade uses.

Resource use efficiency refers to better/more efficient use of available resources—that is, achieving more with the same or even lower inputs. The most relevant biochar applications in this context can be found in agriculture, where multiple inputs are required, such as water and nutrients [8]. Biochar has been shown to affect the efficiency of use of nutrients supplied in the form of fertilizers, and also to improve soil water management, both of which contribute to increased crop yields. For example, a number of studies have shown higher nitrogen use efficiency in agricultural production with the application of biochar to soil [9–11]. This not only reduces the required fertilizer application rates, but also reduces the potential negative environmental effects related to nitrogen leaching and run-off.

Similar observations have been made with respect to phosphorus use efficiency. Phosphorus is a limited resource and therefore it needs to be managed responsibly and where possible reduced application rates, recycling, or utilization of secondary sources should be practiced [12, 13]. Biochar can play an important role in this respect, both as a means to improve phosphorus use efficiency [14, 15] and also as a means to recycle phosphorus from other secondary sources, such as organic residues, waste, and effluent streams (for more information, see the following section). In addition to nutrient use efficiency, biochar can also have a significant effect on soil water management and water use efficiency [16]. This is a critical aspect as water shortages are a serious limiting factor in agricultural productivity and can be expected to become more severe in many parts of the world due to climate change [17–19].

7.3 Waste recycling through biochar production and utilization

A recent study of the European Union estimated that about 50 million cubic meters of wood waste are generated each year [20]. In the United States, about 30 million tonnes of the annual construction and demolition (C&D) waste is composed of wood [21]. It was estimated that around 19% of the total quantity of wood waste worldwide is in the form of clean waste wood streams (for example, discarded wooden pallets) [22]. This clean source is a potentially rich source of feedstock for biochar with a low level of toxic substances.

The remaining global wood waste may contain certain level of toxic materials. For example, large quantities of construction wood may contain volatile organic compounds, such as formaldehydes, which will be emitted from the wood waste during the production of biochar in the pyrolysis or gasification [23]. In these cases, as long as the exhaust gas can be safely and economically treated, biochar production is potentially a reasonably viable method of utilizing discarded wood. However, if heat processing is insufficient or inefficient in removing certain contaminants, such as certain heavy metals, the biochar produced from the wood will have limited applications in the building industry or for agriculture, if at all (see chapter 2, section 2.3.2, for more information on potential contamination issues). Even if other non-wood based organic waste streams are used as feedstock for biochar, potential contamination risks need to be appropriately mitigated as well.

7.4 Upcycling of residues via biochar as an additive in construction materials

The entire process of pyrolysis and char production has been shown to be potentially carbon-negative [24–26]. It is estimated that, depending on feedstock and preparation conditions, biochar application can reduce greenhouse gas emissions by 870 kg CO_2-eq per tonne of dry biomass feedstock, of which 62%–66% can be contributed by the carbon sequestered in the biochar structure inherited from the parent biomass [26]. While application of biochar to soil is the most commonly considered method for the long-term storage of the carbon locked in biochar, it is not the only option.

Application of biochar as an additive to construction materials is a potentially attractive alternative, because biochar is a stable carbon store and it also offers the potential to reduce demand for sand or/and cement, while concurrently improving the physical properties of construction materials. That is, applying biochar as construction materials is a good way to upcycle—not just recycle—biomass waste.

Gupta *et al* [27–32] has extensively investigated the application of biochar as a supplementary cementitious admixture to reduce the demand for cement and silica fumes in cement-based construction materials. In one of their studies, biochar was prepared from wood waste, rice husk, and food waste between temperatures of 300 °C–500 °C. The findings showed that micro-pores of biochar contribute to moisture redistribution and enhance the degree of hydration in biochar–cement composites. Fine biochar also induces a filler effect, leading to a 20%–30% improvement in strength and a 40%–50% reduction in water permeability of biochar–cement and ++biochar–concrete composites.

Similar findings were reported by Choi *et al* [33]—the addition of biochar at 5% by mass of cement led to a 12% improvement in compressive strength. Restuccia and Ferro [34] observed 50%–60% improvement in strength and 60%–70% improvement in fracture toughness of cement composites with 0.80–1 wt.% biochar prepared from hazelnut shell and coffee powder. Biochar has also been applied as a viscosity modifier for asphalt and bitumen. Wang *et al* [35] reported improved hydration due to 1% and 2% biochar addition in sediment based construction products using cement as a binder. However, the application of coarse biochar (<250 μm) led to a reduction in compressive strength.

Biochar was reported to improve the rutting resistance of bitumen [36], improve mix consistency, and reduce water penetration in bitumen [37]. These findings suggest that incorporation of biochar as an admixture in cementitious and bitumen-based construction materials has the potential to improve the strength and durability during their service life. In many of these studies, the filler effect afforded by fine biochar particles was cited as the key reason that resulted in improvement in the mechanical properties and water tightness of the mortar/concrete containing these particles.

These and many other examples show that biochar produced from organic residues can be an effective way for upcycling of waste materials that has little use other than being combusted for energy recovery; biochar is a marketable product with wide-ranging applications in different areas of the construction industry, and biochar mortar was shown to be able to reduce GHG emissions by about 1.2 kg CO_2-eq for every kilogram of mortar.

7.5 Cascade/sequential uses of biochar

The phrase that is sometimes used, ‘cascade uses of biochar’, refers to a sequence of individual applications of biochar that are connected to one another. The term ‘cascade’ can, however, be somewhat misleading, as it implies progressively lower values of uses, as biochar moves through the cascade, but this picture does not necessarily reflect the truth in many cases. Hence, the phrase ‘sequential uses’ is preferred instead.

This concept, although initially proposed nearly a decade ago, is still in a relatively early stage of development. Despite this, several promising examples have already been identified and more will follow, as biochar research in different areas of application progresses and synergistic links between these areas become easier to identify [38]. The attraction of the sequential biochar concept is two-fold:

1. The use of biochar produced from a starting biomass material in more than one application increases the efficiency of the use of this material. This will lead to more positive environmental implications, as it is more likely to replace the use of other materials in individual applications. Consequently, it will also have cost implications, leading to reductions in the need for input materials and their end of life management.
2. The cost of biochar is related to many factors, including the cost of feedstock, transportation, and pyrolysis. When used in a sequence of applications, this cost can be distributed among these applications, where biochar offsets the costs of other materials (for example, activated carbon) or processes, and this makes it more affordable. As the sequence of uses does not diminish biochar's potential to store carbon, even up to the final step in the sequence, it is not difficult to see how a sequence of even just two to three applications can cover the cost of biochar, thus making the cost of carbon sequestration in the form of biochar cost-free. Carbon sequestration with zero cost is a target that is not achievable using any other currently proposed technology.

There are many potential combinations of biochar applications that can yield viable sequences, but the ideal sequences would have one feature in common. Each step/application in the sequence would yield biochar that is not only suitable for the following step/application, but actually yields biochar with improved properties with respect to the subsequent application [38]. Applications potentially suitable for forming synergistic sequences include the removal of nutrients (N, P) from waste water and effluents, use in composting or anaerobic digestion, use in animal feed, or use in gas cleaning.

In addition to the above listed advantages of the sequential biochar system, it also presents a number of practical challenges that need to be taken into consideration when designing such systems. These include issues such as potential contamination during one application that renders the biochar unsuitable for subsequent applications. Ownership of the material as it passes through the sequence is another potential concern that needs to be properly addressed. These and many other aspects of the sequential biochar use concept require attention and careful consideration, and are therefore fertile ground for future research and development.

7.6 Beyond technologies

7.6.1 Enabling policies to accelerate the development of biochar

The crux of building a successful CE ecosystem based on biochar goes beyond getting the science and engineering right. Kapoor *et al* [39] highlighted a need for an integrated approach for bio-waste management and policy to support the successful implementation of CE technologies for resource recycling and recovery. One may view management practices and policies as providing the right sets of conditions or environments for realizing the full potential of these CE technologies and technological strategies. Common management and policy gaps and hindrances have been highlighted in the literature, including the lacunae in local/global supply chains, a lack of technical knowledge of the process of recycling/recovery of bio-wastes, and lack of market incentives (either real or perceived) to develop technical capabilities in the supply chains. For example, if the collection and segregation of biological waste translates into huge capital and operational costs for companies that are not provided with appropriate incentives (such as grants from private investors or government), then it is highly likely that the lack of companies performing the crucial role of waste collection and segregation will result in the said gap in the supply chain, thus failing any effort to establish a holistic and sustainable waste recycling ecosystem.

An example of a lack of market incentive is the lack of a strong 'want signal' sent out by consumers, which can in turn be due to a range of different reasons, including the presence of incumbent competing products that are better known in the market, the lack of consumer confidence in the products that are derived from waste, a lack of enforceable and strict standards that prevent the emergence of sub-standard 'imitation goods' that undercut the authentic or high quality products, and the fact that consumers may be unfamiliar with the products.

In this light, management and policy strategies that can promote CE will be those that address these gaps and barriers. For example, subsidies providing sufficient incentives for start-up firms to play key roles in the supply chain (for example, waste collection) can help to kick start local CE ecosystems. More specifically, these subsidies can come in the form of the preferential leasing of tools, machinery, or equipment, a discounted fee to attend educational programs on financial management or business development for small businesses, and initial discounts for customers who purchase products derived from recycled or recovered wastes.

7.6.2 Developing biochar within an industrial symbiotic network

On the management aspect, special meta-industrial master-planning can be undertaken to deliberately match-make industries or companies to form a local ecosystem to support the development of a network in which resources are shared amongst companies. Such a network is popularly known as an industrial symbiosis. In this network, the resources shared can either be raw materials or waste. One of the best examples of an industrial symbiotic network is the Kalundborg Eco-Industrial Park, which comprises companies in the region collaborating with one another by sharing resources and/or passing waste from one company to another that will utilize it as raw materials for further processing.

Understanding the industrial symbiotic network developed in the Kalundborg Eco-Industrial Park provides an idea of how a local bio-CE based on biochar may be developed, either with or without governmental interventions. In Kalundborg

Eco-Industrial Park, the Asnæs Power Station produces fly ash from the combustion of pulverized coal in its coal-fired boilers. Instead of discarding this residue, the fly ash is passed to the local cement company to be used as a pozzolanic material. For this to happen, the cement company has to change the conventional mindset and expand its product line to include cement-replacing materials as well. The local pharmaceutical company Novo Nordisk passes the biomass and yeast slurry it produces to the local farm as fertilizers or soil conditioners. Finally, Asnæs Power Station and Novo Nordisk are linked by the latter's use of steam that is generated from the heat of the power station. All in all, the highly elaborated material and energy exchange pathways in the industrial symbiotic network of the Kalundborg Eco-Industrial Park were set up to maximize the conversion of the 'waste of one entity into food of another'. Interpreted more widely, the success of an industrial symbiotic network can potentially be a strong driver for a successful bio-CE, in which 'waste' remains in the economy as cyclic flows, and generating economic values within the economy for as long as possible.

A biochar-based CE can be built on the principles of industrial symbiosis. This industrial symbiotic network can be created based on the idea of sequential use. For example, within the network, after biochar completes its 'first use' to remove nutrients (N, P) from waste water, it can be passed to the local agricultural sector that will use the biochar in compost. In other words, forming this biochar-exchange network within the CE helps to increase the overall efficiency in circulating the biochar through the different 'stations of use'.

7.7 Conclusions

The roles of biochar in CE can be related to resource use efficiency, material recycling/upcycling and cascade (sequential) uses. Creating an industrial symbiotic network, in which biochar is circulated within the network for different uses and functions, is argued to be a good way to keep increasing biochar value within the CE. However, the sustainability of this CE will require the creation and implementation of appropriate policies that enable all the relevant entities in the network to benefit from their respective role in circulating biochar to generate values for the network. Finally, it is hoped that the different concepts discussed in this work will motivate more research on these topics in the future.

References

[1] Murray A, Skene K and Haynes K 2017 The circular economy: an interdisciplinary exploration of the concept and application in a global context *J. Bus. Ethics* **140** 369–80

[2] Babbitt C W, Gaustad G, Fisher A, Chen W-Q and Liu G 2018 Closing the loop on circular economy research: from theory to practice and back again *Resour. Conserv. Recycl.* **135** 1–2

[3] Hofmann F 2019 Circular business models: business approach as driver or obstructer of sustainability transitions? *J. Clean. Prod.* **224** 361–74

[4] Morseletto P 2020 Targets for a circular economy *Resour. Conserv. Recycl.* **153** 104553

[5] Potting J, Hekkert M, Worrell E and Hanemaaijer A 2017 *Circular Economy: Measuring Innovation in the Product Chain* (The Hague: PBL Publishers)

[6] Leal Filho W 2018 Bioeconomy meets the circular economy: the RESYNTEX and FORCE projects *Towards a Sustainable Bioeconomy: Principles, Challenges and Perspectives* (Cham: Springer) pp 567–75
[7] Kumar S S, Kumar V, Kumar R, Malyan S K and Pugazhendhi A 2019 Microbial fuel cells as a sustainable platform technology for bioenergy, biosensing, environmental monitoring, and other low power device applications *Fuel* **255** 115682
[8] de Wit C T 1992 Resource use efficiency in agriculture *Agric. Syst.* **40** 125–51
[9] Liao J, Liu X, Hu A, Song H, Chen X and Zhang Z 2020 Effects of biochar-based controlled release nitrogen fertilizer on nitrogen-use efficiency of oilseed rape (*Brassica napus* L.) *Sci. Rep.* **10** 11063
[10] Xia H, Riaz M, Zhang M, Liu B, El-Desouki Z and Jiang C 2020 Biochar increases nitrogen use efficiency of maize by relieving aluminum toxicity and improving soil quality in acidic soil *Ecotoxicol. Environ. Saf.* **196** 110531
[11] Puga A P, Grutzmacher P, Cerri C E P, Ribeirinho V S and de Andrade C A 2020 Biochar-based nitrogen fertilizers: greenhouse gas emissions, use efficiency, and maize yield in tropical soils *Sci. Total Environ.* **704** 135375
[12] Elser J and Bennett E 2011 A broken biogeochemical cycle *Nature* **478** 29–31
[13] Withers P J A *et al* 2018 Transitions to sustainable management of phosphorus in Brazilian agriculture *Sci. Rep.* **8** 2537
[14] dos Santos S R, Lustosa Filho J F, Vergütz L and Melo L C A 2019 Biochar association with phosphate fertilizer and its influence on phosphorus use efficiency by maize *Ciência Agrotecnol.* **43** e025718
[15] Borges B M M N, Strauss M, Camelo P A, Sohi S P and Franco H C J 2020 Re-use of sugarcane residue as a novel biochar fertiliser—increased phosphorus use efficiency and plant yield *J. Clean. Prod.* **262** 121406
[16] Yeboah E, Asamoah G, Kofi B and Abunyewa A A 2016 Effect of biochar type and rate of application on maize yield indices and water use efficiency on an ultisol in Ghana *Energy Procedia* **93** 14–8
[17] Misra A K 2014 Climate change and challenges of water and food security *Int. J. Sustain. Built Environ.* **3** 153–65
[18] Amanambu A C *et al* 2020 Groundwater system and climate change: present status and future considerations *J. Hydrol.* **589** 125163
[19] Lu S, Bai X, Li W and Wang N 2019 Impacts of climate change on water resources and grain production *Technol. Forecast. Soc. Change* **143** 76–84
[20] Bergeron F C 2016 Energy and climate impact assessment of waste wood recovery in Switzerland *Biomass Bioenergy* **94** 245–57
[21] Falk B 1997 Opportunities for the woodwaste resource *For. Prod. J.* **47** 17–22
[22] Bush R J, Reddy V S and Araman P A 1997 Pallets: a growing source of recycled wood *Proc., The Use of Recycled Wood and Paper in Building Applications* pp 24–8
[23] Sivrikaya H, Tesařová D, Jeřábková E and Can A 2019 Color change and emission of volatile organic compounds from Scots pine exposed to heat and vacuum-heat treatment *J. Build. Eng.* **26** 100918
[24] Lehmann J and Joseph S (ed) 2015 *Biochar for Environmental Management* (London: Taylor and Francis)
[25] Lehmann J 2007 Bio-energy in the black *Front. Ecol. Environ.* **5** 381–87

[26] Roberts K G, Gloy B A, Joseph S, Scott N R and Lehmann J 2010 Life cycle assessment of biochar systems: estimating the energetic, economic, and climate change potential *Environ. Sci. Technol.* **44** 827–33

[27] Gupta S, Kua H W and Koh H J 2018 Application of biochar from food and wood waste as green admixture for cement mortar *Sci. Total Environ.* **619–620** 419–35

[28] Gupta S, Kua H W and Pang S D 2018 Biochar–mortar composite: manufacturing, evaluation of physical properties and economic viability *Constr. Build. Mater.* **167** 874–89

[29] Gupta S and Kua H W 2019 Carbonaceous micro-filler for cement: effect of particle size and dosage of biochar on fresh and hardened properties of cement mortar *Sci. Total Environ.* **662** 952–62

[30] Gupta S and Kua H W 2018 Effect of water entrainment by pre-soaked biochar particles on strength and permeability of cement mortar *Constr. Build. Mater.* **159** 107–25

[31] Gupta S and Kua H W 2020 Combination of biochar and silica fume as partial cement replacement in mortar: performance evaluation under normal and elevated temperature *Waste Biomass Valorization* **11** 2807–24

[32] Gupta S, Kua H W and Pang S D 2020 Effect of biochar on mechanical and permeability properties of concrete exposed to elevated temperature *Constr. Build. Mater.* **234** 117338

[33] Choi W C, Do Yun H and Lee J Y 2012 Mechanical properties of mortar containing bio-char from pyrolysis *J. Korea Inst. Struct. Maint. Insp.* **16** 67–74

[34] Restuccia L and Ferro G A 2016 Nanoparticles from food waste: a 'green' future for traditional building materials *Proc. 9th Int. Conf. on Fracture Mechanics of Concrete and Concrete Structures*

[35] Wang L *et al* 2019 The roles of biochar as green admixture for sediment-based construction products *Cem. Concr. Compos.* **104** 103348

[36] Zhao S, Huang B, Ye X P, Shu X and Jia X 2014 Utilizing bio-char as a bio-modifier for asphalt cement: a sustainable application of bio-fuel by-product *Fuel* **133** 52–62

[37] Chebil S, Chaala A and Roy C 2000 Use of softwood bark charcoal as a modifier for road bitumen *Fuel* **79** 671–83

[38] Wurzer C, Sohi S P and Mašek O 2019 Synergies in sequential biochar system *Advanced Carbon Materials from Biomass: An Overview, Zenodo* (Geneva: CERN) pp 147–59

[39] Kapoor R, Ghosh P, Kumar M and Vijay V K 2019 Evaluation of biogas upgrading technologies and future perspectives: a review *Environ. Sci. Pollut. Res.* **26** 11631–61

Part II

Biochar: a filler for composites

Chapter 8

Shielding effectiveness of biochar composites at microwave frequency

Muhammad Yasir and Patrizia Savi

The demand for eco-friendly materials is increasing due to rapid population growth and increasing emissions of greenhouse gases. In order to pave the way for widespread use of eco-friendly materials for novel applications, there is a growing need for morphological, mechanical, electrical, and microwave characterization. A number of studies have been performed on the use of carbon based materials, specifically biochar, in electromagnetic interference/shielding effectiveness applications. Biochar (extracted from plants, sewage sludge, or other sources) can be used as a filler in epoxy resin or cement composites. With an increased interest in the use of eco-friendly materials, novel methods of characterization are required for accurate detection of the material properties and the proposal of relevant applications. There are several measurement techniques for finding the microwave properties of materials. The measurement techniques can be broadly divided into two categories based on the measurement of the transmission/reflection in an enclosed microwave device, e.g. a waveguide, coaxial, resonant cavity, etc, and the measurement of the transmission/reflection in a free space.

In this chapter, the transmission, reflection, and absorption properties of sewage sludge biochar in the microwave frequency band are studied. Samples of epoxy resin filled with sewage sludge are fabricated in order to fit rectangular waveguides in the C-band and X-band. Transmission and reflection coefficients are measured on samples of 4 mm thickness filled with 10 wt.% and 20 wt.% of biochar. Shielding effectiveness is also evaluated from the measured transmission coefficients.

8.1 Introduction

The demand for eco-friendly materials is increasing due to rapid population growth and increasing emissions of greenhouse gases. In order to pave the way for widespread use of eco-friendly materials for novel applications, there is a growing

doi:10.1088/978-0-7503-2660-5ch8

need for morphological, mechanical, electrical, and microwave characterization. A number of studies have been performed on the use of carbon based materials, specifically biochar, in electromagnetic interference/shielding effectiveness applications. Biochar (extracted from plants, sewage sludge, or other sources) can be used as a filler in epoxy resin or cement composites.

There is a high demand for an increase in the data rate and available spectrum for realizing 5G cellular technology [1]. This makes the use of higher frequencies inevitable [2, 3]. At the micro- and millimetre wave frequencies, transmission media such as coaxial and microstrip lines suffer high insertion loss [4, 5]. Waveguide technology is an attractive alternative for use at higher frequencies since it possesses lower insertion loss (see e.g. [6]). The growing interest in the use of waveguide technology calls for the increased use of such technology in various applications, e.g. wireless communications (see e.g. [7]), power transfer (see e.g. [8]), and material and film characterization [9, 10].

With an increased interest in the use of eco-friendly materials, novel methods of characterization are required for accurate detection of the material properties and the proposal of relevant applications. The use of eco-friendly materials has been growing due to the excess production of pollutants that has aggravated the problems of climate change and global warming (see e.g. [11]). The treatment of waste water produces sewage sludge, which contains a number of pollutants including heavy metals and pathogens. With a rapid increase in human population, there has been a considerable increase in the creation of sewage [12, 13]. Sludge derived from sewage is used in agricultural applications as a manure. Due to the ever-increasing production of sewage sludge and limited opportunities of disposal and productive use, there is an imminent need for using it in innovative applications. This will result in value addition to the product and also help in recycling unwanted and environmentally damaging material. The use of carbon negative materials in innovative applications can be effective in reducing the carbon footprint. A number of studies have been performed for the use of carbon based materials (see e.g. [14–16]) specifically biochar in electromagnetic interference/shielding effectiveness applications [17, 18].

There are several measurement techniques for finding the microwave properties of materials [19, 20]. The measurement techniques can be broadly divided into two categories based on the measurement of the transmission/reflection in an enclosed microwave device, e.g. waveguide [21, 22], coaxial [23], resonant cavity [24], etc, and the measurement of the transmission/reflection in a free space. The measurements of transmission reflection in free space can be performed in air [25], inside a reverbation chamber [26], or using a shielded box/room method [27]. In measurements performed in an enclosed microwave device the samples need to be of a particular shape and size in order to fit the microwave device. In the free space method, the sample needs to be placed in between two antennas. There is less constraint on the material shape in the free space measurement techniques.

In this chapter, the transmission, reflection and absorption properties of biochar in the microwave frequency band are studied. Samples of epoxy resin filled with sewage sludge biochar are considered. The structural properties of the biochar used in this chapter have been investigated in [17, 18] using RAMAN and FESEM analysis. In section 8.2, the transmission, reflection, and absorption properties are introduced. In section 8.3, the transmission coefficient, reflection coefficient, and shielding effectiveness of samples filled with 10 wt.% and 20 wt.% of biochar in the C-band and X-band frequencies are shown.

8.2 Transmission, reflection, and absorption

The two main categories of methods used for the measurement of transmission, reflection, and absorption of microwaves in materials are the transmission/reflection methods inside a waveguide or coaxial air-line and free space measurement methods. In both the methods, the S-parameters are acquired from two-port measurements by a vector network analyser (VNA).

The transmission/reflection methods involve the placement of a slab of the material under test (MUT) inside a waveguide or coaxial line and measuring the S-parameters by the help of a VNA. In this case, the MUT should be shaped to fit the inside of a waveguide or coaxial air-line, usually with the help of an MUT holder. The VNA is calibrated at the interconnection between the waveguide and the MUT holder. Usually, the calibration standard used is the TRL calibration.

The free space method involves the placement of the MUT between two antennas connected to a VNA. The antennas are placed facing each other with a flat MUT in the middle. This method also involves the calibration of the VNA prior to the measurements. The S-parameters are measured once the system is calibrated.

The four S-parameters S_{11}, S_{21}, S_{12}, and S_{22} of a two-port system are defined as follows:

$$S_{11} = \left.\frac{b_1}{a_1}\right|_{a2=0}, \quad S_{21} = \left.\frac{b_2}{a_1}\right|_{a2=0}, \quad S_{12} = \left.\frac{b_1}{a_2}\right|_{a1=0}, \quad S_{22} = \left.\frac{b_2}{a_2}\right|_{a1=0}, \tag{8.1}$$

where a_i is the incident signal at the port i and b_i is the reflected signal at the port i (see figure 8.1). When computing S_{11} and S_{21}, the signal $a_2 = 0$ while when computing S_{12} and S_{22}, $a_1 = 0$. The terms a_i and b_i can be defined in terms of forward travelling voltage wave (V_i^+) and backward travelling voltage wave (V_i^-) as follows:

$$a_i = \frac{V_i^+}{\sqrt{Z_0}}, \; b_i = \frac{V_i^-}{\sqrt{Z_0}}, \tag{8.2}$$

where Z_0 is the reference impedance at the port.

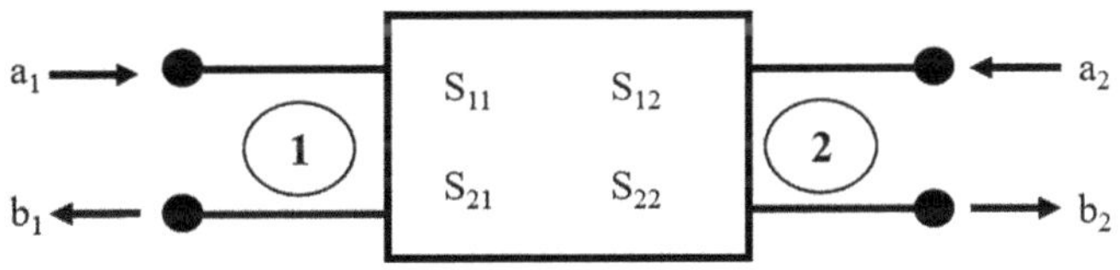

Figure 8.1. Definition of scattering parameters.

There are many useful parameters that are representative of a material's microwave properties that can be derived from the S-parameter, one of which is the shielding effectiveness (SE) defined in terms of transmission scattering as follows:

$$\mathrm{SE} = |S_{21}|.$$

The transmission loss or shielding effectiveness of a material is composed of dissipation loss, L_D, and mismatch loss, L_M [15, 16]:

$$\mathrm{SE} = L_D + L_M. \tag{8.3}$$

The mismatch loss, L_M, can be calculated from the reflection loss, S_{11}, by

$$L_M = -10\log_{10}(1 - |S_{11}|^2). \tag{8.4}$$

The dissipation loss, L_D, can be expressed in terms of reflection loss, S_{11}, and insertion loss, S_{21}, as follows:

$$L_D = -10\log_{10}\left(\frac{|S_{21}|^2}{1 - |S_{11}|^2}\right). \tag{8.5}$$

These parameters can be evaluated by realizing a two-port system by the help of waveguides, which will be performed in section 8.3.

8.3 Waveguide method for transmission reflection evaluation

Samples of standard epoxy and sewage sludge biochar composites are fabricated and measured in a waveguide structure in the C- and X-bands, which covers the frequency bandwidth of 5–12 GHz. Samples of adequate dimensions to fit in the waveguide cross section are fabricated. The measurements of the scattering parameters are performed in WR137 waveguide (5–8 GHz) and the WR90 waveguide (8–12 GHz).

8.3.1 Sample fabrication

The biochar used in the fabrication of the composites is sewage sludge biochar. It is mainly composed of human excrement acquired through the sewerage system. It is then processed and converted into a class B biosolid by anaerobic digestion. Further processing is performed for the conversion of the class B biosolids to class A biosolids by the help of a BioDryer. This is a process developed by the Bioforcetech Corporation that reduces the weight of biodegradable waste by heating it in the initial stages of composting. This is a sustainable process that does not involve any external heat. The heat is a kind of biological heat auto-generated by aerobic degradation. The biosolids are then converted into biochar through pyrolysis. In the absence of oxygen, the biosolids are heated for 20 min at a temperature of 650 °C. In this whole process, no any harmful gases or carbon are left in the environment. A vertical furnace and a quartz reactor are used for the annealing of the biochar acquired from the Bioforcetech Corporation. The annealing is performed at a temperature of 1000 °C for 30 min in argon atmosphere in order to make it more conductive. The heating rate adopted is 15 °C min^{-1}.

In order to make composites of the sewage sludge biochar (Bioforcetech Corporation), a pre-weighed quantity of the biochar is mechanically mixed in an epoxy resin (Hexion, Infusion Resin RIMR-135). A hardener (RIMR-134) is then added to polymerize the mixture. The mixture of the biochar, epoxy resin, and hardener was then transferred to moulds of specific shape. The detailed procedure of the composite preparation is described in [17, 18].

Scattering parameters in the C-band and X-band are measured in waveguides which require cuboid shaped samples to fill the interior of the particular waveguide. In order to fabricate samples of precise dimensions, a master mould is 3D printed. Liquid silicone is used to fabricate reusable moulds. The shape and size of the 3D printed master moulds and the silicone moulds can be varied according to the shape of the waveguides of different dimensions working in different frequency bands. The elasticity of silicone facilitates easy extraction of the biochar samples once they have polymerized. The 3D printed master mould and silicone mould spacer for the WR137 waveguide are shown in figure 8.2(a) and that of the WR90 waveguide are shown in figure 8.2(b).

After the fabrication of the samples, they are placed in a metal holder that is placed between two flanges of the waveguide, as shown in figure 8.3. The dimensions of the WR137 are 34.84 mm × 15.79 mm and the frequency band of operation is from 5 to 8 GHz. The dimensions of the WR90 waveguide are 22.86 × 10.16 mm,

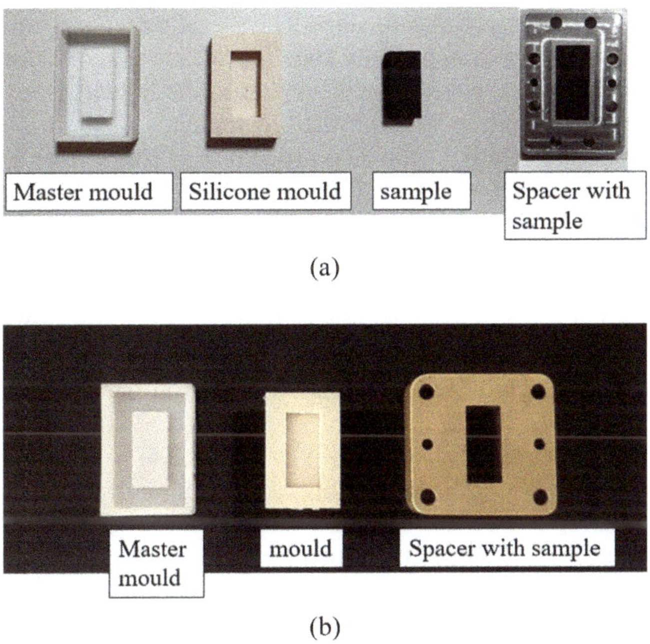

Figure 8.2. Waveguide sample fabrication with a master mould, mould, and spacer: (a) C-band and (b) X-band [18].

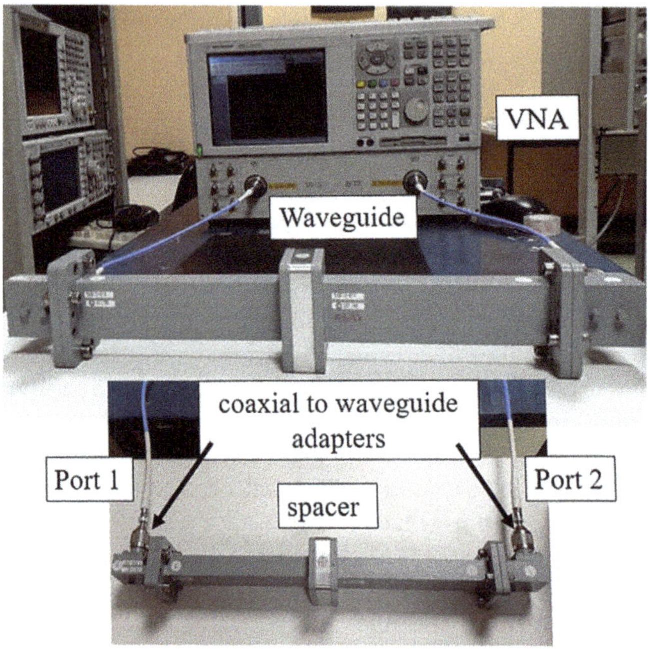

Figure 8.3. Waveguide measurements using VNA. The inset shows the waveguide to coaxial adapters and the spacer connected to the ends of a waveguide.

and the respective frequency band is from 8 to 12 GHz. A vector network analyser is used to measure the two-port S-parameters. The measurement set-up for the WR137 waveguide is shown in figure 8.3. Note that the measurement set-ups for the two waveguides are identical except for the waveguides, the holder, and the sample size.

The vector network analyser is calibrated at the ends of the waveguide sections where the spacer is connected. A standard procedure of transmission, reflection, and isolation is adopted as thru, short, and load as a part of the calibration procedure.

Since the waveguide set-up is calibrated at the point where the spacer is connected and therefore takes into account all the losses incurred in the transmission up to the spacer, the empty waveguide shows almost lossless transmission. Afterwards the spacer is fitted with a pristine resin sample, a 10 wt.% biochar sample and a 20 wt.% biochar sample. The transmission (S_{21}) is measured for each of the mentioned cases. The shielding effectiveness can be calculated from the transmission scattering as $\mathrm{SE} = -10\log(S_{21})$.

In the WR137 waveguide set-up shown in figure 8.4, the measurements of the transmission and reflection scattering shows a transmission loss of around −10 dB for the 20 wt.% sample. For the 10 wt.% sample, the transmission loss is almost −7 dB. For the pristine resin, this value is almost −2 dB. The thickness of the samples for these measurements is 4 mm.

When classified into the two different losses, the total insertion loss and the dissipative loss for the 10 wt.% (figure 8.5 left) and the 20 wt.% (figure 8.5 right)

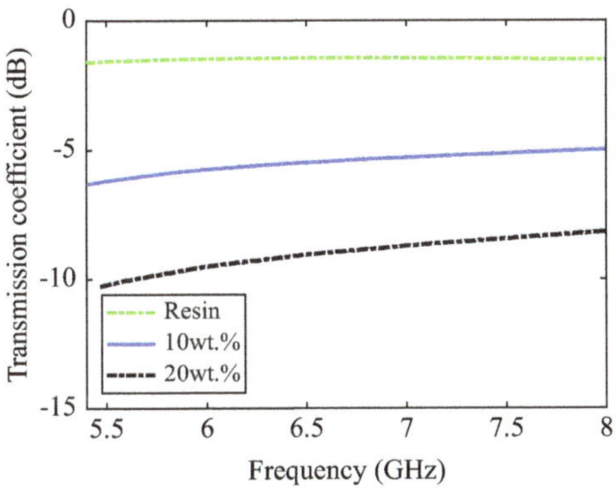

Figure 8.4. Transmission loss of spacer, pristine resin, 10 wt.% and 20 wt.% in a WR137 waveguide set-up [17].

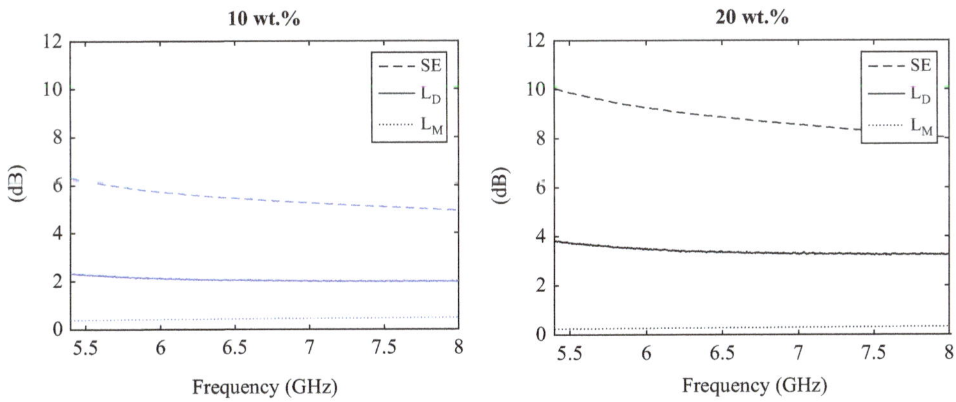

Figure 8.5. Transmission loss and dissipation loss for the 10 wt.% and 20 wt.% samples [17].

samples are shown in figure 8.5. A major portion of the total insertion loss comprised of dissipative loss.

Identical to the analysis performed for the WR137 waveguide, an analysis of the samples is performed with the WR90 waveguide. In this case, the transmission loss in an empty waveguide, the pristine resin, and 10 wt.% and 20 wt.% samples are shown in figure 8.6. The empty waveguide shows almost lossless transmission while the resin causes a transmission loss of almost −3 dB. The 10 wt.% biochar produces less than −5 dB transmission loss, while the 10 wt.% samples cause a transmission loss of less than −10 dB.

A classification of the shielding effectiveness between the reflective and dissipative losses is performed in the case of the WR90 waveguide. In figure 8.7, the classification of the total transmission loss performed according to (1)–(3) is reported. A significant portion of the transmission loss is due to the dissipative loss, which increases with an increase in the filler percentage.

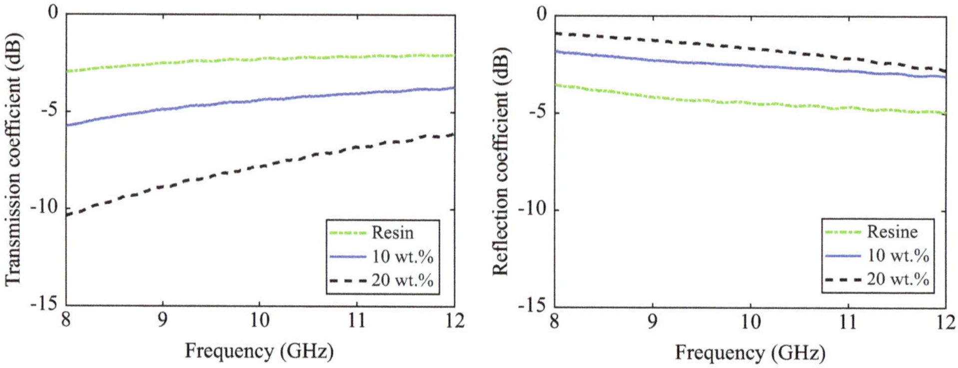

Figure 8.6. Transmission loss (S_{21}) versus frequency of the empty waveguide, pristine resin, and 10 wt.% and 20 wt.% samples [18].

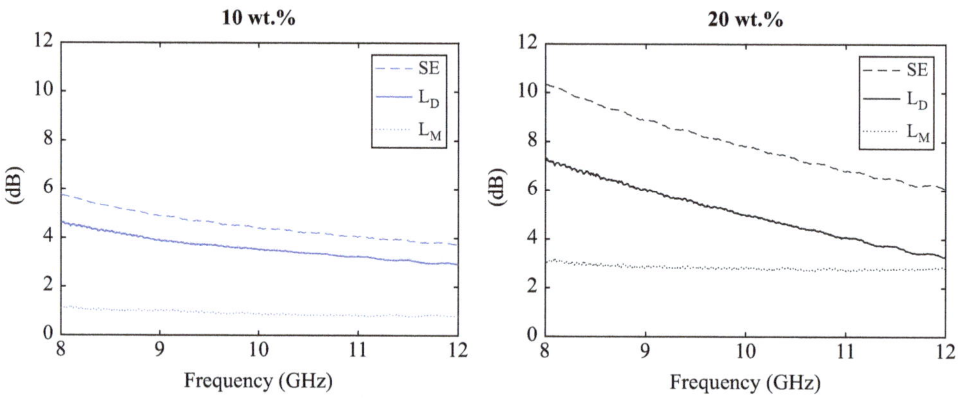

Figure 8.7. Shielding effectiveness classified into dissipation loss and mismatch loss for the 10 wt.% sample (top) and 20 wt.% sample (bottom) [18].

8.4 Conclusions

Measurements of the transmission and reflection scattering parameters is crucial for understanding the microwave properties of novel materials such as biochar. The waveguide method is effective in measurements of the scattering parameters and the shielding effectiveness values of composites filled with biochar being a method covering a sufficiently large bandwidth with reasonable accuracy. Its main drawback is that it requires samples to be of a specific dimension and shape in order to fit the interior of the waveguide. Measurement results have shown that the increase of percentage of biochar filler in composites considerably increases the transmission loss. Classification of the total transmission loss into dissipative and mismatch loss shows that a major portion of the total transmission loss is dissipative in nature.

Acknowledgements

The authors would like to thank Matteo Longo of Bioforcetech for providing the sludge biochar and Dr Giuseppe Virone for the WR137 waveguide set-up.

References

[1] Chen S and Zhao J 2014 The requirements, challenges, and technologies for 5G of terrestrial mobile telecommunications *IEEE Commun. Mag.* **52** 36–43

[2] Al-Yasir Y I A, Parchin N O, Abd-Alhameed R A, Abdulkhaleq A M and Noras J M 2019 Recent progress in the design of 4G/5G reconfigurable filters *Electronics* **8** 114

[3] Hong W *et al* 2017 Multibeam antenna technologies for 5G wireless communications *IEEE Trans. Antennas Propag.* **65** 6231–49

[4] Yoneyama T and Nishida S 1981 Nonradiative dielectric waveguide for millimeter-wave integrated circuits *IEEE Trans. Microwave Theory Tech.* **29** 1188–92

[5] Sisson M J, Brigginshaw P M, Wood P N, Brown P R, Hansom A M and Nicholls M R 1982 Microstrip devices for millimetric frequencies *IEEE MTT-S Int. Microwave Symp. Digest (Dallas, TX)* pp 212–14

[6] Itoh T 1976 Inverted strip dielectric waveguide for millimeter-wave integrated circuits *IEEE Trans. Microwave Theory Tech.* **24** 821–27

[7] Rappaport T S, Murdock J N and Gutierrez F 2011 State of the art in 60-GHz integrated circuits and systems for wireless communications *Proc. IEEE* **99** 1390–436

[8] Rakotomalala I, Lemaitre-Auger P and Tedjini S 2019 UHF near-field wireless RFID power transfer through two distant rectangular waveguides *49th European Microwave Conf. (EuMC) (Paris, France)* pp 547–50

[9] Lee C-K *et al* 2019 Evaluation of microwave characterization methods for additively manufactured materials *Designs* **3** 1–17

[10] Costa F, Borgese M, Degiorgi M and Monorchio A 2017 Electromagnetic characterisation of materials by using transmission/reflection (T/R) devices *Electronics* **6** 1–27

[11] Bezama A and Agamuthu P 2019 *Addressing the Big Issues in Waste Management* (London: SAGE)

[12] Yang G, Zhang G M and Wang H C 2015 Current state of sludge production, management, treatment and disposal in China *Water Res.* **78** 60–73

[13] Giorcelli M, Savi P, Delogu A, Miscuglio M, Yahya M H and Tagliaferro A 2013 Microwave absorption properties in epoxy resin multi walled carbon nanotubes composites *Int. Conf. Electromagnetics in Advanced Applications (Torino, Italy)*

[14] Giorcelli M, Savi P, Miscuglio M, Yahya M H and Tagliaferro A 2014 Analysis of MWCNT/epoxy composites at microwave frequency: reproducibility investigation *Nanoscale Res. Lett.* **9** 168

[15] Giorcelli M, Savi P, Yasir M, Miscuglio M, Yahya M H and Tagliaferro A 2015 Investigation of epoxy resin/multiwalled carbon nanotube nanocomposite behavior at low frequency *J. Mater. Res.* **30** 101–7

[16] Savi P, Yasir M, Giorcelli M and Tagliaferro A 2017 The effect of carbon nanotubes concentration on complex permittivity of nanocomposites *Prog. Electromagn. Res. M.* **55** 203–9

[17] Savi P and Yasir M 2020 Waveguide measurements of biochar derived from sewage sludge *IET Electronics Lett.* **56** 335–7

[18] Savi P, Yasir M, Bartoli M, Giorcelli M and Longo M 2020 Electrical and microwave characterization of thermal annealed sewage sludge derived biochar composites *Appl. Sci.* **10** 1334

[19] Jagadeesh Chandra R B *et al* 2019 Hybrid polymer composites for EMI shielding application —a review *Mater. Res. Express* **6** 082008

[20] Ibrar Ahmed *et al* 2020 Graphene-ferrites interaction for enhanced EMI shielding effectiveness of hybrid polymer composites *Mater. Res. Express* **7** 016304

[21] Rudd M, Baum T C and Ghorbani K Jan. 2020 Determining high-frequency conductivity based on shielding effectiveness measurement using rectangular waveguides *IEEE Trans. Instrum. Meas.* **69** 155–62

[22] Marra F, Lecini J, Tamburrano A, Pisu L and Sarto M S 2020 Broadband electromagnetic absorbing structures made of graphene/glass-fiber/epoxy composite *IEEE Trans. Microwave Theory Tech.* **68** 590–601

[23] Diaz Angulo L M *et al* Dec. 2019 Modeling and measuring the shielding effectiveness of carbon fiber composites *IEEE J. Multiscale Multiphys. Comput. Tech.* **4** 207–13

[24] Massoni E, Siciliano G, Bozzi M and Perregrini L 2018 Enhanced cavity sensor in SIW technology for material characterization *IEEE Microwave Wireless Compon. Lett.* **28** 948–50

[25] Neo C P and Varadan V K Feb. 2004 Optimization of carbon fiber composite for microwave absorber *IEEE Trans. Electromagn. Comput.* **46** 102–6

[26] Gifuni A *et al* 2020 Latest developments on the shielding effectiveness measurements of materials and gaskets in reverberation chambers *IET Sci. Meas. Technol.* **14** 435–45

[27] Jiajia Duan *et al* 2019 Effect of double-layer composite absorbing coating on shielding effectiveness of electromagnetic shielding fabric *Mater. Res. Express* **6** 086109

Chapter 9

Flame retardant polymer systems containing biochar: current state-of-the-art and perspectives

Samuele Matta, Mattia Bartoli and Giulio Malucelli

Both in the academic and industrial worlds, the field of flame retardance of plastics and textile materials is currently experiencing the use of bio-sourced and low environmental impact products for the design of effective and 'green' alternatives to common flame retardants. Among them, biochar, a solid product obtained from the thermo-chemical conversion of biomasses in an oxygen-limited environment, is gathering great interest as a cheap carbon source, suitable, in combination with standard flame retardant additives, for enhancing the resistance of different polymer-based systems to a flame application or to the exposure of an irradiative heat source. This chapter will thoroughly describe the novel current state-of-the-art concerning the use of this carbon material as a flame retardant, discussing the recent advances and highlighting some future possible developments.

9.1 Introduction

The ease of flammability is one of the most severe issues for many plastics and textiles. In fact, if not inherently flame retardant, these materials burn very vigorously when exposed to an irradiative heat source or to a flame. In addition, during the combustion smoke is generated, which can be toxic, hence increasing the hazards of fire [1–3].

To overcome this problem, more than sixty years ago academic and industrial scientists started designing, developing, and producing so-called flame retardants (FRs), i.e. chemical products that are able to interfere with flame propagation, slowing down or even stopping combustion processes [4–6]. In addition, the efficiency of these products has been widely assessed, tailoring their chemical structure and composition and the related flame retardant mechanisms to the polymer matrices to be protected.

doi:10.1088/978-0-7503-2660-5ch9

The first family of very efficient flame retardants comprised halogen-based products (mainly chlorinated and brominated structures). Despite their outstanding efficacy in protecting the polymer matrices, which they are embedded in from fire, some of them (namely, decabromodiphenyl or pentabromodiphenyl ethers and polychlorinated biphenyls) showed high toxicity or even mutagenic features, prompting the European Community to ban them from the market [7–9].

This concern inspired the scientific and industrial world toward the design of less toxic products, i.e. with low environmental impact. Within this effort, phosphorus-based (also in combination with nitrogen) flame retardants have been recognized as suitable replacements for their halogenated counterparts [10–12].

Apart from the quite recent exploitation of nanotechnology and nanomaterials (in particular in the form of nanoparticles or nanolamellae) as flame retardant fillers for different thermoplastics and thermosets [13–16], at present, the scientific community is moving toward the design of bio-sourced products. In this way it is possible to combine the 'green' character of these materials with the current need to provide wastes, by-products, and crops with further added value within the circular economy concept. In the literature, there are several examples that prove the efficiency of utilizing biomacromolecules (such as selected proteins and nucleic acids, alone or in combination with standard flame retardants) or bio-sourced compounds (i.e. tannins, phytic acid, banana pseudostem sap, lignins, pomegranate rind extract, to mention only a few) as low environmental impact FRs [17–22]. Indeed, these biomacromolecules and bio-sourced products possess a chemical composition that comprises those elements (namely phosphorus, nitrogen, and/or sulfur) usually required for designing effective flame retardants [23, 24].

In view of sustainability, biochar, a solid product obtained from the thermo-chemical conversion of wastes/biomasses in an oxygen-limited environment, represents a renewable/bio-sourced material that has been widely utilized in environmental management and agriculture. In addition, it has been exploited as a low-cost carbon sequester and a natural adsorbent [25, 26], as well as in soil remediation and amendment, thanks to its remarkably stable honeycomb-like carbonaceous structure [27, 28]. All these peculiarities suggest the possibility of using biochar either as a filler in polymer composites with enhanced thermal and flame retardant properties [29], or as a component of flame retardant coatings for fabrics [30], hence further widening its potential uses. This very recent application will be discussed in the following, providing the reader with the most significant examples discussed in the scientific literature so far.

9.2 Flame retarded systems containing biochar

The first pioneering works on the use of biochar as flame retardant filler in thermoplastics dates back to 2016. In particular, Das *et al* [31] prepared polypropylene-biochar composites at four loading levels (namely, 15, 25, 30, and 35 wt.%) by melt compounding and injection molding. The biochar was derived from pine wood feedstock that was subjected to pyrolysis at 500 °C and subsequent activation at 900 °C.

The incorporation of increasing filler loadings in the polymer matrix turned out to enhance the mechanical behavior of polypropylene, in terms of tensile modulus and flexural strength/modulus. Interestingly, as assessed by cone calorimetry tests, the presence of the carbon filler promoted a remarkable decrease of the peak of heat release rate (from 1054-unfilled polymer, to 477 kW m^{-2} for the composite containing 35 wt.% of biochar).

Pursuing this research, Ikram *et al* [32] further demonstrated the efficiency of this carbon filler in polypropylene. The composites containing 36 wt.% of biochar showed an enhanced thermal stability, as assessed by thermogravimetric analyses; furthermore, forced combustion tests revealed a significant decrease of the peak of heat release rate (almost −50% with respect to unfilled polypropylene), hence demonstrating the flame retardance provided by the carbon filler. This finding was ascribed to the formation of a char layer on the top surface of the composites, upon exposure to the irradiative heat flux: the layer behaved as a steady barrier, delaying the transfer of heat, oxygen, and combustible volatiles between the polymer and the radiant heat source.

Pursuing this research, the same group [33] demonstrated the occurrence of synergistic effects between biochar and wool fibers, when both were compounded with polypropylene, in the presence of a standard intumescent flame retardant (i.e. ammonium polyphosphate at 20 wt.% loading). The obtained composites showed enhanced flame retardant features, as witnessed by the increase of limiting oxygen values (from 18% to 23%, for unfilled polypropylene and for the composite containing 25 wt.% of biochar and 10 wt.% of wool fibers, respectively) and the decrease of peak of heat release rate values (from 1054 to 318 kW m^{-2} for unfilled polypropylene and for the composite, respectively).

Then biochar derived from pine wood was compounded with polypropylene in the presence of two different common flame retardants (i.e. ammonium polyphosphate or magnesium hydroxide, each at 20 wt.% loading). In addition, the feedstock for the biochar (i.e. pine wood) was also employed as biomass to be incorporated into biochar–polypropylene composites [34]. In particular, the resulting biochar–pine wood–polypropylene batches had similar FRs loadings, but differed as far as the amounts of the wood and biochar are concerned. More specifically, one batch had less wood (i.e. 10 wt.%) and more biochar (24 wt.%), while the second one had more wood (30 wt.%) and less biochar (4 wt.%). In all cases, the total amount of wood + biochar was set at 34 wt.%, keeping constant the polymer loading (at 42 wt.%) and also embedding 4 wt.% of maleic anhydride grafted polypropylene as compatibilizer. As assessed by cone calorimetry tests performed at 50 kW m^{-2}, all the composites containing ammonium polyphosphate, irrespective of their composition, showed peak values of the heat release rate that were 60%–65% lower than unfilled polypropylene. In addition, using more biochar and less wood in the composite formulation allowed achieving the lowest peak of the heat release rate (namely 376 kW m^{-2}).

A similar behavior was found for the composites containing magnesium hydroxide, although the fire performance was slightly lower with respect to ammonium phosphate because of the different flame retardant mechanisms involved (i.e. char formation and intumescence for ammonium polyphosphate vs. flame dilution through dehydration for magnesium hydroxide). In any case, the thermally stable biochar contributed to the creation of a protective char layer, capable of limiting the transfer of oxygen to the burning polymer and of the volatile combustible gases derived from the degradation of polypropylene to the surroundings. Finally, the mechanical behavior of the prepared composites was not significantly worsened by the presence of the FR particles that are likely to lower the effectiveness of biochar pore infiltration by the polymer, as a consequence of their trapping inside the filler pores.

Further, Zhang *et al* [35] prepared a series of high-density polyethylene/biochar/wood flour composites, using a biochar obtained from the pyrolysis of corn stalk at 500 °C. In particular, the polymer content in the composites was kept at 40 wt.%, while biochar and wood flour varied between 10 and 60 wt.%, keeping the total filler loading at 60 wt.%. In addition, the composite containing high-density polyethylene/biochar/wood flour at 40/30/30 (weight percentages) was further compounded with $Mg(OH)_2$ or $Al(OH)_3$ fillers, which are very well-known as 'heat-sink' flame retardants. The best flame retardant properties were found in the presence of Mg $(OH)_2$ at 40 wt.% loading, as assessed by limiting oxygen index measurements. For this composition, this parameter achieved 31.9%, further improving that of the whole composite without flame retardant (equal to 23.9%).

The first example on the use of biochar as a flame retardant for a thermosetting system was recently reported by Jagdale *et al* [36]. In particular, the effect of biochar, derived from bamboo pyrolyzed at 800 °C for 1 h in argon atmosphere, on the flame retardance of polydimethylsiloxane was thoroughly investigated and compared to that provided by incorporating other carbon fillers (namely, commercial graphitized, –COOH functionalized or pristine multiwall carbon nanotubes, as well as carbon soot). As assessed by cone calorimetry tests, the presence of 3 wt.% of biochar decreased both the time to ignition of polydimethylsiloxane (by about 35%) and the peak of heat release rate (by 7%), but was detrimental as far as the smoke production is considered. In fact, the specific extinction area and total smoke release values significantly increased in the presence of the carbon filler.

Very recently, Zhang *et al* [37] prepared a series of high-density polyethylene-biochar composites, using rice husk as a feedstock to produce carbon filler at 600 °C. The biochar loadings were between 10 and 70 wt.%. Notwithstanding an increase of the tensile modulus with increasing filler content (up to 60 wt.% loading of biochar), this latter was also capable of enhancing the flame retardance of the polymer, as witnessed by the increased limiting oxygen index values (from 18.2 for the unfilled polymer matrix up to 25.1 for the composite containing 70 wt.% of carbon filler).

Then, the use of biochar as a component of a flame retardant treatment for cotton fabrics was reported in a recent work by Barbalini *et al* [30], who exploited the combination of this carbon filler (derived from exhausted coffee powder dried at 105 °C and then subjected to pyrolysis using a vertical furnace and a quartz reactor up to 800 °C in argon atmosphere) with phytic acid, a molecule bearing six phosphate groups (corresponding to 28 wt.% of phosphorus based on the molecular weight) extracted from different plant tissues, such as soy beans, cereal grains, and oil seeds [18]. The cellulosic fabrics were treated using an aqueous dispersion of the two components, keeping a 1:1 weight ratio, and achieving a final dry add-on from 4 to 10 wt.%. As a reference, the fabrics were also treated with single components (achieved final dry add-on: 8 wt.%). Vertical and horizontal flame spread tests revealed that biochar alone was not able to stop the flame propagation. Conversely, the concurrent presence of biochar and phytic acid, at least at 6 wt.%, provided self-extinction to cotton, as clearly shown in figure 9.1.

Furthermore, the self-extinguishing treated fabrics did not ignite when exposed to a 35 kW m^{-2} irradiative heat source in forced combustion tests, thus confirming the high flame retardant performances provided by the designed treatment. Conversely, the washing fastness of the treated fabrics was not satisfactory, as the observed flame retardant properties were notably lost after few washing cycles. However, it is worth noting that the washed flame retarded fabrics still exhibited higher flame retardant performances compared to the neat cellulosic substrate.

9.3 Conclusions and perspectives

This chapter has tried to demonstrate the suitability and feasibility of using biochar, derived from different natural sources, as a component for improving the flame retardance of mainly thermoplastics-based systems. In fact, the high thermal stability and the peculiar morphology of this filler can be successfully exploited for the design of flame retarded polymer composites. This utilization of biochar is really new, as documented by the limited number of publications in the current scientific literature. However, it is expected that the research on this specific application of biochar will grow in the forthcoming years. Indeed, it is a low-cost filler, with a porous structure, as well as a tunable chemical composition (on the basis of the feedstock the biochar is obtained from). In addition, it can be exploited as a carbon source derived from wastes and by-products in combination with 'standard' flame retardants, hence further improving the overall fire performances

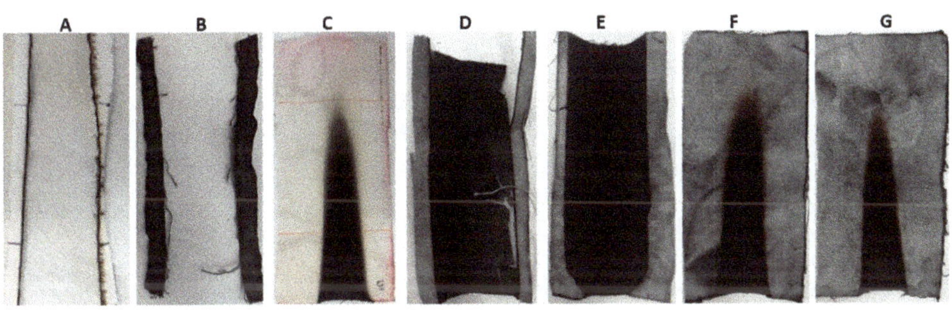

Figure 9.1. Residues after vertical flame spread tests: (a) neat cotton, (b) cotton treated with biochar (8 wt.%), (c) cotton treated with phytic acid (8 wt.%), (d) cotton treated with 4 wt.% phytic acid/biochar (ratio 1:1), (e) cotton treated with 6 wt.% phytic acid/biochar (ratio 1:1), (f) cotton treated with 8 wt.% phytic acid/biochar (ratio 1:1), and (g) cotton treated with 10 wt.% phytic acid/biochar (ratio 1:1), as reported by Barbalini *et al* [30]. Reprinted with permission from MDPI.

on the basis of synergistic effects taking place between the components of the flame retardant recipe.

In addition, apart from its use for bulky polymer systems, very recently biochar has started to be utilized as a flame retardant for fabrics (namely, cotton). Here the results are very promising, as described in the present chapter and deserve further experimental investigation.

All these findings clearly show the potential of this carbon material in an application sector that is quite far from its standard uses. However, at present there are still some challenging issues that are limiting its complete exploitation in flame retardance. In particular, its dispersion in bulky polymer systems shows some limitations as biochar tends to aggregate at high loadings, hence possibly reducing its overall (mechanical, thermal, and fire) performances. In addition, when exploited for flame retardant cotton, biochar shows a limited durability, i.e. it comes off the fabrics during the laundry treatments, fabrics are usually subjected to.

These are all challenging issues that may be overcome in the next future, using new technological approaches and fine-tuning the designed flame retardant systems.

References

[1] Weil E D and Levchik S V 2009 *Flame Retardants for Plastics and Textiles: Practical Applications* (München: Hanser)

[2] Morgan A B, Wilkie C A and Nelson G L 2012 *Fire and Polymers VI: New Advances in Flame Retardant Chemistry and Science* (Washington, DC: American Chemical Society)

[3] Ray S S and Kuruma M 2020 *Halogen-free Flame-retardant Polymers: Next-Generation Fillers for Polymer Nanocomposite Applications* vol 294 (Berlin: Springer)

[4] Horrocks A R, Price D and Price D 2001 *Fire Retardant Materials* (Cambridge: Woodhead Publishing)

[5] Neisius M, Stelzig T, Liang S and Gaan S 2014 Flame retardant finishes for textiles *Functional Finishes for Textiles: Improving Comfort, Performance and Protection* (Cambridge: Woodhead) p 429

[6] Horrocks A R and Price D 2008 *Advances in Fire Retardant Materials* (Amsterdam: Elsevier)

[7] Weijs L, Dirtu A C, Malarvannan G and Covaci A 2015 *Comprehensive Analytical Chemistry* (Amsterdam: Elsevier) p 433–91

[8] Howard P H and Muir D C 2010 *Identifying New Persistent and Bioaccumulative Organics Among Chemicals in Commerce* (Washington, DC: American Chemical Society)

[9] Iqbal M, Syed J H, Katsoyiannis A, Malik R N, Farooqi A, Butt A, Li J, Zhang G, Cincinelli A and Jones K C 2017 Legacy and emerging flame retardants (FRs) in the freshwater ecosystem: a review *Environ. Res.* **152** 26–42

[10] Hull T, Law R and Bergman Å 2014 *Polymer Green Flame Retardants* (Amsterdam: Elsevier) pp 119–79

[11] Hörold S 2014 *Polymer Green Flame Retardants* (Amsterdam: Elsevier) pp 221–54

[12] Salmeia K A, Gaan S and Malucelli G 2016 Recent advances for flame retardancy of textiles based on phosphorus chemistry *Polymers* **8** 319

[13] Haurie L, Fernández A I, Velasco J I, Chimenos J M, Cuesta J-M L and Espiell F 2007 Thermal stability and flame retardancy of LDPE/EVA blends filled with synthetic

hydromagnesite/aluminium hydroxide/montmorillonite and magnesium hydroxide/aluminium hydroxide/montmorillonite mixtures *Polym. Degrad. Stab.* **92** 1082–7
[14] Goodarzi V, Jafari S H, Khonakdar H A, Monemian S A and Mortazavi M 2010 An assessment of the role of morphology in thermal/thermo-oxidative degradation mechanism of PP/EVA/clay nanocomposites *Polym. Degrad. Stab.* **95** 859–69
[15] Kiliaris P and Papaspyrides C 2010 Polymer/layered silicate (clay) nanocomposites: an overview of flame retardancy *Prog. Polym. Sci.* **35** 902–58
[16] Laoutid F, Bonnaud L, Alexandre M, Lopez-Cuesta J-M and Dubois P 2009 New prospects in flame retardant polymer materials: from fundamentals to nanocomposites *Mater. Sci. Eng.* R **63** 100–25
[17] Basak S and Ali S W 2016 Sustainable fire retardancy of textiles using bio-macromolecules *Polym. Degrad. Stab.* **133** 47–64
[18] Costes L, Laoutid F, Brohez S and Dubois P 2017 Bio-based flame retardants: when nature meets fire protection *Mater. Sci. Eng.* R **117** 1–25
[19] Cheng X-W, Liang C-X, Guan J-P, Yang X-H and Tang R-C 2018 Flame retardant and hydrophobic properties of novel sol–gel derived phytic acid/silica hybrid organic–inorganic coatings for silk fabric *Appl. Surf. Sci.* **427** 69–80
[20] Basak S, Samanta K K, Chattopadhyay S K, Narkar R S and Mahangade R 2015 Flame retardant cellulosic textile using banana pseudostem sap *Int. J. Clothing Sci. Technol.* **27** 247–61
[21] Basak S and Ali S W 2017 Leveraging flame retardant efficacy of pomegranate rind extract, a novel biomolecule, on ligno-cellulosic materials *Polym. Degrad. Stab.* **144** 83–92
[22] Bosco F, Casale A, Gribaudo G, Mollea C and Malucelli G 2017 Nucleic acids from agro-industrial wastes: a green recovery method for fire retardant applications *Ind. Crops Prod.* **108** 208–18
[23] Alongi J, Bosco F, Carosio F, Di Blasio A and Malucelli G 2014 A new era for flame retardant materials? *Mater. Today* **17** 152–3
[24] Malucelli G, Bosco F, Alongi J, Carosio F, Di Blasio A, Mollea C, Cuttica F and Casale A 2014 Biomacromolecules as novel green flame retardant systems for textiles: an overview *RSC Adv.* **4** 46024–39
[25] Ghaedi M, Mazaheri H, Khodadoust S, Hajati S and Purkait M 2015 Application of central composite design for simultaneous removal of methylene blue and Pb^{2+} ions by walnut wood activated carbon *Spectrochim. Acta* A **135** 479–90
[26] Ghaedi M, Nasab A G, Khodadoust S, Rajabi M and Azizian S 2014 Application of activated carbon as adsorbents for efficient removal of methylene blue: kinetics and equilibrium study *J. Ind. Eng. Chem.* **20** 2317–24
[27] Zhang X, Wang H, He L, Lu K, Sarmah A, Li J, Bolan N S, Pei J and Huang H 2013 Using biochar for remediation of soils contaminated with heavy metals and organic pollutants *Environ. Sci. Pollut. Res.* **20** 8472–83
[28] Nan N, DeVallance D B, Xie X and Wang J 2016 The effect of bio-carbon addition on the electrical, mechanical, and thermal properties of polyvinyl alcohol/biochar composites *J. Compos. Mater.* **50** 1161–8
[29] Liu Z, Fei B and Jiang Z 2014 Combustion characteristics of bamboo-biochars *Bioresour. Technol.* **167** 94–9
[30] Barbalini M, Bartoli M, Tagliaferro A and Malucelli G 2020 Phytic acid and biochar: an effective all bio-sourced flame retardant formulation for cotton fabrics *Polymers* **12** 811

[31] Das O, Bhattacharyya D, Hui D and Lau K-T 2016 Mechanical and flammability characterisations of biochar/polypropylene biocomposites *Composites* B **106** 120–28
[32] Ikram S, Das O and Bhattacharyya D 2016 A parametric study of mechanical and flammability properties of biochar reinforced polypropylene composites *Composites* A **91** 177–88
[33] Das O, Kim N K, Sarmah A K and Bhattacharyya D 2017 Development of waste based biochar/wool hybrid biocomposites: flammability characteristics and mechanical properties *J. Clean. Prod.* **144** 79–89
[34] Das O, Kim N K, Kalamkarov A L, Sarmah A K and Bhattacharyya D 2017 Biochar to the rescue: balancing the fire performance and mechanical properties of polypropylene composites *Polym. Degrad. Stab.* **144** 485–96
[35] Zhang Q, Cai H, Yang K and Yi W 2017 Effect of biochar on mechanical and flame retardant properties of wood–plastic composites *Results Phys.* **7** 2391–5
[36] Jagdale P, Salimpour S, Islam M H, Cuttica F, Hernandez F C R, Tagliaferro A and Frache A 2018 Flame retardant effect of nano fillers on polydimethylsiloxane composites *J. Nanosci. Nanotechnol.* **18** 1468–73
[37] Zhang Q, Zhang D, Xu H, Lu W, Ren X, Cai H, Lei H, Huo E, Zhao Y and Qian M 2020 Biochar filled high-density polyethylene composites with excellent properties: towards maximizing the utilization of agricultural wastes *Ind. Crops Prod.* **146** 112185

Chapter 10

Review of biochar as a sustainable mortar admixture and evaluation of its potential as coating for PVA fibers in mortar

Harn Wei Kua, Souradeep Gupta and Sek Teng Koh

The idea of applying biochar–cement coating on to the surface of polyvinyl alcohol (PVA) fibers was born out of the need to reduce the air voids at the fiber–matrix interface. This study found that while adding biochar–cement coating to the fibers could increase the flexural and split-tensile strengths of the mortar, the coating led to a lower compressive strength. The coating also decreased the water permeability of the mortar.

These results indicate that while biochar particles may reduce the porosity at the fiber–coating interface, proper curing of this coating before the coated fibers are incorporated in the mixture is crucial to avoid or reduce the formation of air voids at the coating–matrix interface, which is deduced to be the main reason for the weakening of the mortar. Furthermore, modifying the surface chemistry of the biochar is required to reduce the hydrophobicity of the biochar in order to avoid the biochar–matrix interface from becoming the weak zone that enhances crack propagation. Further research is needed to optimize the properties of biochar in enhancing the capability of synthetic fibers in reinforcing mortar and concrete.

10.1 Introduction—the need to improve the fiber reinforcement of cementitious composites

Cement is the most widely used construction material in the world [1]. Currently, the world produces 4.4 billion tons of concrete annually, and this quantity is expected to rise to over 5.5 billion tons by 2050. It is a well-known fact that cement is strong in compression but weak in tension. To increase the tensile strength of cementitious composites, attempts have been made to use different types of fibers as a reinforcement in mortar and concrete.

doi:10.1088/978-0-7503-2660-5ch10

When cracks that are initiated by loading advance and reach the fiber–matrix interface, de-bonding at this interface may occur or the fibers themselves may be stressed. This dissipates the energy and decreases the concentration of stress at the tip of the cracks, thus preventing the crack from further propagation. If local de-bonding at the fiber–matrix interface occurs, the crack will continue to propagate through the matrix. As long as the fibers remain intact, they can help transfer loads across the cracks, thereby increasing the overall strength of the composite.

The effectiveness of fiber reinforcement in concrete depends on several factors, including the types of fibers, the fibers' age, the water–cement (w/c) ratio, particle grading, chemical admixtures, and concrete quality. The use of polyvinyl alcohol (PVA) fibers as reinforcement is receiving increasing attention in the industry. PVA is selected because of its stability and durability in the alkaline environment of a cementitious matrix. PVA fibers are high in tensile strength, and have a high modulus of elasticity, high chemical resistance to an alkaline environment, and high affinity to water. PVA also poses no adverse health risk [2]. The hydrophilic surface of PVA fibers helps it to bond easily with the cement matrix, thus enhancing the bending strength and other mechanical properties of the fiber-reinforced concrete [3].

The effectiveness of reinforcement fibers in increasing concrete's tensile strength depends on the quantity of fibers present in the concrete. For example, Noushini *et al* [4] investigated the effect of adding 0.125 vol.%, 0.25 vol.%, 0.375 vol.%, and 0.5 vol.% (by volume) of fibers in concrete samples and found that 0.25 vol.% yields the best results—the ultimate strength and ductility of the fiber-reinforced concrete was significantly improved. Similarly, Erdogmus [5] confirmed that a lower fiber percentage led to higher flexural strength, toughness, energy absorption ability, and ductility compared to the control. Xu *et al* [6] also verified that a low volume fraction of PVA at 0.6 vol.% was sufficient to increase the cement composite's capability to resist load. Noushini *et al* [7] showed that the highest compressive strength in the composite was observed when only 0.25 vol.% volume of fibers were incorporated, which resulted in a 12% improvement in strength.

Adding reinforcement fibers may alter the optimum w/c ratios in the mixture. Atahan *et al* [8] studied the effect of different w/c ratios on cement composites containing 0.5 vol.%, 1 vol.%, 1.5 vol.%, and 2 vol.% of PVA fibers. They discovered that, in general, the higher the w/c ratio, the higher the energy absorption during static and impact loading on the composites containing a 0.5 vol.% volume fraction of PVA fibers. At a higher dosage, PVA fibers tend to 'ball' up by forming agglomerates and affecting the workability of cementitious mixes. This is also supported by Yew *et al* [9]. Therefore, a volume fraction of fiber ranging from 0.10% and 0.30% was recommended [2].

However, the addition of fibers may result in a reduction of the mechanical strength of the composites. Noushini *et al* [7] studied the compressive strength of fiber-reinforced cement composite and found that the addition of fibers increases the composite's compressive strength, but it may also decrease the compressive strength as the vol.% of fibers increases. This agrees with the conclusions drawn by Li [10], who found that increasing fiber content beyond a certain point in a cementitious composite may contribute to compressive strength degradation.

Why is it possible for fibers to reduce the strength of the composites? One possible reason is that adding fibers may increase the total volume of air voids in the composites (in particular those at the fiber–matrix interface) or an increase the density of micro-cracks at the fiber–matrix interface [7]. These air gaps may be caused by poor fiber–matrix bonding and/or a lack of effective adhesion between the fibers and matrix [11]. Air gaps can also be formed due to ineffective or insufficient compaction of the composite [12].

The objective of this study is to evaluate the effectiveness of fillers in reducing air voids in mortar that is reinforced with PVA fibers. Specifically, the combined effect of the fillers and PVA fibers on the mechanical strength and water permeability of mortar was evaluated. A vol.% value of less than 0.2 was considered because the emphasis was to use as few PVA fibers as possible, in order to reduce the material requirement for strengthening the mortar specimens. The novelty of our approach lies in the use of biochar–cement as the coating material for PVA fiber in order to modify the adhesion of fibers to the matrix. For fillers, biochar was chosen because it has a high potential to be a carbon-negative material throughout its whole life cycle.

10.2 A review of the state-of-the-art of biochar as a supplementary admixture in cementitious composites

Biochar is a carbon-rich solid produced by thermo-chemical conversion of waste biomass, including agricultural waste, forest waste, and domestic waste, under a controlled set of preparation conditions. Pyrolysis and gasification are technologies commonly used for this purpose, which involves heating of biomass in the absence or with a limited supply of oxygen to yield three products—bio-oil, biogas, and biochar. If all three products are utilized to offset the usage of natural resources and fossil fuels, the entire process of pyrolysis and char production would be effectively carbon-negative [13–15]. It is estimated that depending on the feedstock and preparation conditions, biochar application can reduce greenhouse gas emission by 870 kg CO_2-eq per kg of biomass, of which 62%–66% can be contributed by the carbon sequestered in the biochar structure inherited from the parent biomass [15]. Nevertheless, the application of biochar to reduce the demand for cement and improve the physical properties of construction materials could be an effective means of waste recycling and valorization.

10.2.1 Biochar as an additive in cementitious composites

Gupta *et al* [16–20] investigated extensively the application of biochar from wood waste and mixed food waste, prepared by slow pyrolysis at 500 °C, to replace 2%–8% of cement in mortar. The findings showed that the addition of biochar in the size range of 1–100 μm results in a 20%–30% improvement in the 1 d and 7 d strength of mortar compared to a control. The addition of biochar at 1%–2% by weight of cement accelerated hydration, which is confirmed from the reduction in the setting time of fresh cement paste by 45 min to 1 h compared to the control. Moreover, the pores and surface of biochar offer sites for the growth of hydration products, which contribute to improvement in hydration and early densification. Similar results were obtained by Dixit *et al* [21]—the application of 5% biochar, prepared from sawdust at 500 °C, accelerated the hydration kinetics by 2–6 h depending on dosage, and increased the degree of hydration by 10%–17%.

Biochar with a finer particle size compared to cement and aggregates induces a filler effect by reducing the pores between solid particles. This has a positive effect on packing, which improves the compactness of the cementitious matrix and increases early strength development [16]. Pores of different size ranges in biochar also retain water during the preparation of cementitious mixes, which reduces the local effective water–cement ratio around biochar particles. This mechanism leads to initial densification, contributing to a higher early strength of biochar–mortar [20]. However, the stored water in the pores of biochar is later released, which leads to an internal curing effect and improved hydration in the cementitious matrix by 10%–17%, depending on the dosage and biochar size [20, 21]. The addition of biochar from wood and food waste led to a reduction in water absorption and the depth of water penetration by 30%–50% compared to the control. This indicates the potential of biochar in improving water tightness and reducing seepage in cement based construction materials, which is important to prevent the ingress of detrimental fluids into the structure. Gupta and Kua [18] found that a combination of biochar (4 wt.% of cement) and silica fume (6 wt.% of cement) enhance the compressive strength by 20% compared to the control, while the sorptivity and rate of absorption can be reduced by 50%–60% compared to the control mortar.

For reliable and durable constructions, the addition of biochar must not have any adverse effects on the cementitious matrix over the long term. Gupta *et al* [22] investigated the influence of biochar from sorghum, cotton stalk, dairy manure, wood, and algae, prepared under different conditions on the compressive strength of mortar at an age of two years. The study found that the addition of 2% biochar from wood waste contributes to a 10%–15% higher compressive strength of biochar–mortar at an age of two years compared to the control, which reflects the stability of biochar in a cementitious matrix at a relatively advanced stage. However, biochar with a high ash content, for example dairy manure, and those prepared at low temperatures (below 300 °C) have an adverse effect on strength development, due to undecomposed hemicellulose and unburnt particles of raw biomass.

Gupta and Kua [17] found that meso- and micro-pores in ball milled biochar (a particle size between 0.10 and 4 μm), prepared from sawdust at 500 °C, are more effective in retaining moisture which leads to 10%–16% higher compressive strength after 28 d of dry curing (65% relative humidity and 30 °C), compared to mortar with coarser and macro-porous biochar with particle sizes between 1 and 100 μm. While the control (without biochar) showed a reduction in strength by 25% as the curing condition was changed from wet to dry, the mortar with biochar suffered only a 7%–10% loss in strength under similar conditions [17]. These findings suggest that the addition of ground biochar at 1%–2% by weight of cement can lead to higher strength retention even when external wet curing is not consistently available.

A positive influence of biochar's filler effect in cement has also been demonstrated by Ahmad *et al* [23], who reported an increase in the compressive strength and fracture toughness of cement paste by 25% and 76%, respectively, due to inclusion of

0.08 wt.% biochar derived from bamboo at 850 °C. Restuccia and Ferro [24] observed a 50%–60% improvement in strength and a 60%–70% improvement in fracture toughness of cement composites with 0.80–1 wt.% biochar prepared from hazelnut shell and coffee powder. In many of these studies, the filler effect afforded by fine biochar particles was cited as the key reason for the improvement in the mechanical properties and water tightness of the mortar/concrete containing these particles.

Wang *et al* [25] reported improved hydration due to 1% and 2% biochar addition in sediment based construction products using cement as a binder. However, application of coarse biochar (<250 μm) led to a reduction in compressive strength. However, the pores of biochar offered excellent immobilization of toxic elements and organic contaminants in the sediment.

10.2.2 Biochar as a supplement for self-healing concrete

Gupta *et al* [26] investigated a novel application of biochar as an immobilization medium for special bacteria spores that precipitate calcium carbonate and generate self-healing in pre-cracked cement composites. Although self-healing with bacteria has been widely explored, one of the limitations of this technique is the reduced healing efficiency due to the lower metabolism of bacteria in the alkaline cementitious matrix. The investigation by Gupta *et al* [26, 27] attempted to improve the healing efficiency by immobilizing the bacteria spores in pores of biochar prior to addition in the cement mortar and concrete, respectively. The healing efficiency was investigated through monitoring of crack sealing and recovery of the mechanical and permeability properties. The authors proposed an innovative approach by combining biochar–bacteria with fibers and superabsorbent polymer (SAP) to restrict the crack width and source for internal moisture, respectively, which are critical for effective self-healing. The findings suggested that this technique could completely seal cracks of width 700–800 μm compared to 200–300 μm in the case of the control. A major part of the wider micro-cracks (300–400 μm) in the control is left unsealed. This strategy also led to a reduction in water permeability by 60%–70% and an increase in strength by 38% compared to a mix with directly added bacteria.

In summary, the outcomes suggest that a combination of biochar immobilized bacteria, SAP, and fiber can potentially seal micro-cracks at an early stage, preventing the ingress of external contaminants and restoring the mechanical and durability properties of cementitious composites. This translates into reduced repair and maintenance operations on concrete structures and better performance during the serviceable life.

10.2.3 The role of biochar to modify carbonation potential and enhance the performance of recycled aggregate concrete

The versatility of biochar based cementitious composites is key for a wide range of industrial applications. One aspect of this versatility is the potential to combine biochar with other products from industrial waste-streams. Praneeth *et al* [28] investigated the carbon dioxide uptake of cement blocks with fly ash and biochar from corn stover, added at 2%, 4%, and 6% by weight of cement. The authors reported that the porous nature of biochar and fly ash increased the diffusion of carbon dioxide resulting in increased CO_2 uptake. Depending on the biochar dosage, carbonated cement blocks with biochar showed a 25%–35% increase in compressive strength compared to uncarbonated blocks. This was attributed to the conversion of hydration products to calcium carbonate leading to densification of the cementitious matrix. The findings from this study present a potential means for developing low carbon biochar–cement composites.

Akhtar and Sarmah [29] explored the application of biochar from rice husk and poultry litter as a filler in recycled aggregate concrete to replace cement by 0.10%–0.75% by volume of concrete. The authors found that 0.10% rice husk biochar and poultry litter char increased the compressive strength by 17% and 3%, respectively. Added at 0.10% by volume, the rice husk biochar reduced water absorption by 17% compared to the control, while both biochars reduced the permeable void volume at 0.50% volume addition. This improvement is attributed to the filler effect of biochar densifying interfacial zones around aggregates. Rice husk biochar also has some pozzolanic action, resulting in precipitation of calcium silicate hydrate gel which reduces the porosity of concrete.

10.2.4 Biochar as an additive in concrete and lightweight mortar subjected to elevated temperature

An important consideration for the application of biochar–mortar and biochar–concrete is its thermal resistance at elevated temperatures. The surface temperature of concrete for applications including thermal plants and reactor walls can go up to 300 °C–500 °C [30]. Therefore, the durability and strength retention of biochar–concrete at elevated temperatures is important for such special applications. Gupta *et al* [31] found that the addition of 1–2 wt.% biochar can potentially minimize the internal thermal damage of concrete at 300 °C and 550 °C, which results in 11% and 20% higher strength retention than plain concrete (without biochar) and concrete containing 10% silica fume, respectively. Lower damage to the micro-structure of biochar–concrete contributed to 20%–25% lower permeability compared to the control. This shows that the addition of biochar improves the durability of concrete components by enhancing its resistance to thermal damage.

The findings from the above-mentioned study inspired Gupta and Kua [32] to investigate the influence of rice husk biochar (RHB), added at 40% replacement of silica fume, on the thermal resistance of cenosphere modified lightweight mortar (dry density of 1600–1800 kg m^{-3}) at 450 °C. The results showed that the addition of RHB led to 15%–20% higher strength retention and 20%–25% lower permeability of cenosphere mortar compared to the control (containing silica fume and cenosphere) after thermal exposure.

10.2.5 Biochar based coating for polymer fibers to improve the strength of fiber-reinforced mortar

Polypropylene (PP) fiber has a hydrophobic surface, which weakens the bond strength at the fiber–matrix interface. Adding a biochar–cement coating on the surface of PP fibers can increase its surface roughness, which is important for improving the interaction of the fiber with the cementitious matrix [33]. Biochar from sawdust, prepared at 300 °C, has been employed as coating on PP fiber to improve the bond at the fiber–matrix interface [33]. The authors found that a biochar coating led to improvement in the compressive strength of fiber-reinforced mortar by 28% and 16% at 7 d and 28 d, respectively. This strategy also increased the fiber pull-out response and post-peak ductility of coated PP fiber mortar compared to mortar with uncoated fibers. This motivated the authors to develop a similar coating for PVA fibers, which can increase the flexural strength and tensile performance compared to control (without fiber) and mortar with uncoated PVA fiber.

10.3 Materials and methods

The materials used to produce the fiber-reinforced mortar are shown in table 10.1 and the compositions of the specimens are shown in table 10.2. The specifications for the PVA fibers used are presented in table 10.3. The prepared biochar was manually ground into fine particles and sieved using a 150 μm sieve. The sieved biochar was used in the preparation of the biochar–cement coating for PVA fibers.

For W/F&B specimens, every fiber strand was coated with biochar (production conditions summarized in table 10.4) as homogeneously as possible. This coating process used 28 g of biochar, 28 g of cement, and 70 g of water for every 14 g of PVA fibers. The PVA fibers were separated into individual strands as much as possible. The separated fibers were dipped into the prepared cement paste as fully as possible. The coated fibers were then placed and spaced out on aluminum foil. They were then covered with another piece of foil and placed in an indoor enclosed environment to cure at room temperature for 7 days, before they were added into the W/F&B mix.

For the compression test, the specimens were cast in cylindrical molds measuring 100 mm (diameter)-by-200 mm (height). For the tensile test, 40 mm-by-40 mm-by-

Table 10.1. General descriptions of the materials used.

Materials	Specifications
Cement	Ordinary Portland cement, ASTM type 1, 52.5N
Sand	Regional river sand
Water	Tap water at room temperature
Superplasticizer	ADVA 18IN, W.R. Grace (Singapore)
Fiber (PVA)	Microfiber 8 mm polyvinyl alcohol (PVA)
Biochar	Mixed wood sawdust obtained from Indonesia; produced at 300 °C for 45 min in a Nabertherm furnace

Table 10.2. Compositions of specimens.

Sample	PVA fibers (vol.%)	Biochar (g)	Cement (kg)	Sand (kg)	Water (kg)	Superplasticizer (wt.% of cement) (g)	Density (kg m^{-3})
Control	0	0	1	2.75	0.4	0.3	2291
W/F	0.14						2271
W/F&B		28					2257

Table 10.3. Physical properties of the PVA fibers used.

Properties	Polyvinyl alcohol (PVA)
Average diameter (μm)	~10
Average length (mm)	8
Density (g cm^{-3})	1.29
Melting point (°C)	200

Table 10.4. Production conditions for biochar.

Temperature	Duration (min)	Heating rate (°C min^{-1})	Yield mass (%)
300	45	10	30

160 mm prismatic molds were used instead. After the specimens were cast, to prevent early moisture loss that could lead to drying shrinkage, they were covered with plastic sheets before they were demolded after 24 h. Wet curing was carried out by placing the specimens in a fog room, with a relative humidity (RH) of about 100% and a temperature of 28 ± 2 °C. The specimens were tested after 7 and 28 days of curing.

The produced biochar was characterized for morphology using a scanning electron microscope (SEM). The carbon content of the biochar was determined using semi-quantitative electron dispersive spectra, attached to the SEM. The water absorption of biochar was determined using the filtration method outlined in Gupta *et al* [22]. The presence of organic functional groups in the produced biochar was determined by Fourier transform infrared spectroscopy (FTIR), using the sample preparation method mentioned in Gupta *et al* [18].

Uniaxial compressive strength test on mortar samples (in accordance with EN 12 390–3:2009) was carried out on a Denison Mayes hydraulic compression machine, at a constant loading rate of 157 kN min^{-1}. The tensile strength test was performed in accordance with ASTM C348, using a central-loading method with a 10 kN load cell and at a displacement rate of 0.5 mm min^{-i}. Tests were conducted in triplicate and average values (with their standard deviation) were recorded. To check for the statistical significance of any difference in the strength values of the specimens, the *p*-statistic from *t*-tests at a 90% (*p*-critical = 0.10) confidence interval is used.

10.4 Results and analyses

10.4.1 Characterization of biochar

Figure 10.1(a) shows the morphology of biochar particles. The particles' surfaces show ridges that are typical of biochar derived from woody feedstock. The particles are randomly shaped with a combination of elongated and angular morphologies. There are surface macro-pores (figure 10.1(b)), which are generally inherited from the parent feedstock. These macro-pores play active role in water absorption and retention, which contribute to moisture redistribution in biochar–cementitious matrices.

The carbon content in the biochar is 63%, which is higher than the carbon content in the parent biomass (ranging between 45%–50%) [34]. This is due to the fixation of a major fraction of carbon in the parent biomass during pyrolysis, while other inorganics and volatiles are removed. The water absorption capacity of biochar as determined by a filtration method is 2.50 ± 0.25 g g^{-1} of dry biochar.

As shown in the FTIR spectra (figure 10.2), the band centered around 3446 cm^{-1} is attributed to the presence of –OH functional groups, while the peaks between 2800

Figure 10.1. (a) The morphology of biochar particles and (b) macro-pores on biochar surface.

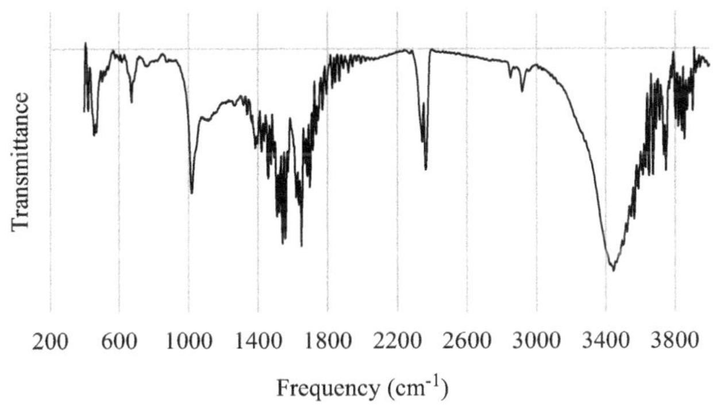

Figure 10.2. FTIR spectra of produced biochar.

and 3000 cm^{-1} is due to the presence of alkyl groups. Peaks around 1420 cm^{-1} indicates presence of aromatic compounds, whereas the bands occurring between 1600 and 1800 cm^{-1} indicate the presence of carbonyl groups, alkanes, and other oxygenated hydrocarbons [35].

10.4.2 Mechanical characterization and influence on mechanical properties

Figures 10.3(a)–(c) show the compressive strength, flexural strength, and split-tensile strength of the mortar specimens. In general, the addition of biochar–cement coated

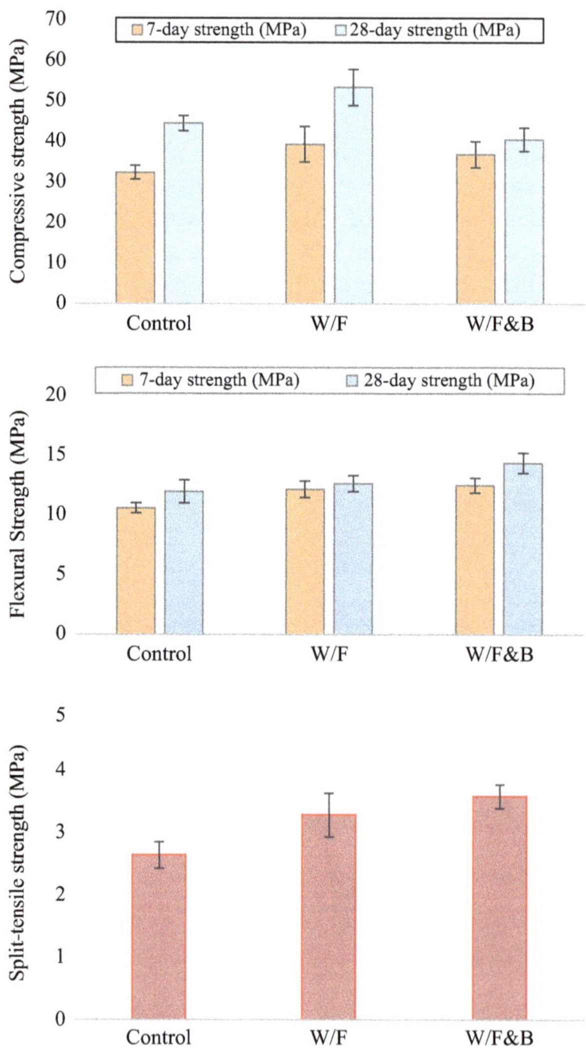

Figure 10.3. Comparisons of the (a) compressive strength, (b) flexural strength, and (c) split-tensile strength of different mortar mixes. The difference in split-tensile strength between W/F and W/F&B is statistically insignificant.

increased the flexural and split-tensile strength but reduced the compressive strength of the mortar.

As shown in figure 10.3(a), both W/F and W/F&B showed higher compressive strength than control at a 7 d age, although only W/F showed higher strength (by 19%) than the control at 28 d age. In fact, W/F shows a 24% higher compressive strength than W/F&B at an age of 28 d, which is statistically significant ($p = 0.025 < 0.100$). This is indicative of the adverse effect of biochar–cement fiber coating on the compressive strength development. The surface of the PVA fibers is hydrophilic due to presence of hydroxyl (–OH) groups on its surface, which enables it to develop strong bonds with the mortar matrix [12, 36]. In contrast, the presence of C–H groups on the surface of the biochar, due to the deposition of aliphatic and aromatic compounds [37], may lead to weak biochar–matrix interfaces. As shown in figure 10.4(a), at 'A' this local weakness may lead to cracks forming at and/or propagating around embedded biochar particles, thus reducing the compressive strength of the mortar. Furthermore, the lower density of the W/F&B mixture compared to W/F (table 10.2) implies that the presence of air voids at the coating–mortar matrix interface ('B' in figure 10.4(a)) and the pores in the biochar itself increases the overall porosity. The formation of these air voids at the coating–matrix interface could be exacerbated by the fact that the coating was cured separately for 7 d before the coated fibers were added into the mortar mix. Water loss from this coating during the 7 d curing process could have reduced the cohesion of the coating with the surrounding mortar. The strength reduction due to these coating–matrix interface air voids could have overridden any strengthening effects due to two other factors—reduction in the air voids at the fiber–coating interface due to the presence of biochar ('C' in figure 10.4(a)), as well as the strong fiber–matrix bonds due to the

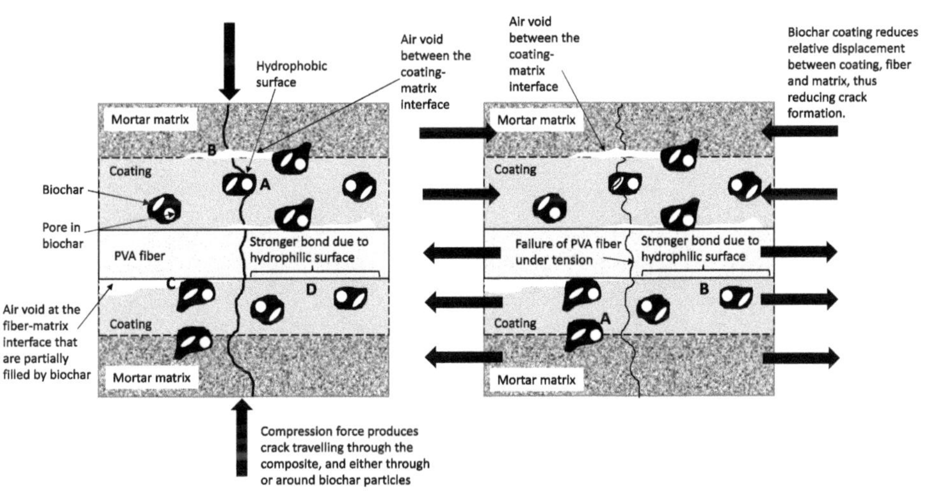

Figure 10.4. A qualitative model that explains (a) the reduction in compressive strength and (b) the contrasting increase in flexural strength. The biochar particles are not drawn to scale and the positions of the arrows in (b) are indicative in nature.

hydrophilicity of the PVA surface ('D' in figure 10.4(a)). By shortening the separate curing duration of the biochar–cement coating, this strength-reducing effect may be controlled.

In contrast, W/F&B exhibits 13.50% higher flexural strength than W/F ($p = 0.072 < 0.100$), and 27% higher flexural strength than the control ($p = 0.049 < 0.100$) at 28 d age, while the split-tensile strength of W/F and W/F&B are similar. As shown in figure 10.4(b), when the composite undergoes flexural tests, relative displacement of the matrix, coating and fiber will accelerate crack formation and failure of the composite. The surface roughness afforded by the biochar particles, which anchor the coating into the matrix at their interface ('A' in figure 10.4(b)) reduces the relative displacement between the coating and matrix, thus reducing the formation and propagation of cracks in the matrix. Furthermore, the hydrophilicity of the fibers strengthens the bond at the fiber–coating interface ('B' in figure 10.4(b)), thus reducing relative displacement at this interface and delay the rupture of the fibers. On the whole, these mechanisms contribute to higher resistance to crack formation.

10.4.3 Influence on permeability

Figure 10.5 shows that the addition of uncoated (W/F) and biochar–cement coated PVA fibers (W/F&B) led to a reduction in the depth of water penetration (compared to the control) by 46 % and 37%, respectively. These differences are significant at a 90% confidence interval ($p = 0.005$ and $0.010 < 0.100$). This is in agreement with findings by Zhao and He [38], which indicates that adding 0.30 vol.% of PVA fibers led to a decrease in permeability. This result shows that although the presence of air voids and biochar pores led to higher porosity (thus lower densities, as indicated in table 10.2), these voids and pores were not connected to a degree that the overall permeability of the specimens was increased. Reduction in water penetration is critical to the durability of cementitious composites, because water ingress must be controlled by the cover concrete (usually 30–40 mm thick in a non-aggressive

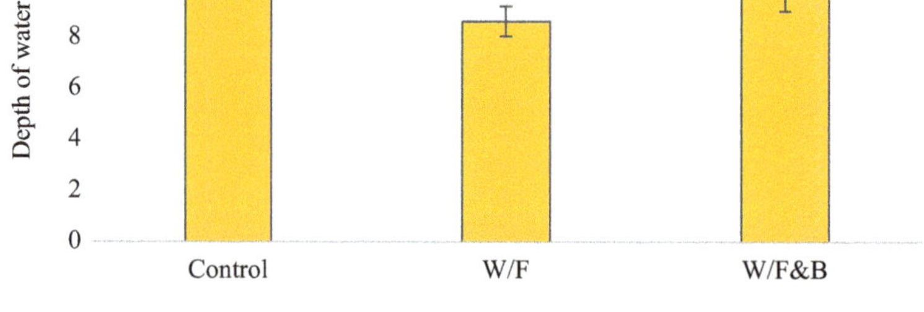

Figure 10.5. Depth of water penetration in hardened mortar at a 28 d age.

environment) before it reaches the level where reinforcement bars are located. Our findings show that PVA fibers at low volume dosage improve the compactness of the matrix and prevent ingress of moisture even under high external pressure.

Although W/F&B shows a slightly higher penetration depth than W/F, the difference is not statistically significant. The higher penetration depth in W/F&B is attributed to air voids in the coating–matrix interface, which contributed to the reduction in compressive strength compared to W/F. However, due to the low dosage of fiber, this difference in air voids at the interface does not result in a significant difference in permeability between W/F and W/F&B.

10.5 Conclusions

The idea of applying a biochar–cement coating on to the surface of PVA fibers was born out of the need to reduce the air voids at the fiber–matrix interface. However, the fact that adding a biochar–cement coating reduced the compressive strength of the composite indicates that proper curing of this coating before the coated fibers are incorporated in the mixture is crucial to avoid or reduce the formation of air voids at the coating–matrix interface. Furthermore, modifying the surface chemistry of the biochar is required to reduce the hydrophobicity of biochar, if the biochar contains a substantial amount of C–H groups. Further research is needed to optimize the properties of biochar in enhancing the capability of synthetic fibers in reinforcing mortar and concrete.

References

[1] Worrell E, Price L, Martin N, Hendriks C and Meida L O 2001 Carbon dioxide emissions from the global cement industry *Annu. Rev. Energy Environ.* **26** 303

[2] Redon C, Li, Victor C, Wu C, Hoshiro H, Saito T and Ogawa A 2001 Measuring and modifying interface properties of PVA fibers in ECC matrix *ASCE J. Mater. Civil Eng.* **13** 399–406

[3] Ogawa A and Hoshiro H 2011 Durability of fibres *Durability of Strain-Hardening Fibre-Reinforced Cement-Based Composites (SHCC)* (Dordrecht: Springer) pp 81–8

[4] Noushini A, Samali B and Vessalas K 2013 Effect of polyvinyl alcohol (PVA) fibre on dynamic and material properties of fibre reinforced concrete *Constr. Build. Mater.* **49** 374–83

[5] Erdogmus E 2015 Use of fiber-reinforced cements in masonry construction and structural rehabilitation *Fibers* **3** 41–63

[6] Xu B, Toutanji H A and Gilbert J 2010 Impact resistance of poly(vinyl alcohol) fiber reinforced high-performance organic aggregate cementitious material *Cem. Concr. Res.* **40** 347–51

[7] Noushini A, Vessalas K and Samali B 2014 Static mechanical properties of polyvinyl alcohol fibre reinforced concrete (PVA-FRC) *Mag. Concr. Res.* **66** 465–83

[8] Atahan H N, Tuncel E Y and Pekmezci B Y 2013 Behavior of PVA fiber-reinforced cementitious composites under static and impact flexural effects *J. Mater. Civ. Eng.* **25** 1438–45

[9] Yew M K, Mahmud Bin H, Ang B C and Yew M C 2015 Effects of low volume fraction of polyvinyl alcohol fibers on the mechanical properties of oil palm shell lightweight concrete *Adv. Mater. Sci. Eng.* **2015** 425236

[10] Li V C 1992 A simplified micromechanical model of compressive strength of fiberreinforced cementitious composites *Cem. Concr. Compos.* **14** 131–41
[11] Sun Z and Xu Q 2009 Microscopic, physical and mechanical analysis of polypropylene fiber reinforced concrete *Mater. Sci. Eng.* A **527** 198–204
[12] Bentur A and Mindess S 2005 *Fibre Reinforced Cementitious Composites* (London: Elsevier)
[13] Lehmann J and Joseph S 2015 *Biochar for Environmental Management: Science, Technology and Implementation* (New York: Routledge)
[14] Lehmann J 2007 Bio-energy in the black *Front. Ecol. Environ.* **5** 381–7
[15] Roberts K G, Gloy B A, Joseph S, Scott N R and Lehmann J 2009 Life cycle assessment of biochar systems: estimating the energetic, economic, and climate change potential *Environ. Sci. Technol.* **44** 827–33
[16] Gupta S and Kua H W 2018 Application of biochar from food and wood waste as green admixture for cement mortar *Sci. Total Environ.* **619-620C** 419–35
[17] Gupta S and Kua H W 2019 Carbonaceous micro-filler for cement: effect of particle size and dosage of biochar on fresh and hardened properties of cement mortar *Sci. Total Environ.* **662** 952–62
[18] Gupta S and Kua H W 2019 Combination of biochar and silica fume as partial cement replacement in mortar: performance evaluation under normal and elevated temperature *Waste Biomass Valorization* **11** 1–18
[19] Gupta S and Kua H W 2018 Effect of water entrainment by pre-soaked biochar particles on strength and permeability of cement mortar *Constr. Build. Mater.* **159** 107–25
[20] Gupta S, Kua H W and Low C Y 2018 Use of biochar as carbon sequestering additive in cement mortar *Cem. Concr. Compos.* **87** 110–29
[21] Dixit A, Gupta S, Pang S D and Kua H W 2019 Waste valorisation using biochar for cement replacement and internal curing in ultra-high performance concrete *J. Clean. Prod.* **238** 117876
[22] Gupta S, Palansooriya K N, Igalavithana A D, Ok Y S and Kua H W 2020 Carbonaceous inserts from lignocellulosic and non-lignocellulosic sources in cement mortar: preparation conditions and its effect on hydration kinetics and physical properties *Constr. Build. Mater.* **264** 120214
[23] Ahmad S, Khushnood R A, Jagdale P, Tulliani J-M and Ferro G A 2015 High performance self-consolidating cementitious composites by using micro carbonized bamboo particles *Mater. Des.* **76** 223–29
[24] Restuccia L and Ferro G A 2016 Nano-particles from food waste: a 'Green' feature for traditional building materials *9th Int. Conf. Fracture Mechanics of Concrete and Concrete Structures* (Berkeley, CA, USA) pp 1–6
[25] Wang L, Chen L, Tsang D C, Kua H W, Yang J, Ok Y S, Ding S, Hou D and Poon C S 2019 The roles of biochar as green admixture for sediment-based construction products *Cem. Concr. Compos.* **104** 103348
[26] Gupta S, Kua H W and Pang S D 2018 Healing cement mortar by immobilization of bacteria in biochar: an integrated approach of self-healing and carbon sequestration *Cem. Concr. Compos.* **86** 238–54
[27] Kua H W, Gupta S, Aday A N and Srubar W V III 2019 Biochar-immobilized bacteria and superabsorbent polymers enable self-healing of fiber-reinforced concrete after multiple damage cycles *Cem. Concr. Compos.* **100** 35–52

[28] Praneeth S, Guo R, Wang T, Dubey B K and Sarmah A K 2020 Accelerated carbonation of biochar reinforced cement-fly ash composites: enhancing and sequestering CO_2 in building materials *Constr. Build. Mater.* **244** 118363

[29] Akhtar A and Sarmah A K 2018 Strength improvement of recycled aggregate concrete through silicon rich char derived from organic waste *J. Clean. Prod.* **196** 411–23

[30] Sakr K and El-Hakim E 2005 Effect of high temperature or fire on heavy weight concrete properties *Cem. Concr. Res.* **35** 590–6

[31] Gupta S, Kua H W and Pang S D 2020 Effect of biochar on mechanical and permeability properties of concrete exposed to elevated temperature *Constr. Build. Mater.* **234** 117338

[32] Gupta S and Kua H W 2020 Application of rice husk biochar as filler in cenosphere modified mortar: preparation, characterization and performance under elevated temperature *Constr. Build. Mater.* **253** 119083

[33] Gupta S, Kua H W and Tan S Y C 2017 Use of biochar-coated polypropylene fibers for carbon sequestration and physical improvement of mortar *Cem. Concr. Compos.* **83** 171–87

[34] Chebil S, Chaala A and Roy C 2000 Use of softwood bark charcoal as a modifier for road bitumen *Fuel* **79** 671–83

[35] Ghani W A W A K, Mohd A, da Silva G, Bachmann R T, Taufiq-Yap, Rashid Y H and Ala'a U H 2013 Biochar production from waste rubber-wood-sawdust and its potential use in C sequestration: chemical and physical characterization *Ind. Crops Prod.* **44** 18–24

[36] Arain M F, Wang M, Chen J and Zhang H 2019 Study on PVA fiber surface modification for strain-hardening cementitious composites (PVA-SHCC) *Constr. Build. Mater.* **197** 107–16

[37] Choi W C, Yun H D and Lee J Y 2012 Mechanical properties of mortar containing bio-char from pyrolysis *J. Korea Inst. Struct. Mainten. Inspec.* **16** 67–74

[38] Zhao X and He X-J 2014 High-toughness and durability performance characterization of concrete reinforced with poly (vinyl alcohol) fibers *Mater. Express* **4** 247–52

Chapter 11

Biochar addition to inorganic binders

Daniele Ziegler, Elisabetta Di Francia, Patrizia Savi and Jean-Marc Tulliani

Cement based composites have a low conductivity and a reduced capacity to absorb electromagnetic waves. This latter feature becomes a serious problem with the multiplication of wireless telecommunication devices in our modern lives. Therefore, many studies aim at enhancing the electromagnetic interference shielding effectiveness of cementitious composites through the addition of conductive carbonaceous nano/microparticles. To this aim, biochar, obtained from different biomasses, proved to be effective in increasing the electromagnetic shielding efficiency of cementitious materials.

Biochar also has the potential to partially replace cement powder in ultra-high-performance concrete (UHPC), enhancing its hydration due its internal curing ability and favouring the nucleation of hydrated cement phases. These effects were also observed on magnesium phosphate cements and in alkali activated slags. Furthermore, when entrapping bacteria in biochar, self-healing of cracks was possible because of calcium carbonate precipitation.

Carbon dioxide emissions in the construction industry are a serious concern and account for 4%–6% of anthropogenic CO_2 emissions yearly. One of the potential solutions to address this issue is to transform building materials into carbon sinks. Depending on the type of feedstock and preparation conditions, biochar has the potential of reducing greenhouse gas emissions by about 870 kg CO_2 equivalent per ton dry feedstock. Thus, biochar can be mixed with cement or other inorganic binders to sequester carbon for a long time. In the pre-cast industry, biochar addition can be associated with a carbon mineralization process to increase the CO_2 of the obtained products.

Therefore, this short review aims to provide a survey of recent research dealing with carbonaceous materials produced from different sources of biochars used for increasing the electromagnetic shielding properties of cementitious materials, as well as to favor cement paste late hydration and to make them act as carbon sinks.

doi:10.1088/978-0-7503-2660-5ch11

11.1 Introduction

Cement based composites have a low conductivity and a reduced capacity to absorb electromagnetic waves. This latter feature becomes a serious problem with the multiplication of wireless telecommunication devices in our modern lives. Therefore, many studies are aimed at enhancing the electromagnetic interference shielding effectiveness of cementitious composites through the addition of conductive carbonaceous nano/microparticles.

Carbon dioxide emissions in the construction industry are a serious concern and account for from 4% to 6% of anthropogenic CO_2 emissions yearly [1]. One of the potential solutions to address this issue is to transform building materials into carbon sinks. Depending on the type of feedstock and the preparation conditions, biochar has the potential of reducing greenhouse gas emissions by about 870 kg CO_2 equivalent per tonne dry feedstock [2]. Thus, biochar can be mixed with cement or other inorganic binders to sequester carbon for a long time. In addition, in the precast industry, biochar incorporation can be associated with carbon mineralization process to increase the CO_2 of the obtained products.

Recently, internal curing has been proposed, which it is based on saturated porous aggregates that constitute reservoirs inside concrete and provide water for concrete curing internally [3]. The same concept was applied to biochar to partially replace cement powder in ultra-high-performance concrete (UHPC), enhancing its hydration due its internal curing ability and favoring nucleation of hydrated cement phases. These effects were also observed in magnesium phosphate cements and in activated slags.

Finally, when entrapping bacteria in biochar, the self-healing of cracks was possible in mortars because of calcium carbonate precipitation.

Therefore, this short review aims to provide a survey of recent studies dealing with carbonaceous materials produced from different sources of biochars used for increasing the electromagnetic shielding properties of cementitious materials, as well as to favor cement paste late hydration and to make them act as carbon sinks. At the very end, a short survey on the presented patents on these topic is illustrated.

11.2 Electromagnetic interference shielding effectiveness

Electromagnetic wave (EMW) pollution is continuously growing because of the modern developments in wireless and communication systems that are contaminating our built environment. In addition, a cement matrix does not offer enough shielding efficiency (SE) against EMWs. Therefore, the shielding effectiveness of cement composites has been increased by conducting element additions [4–8], (figure 11.1).

Although carbon nanotubes (CNTs) and graphene present high specific surface areas, low density, high aspect ratios, and very high electrical conductivity [4, 9], making them ideal candidates to improve the shielding efficiency of cementitious materials, there are several major concerns that limit their use on large scales. One such concern is their still rather high production costs and another is the difficulty of dispersing them in the cement matrix because of the strong van der Waals forces that create hard agglomeration. This latter inconvenience is being overcome through

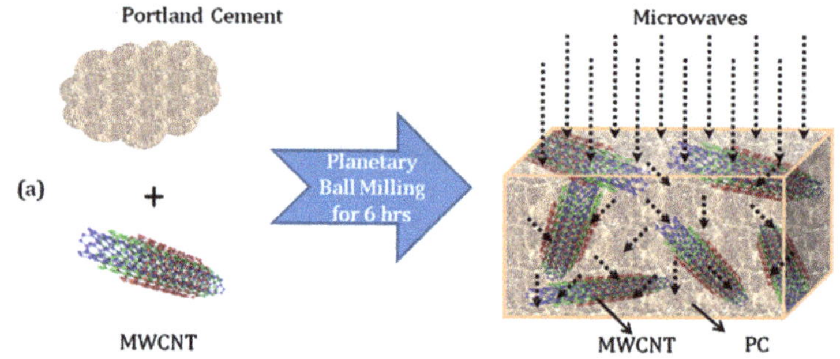

Figure 11.1. Scheme of the addition of multi-walled carbon nanotubes (MWCNTs) to Portland cement for composite manufacturing and their interaction with microwaves [4]. Reproduced with permission from [4]. Copyright 2013 Elsevier.

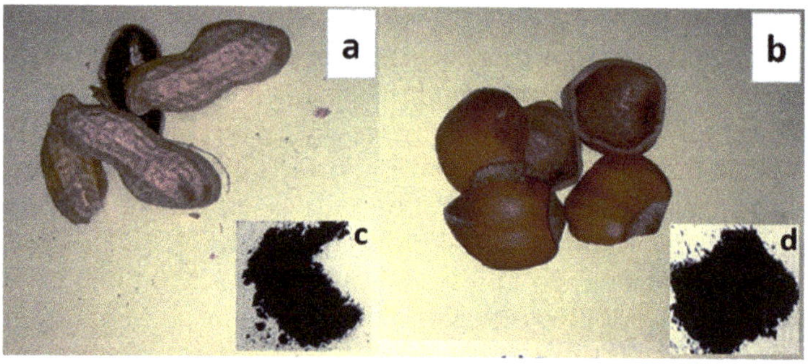

Figure 11.2. Peanut and hazelnut shells: (a) and (b) as received and (c) and (d) after pyrolysis [11]. Reproduced with permission from [11]. Copyright 2015 Elsevier.

surface functionalization and carboxylic group grafting, for example. However, acid functionalization of CNTs can lead to an increase in their toxicity [10]. Thus, carbonized agricultural wastes such as peanut and hazelnut shells were proposed for cement composite preparation and their electromagnetic interference shielding effectiveness was assessed in [11, 12]. The raw peanut shell (PS) and hazelnut shell (HS) were pyrolyzed at 850 °C for 1 h, manually ground in an agate mortar with an agate pestle and then ball milled for 24 h in ethanol, followed by 2 h of attrition milling (figure 11.2).

The amount of carbonized nano/microparticles was 0.2 wt.% and 0.5 wt.% with respect to cement. The relative complex permittivity ($\varepsilon r = \varepsilon' - j\varepsilon''$) was assessed using an open-ended coaxial probe sensor in the frequency range of 0.2 to 10 GHz. The real part refers to the content of energy stored by a material when exposed to an external electric field while the imaginary part represents the measure of dissipated or lost energy. A noticeable increase was observed in the values of real and imaginary permittivity due to the addition of the carbonaceous filler in the composite samples.

Total SE is given by the sum of three major contributions: the reflection loss, which happens due to reflection of EMWs at the material's interfaces, the absorption loss of the wave as it passes through the sample, and the re-reflections and transmission phenomena of EMWs (multiple reflection loss). Total SE along with these three contributions was analytically evaluated from measured values of dielectric constant (ε') and dielectric loss (ε'') considering samples of thickness 10 mm [11]. The total SE of all five different compositions are compared at four different frequencies of 0.94 GHz (GSM mobiles), 1.56 GHz (GPS communication devices), 2.46 GHz (microwave ovens), and 10.0 GHz (radio communication devices) and are reported in table 11.1. With respect to pristine cement, the shielding efficiency increases with the increase of the percentage of filler. A shielding efficiency of 10 dB can be reached with both HS and PS filler (0.5 wt.%). These results can be improved by increasing the percentage of the filler or considering samples with greater thickness.

Furthermore, the addition of 0.5 wt.% HS (PS) led not only to the maximum in electromagnetic SE but also to the maximum in fracture energy in the three-point bending test. For PS, the maximum in fracture energy was reached for a 0.08 wt.% addition, but a higher amount (0.5 wt.%) gave rise to a significant increase in fracture energy as well, with respect to pristine cement. To conclude, the total increase in SE at the four studied frequencies when adding 0.5 wt.% of PS or HS in cement composites was much higher compared to those reported by other researchers using CNTs in the same amount [4, 13].

Recently, cotton stalk and waste tire were pyrolyzed at 500 °C for 30 min and added to cement [14] in the same amounts as PS and HS in [11, 12]. Plates of 16×14 cm^2 with a thickness of 50 mm are measured in the X-band (8 to 12 GHz) with a free-space set-up and present SEs of about 23 and 24 dB on 0.5% optimum addition of carbonized cotton stalk and carbonized waste tire, respectively.

11.3 Internal curing ability

The porous nature of biochar and its water retention ability are known in the literature and contribute to the hydration degree of cementitious mortar through internal curing [15, 16]. Apart from the incorporation of biochar as a carbon sequestering mineral admixture for cement based materials, its adsorption capabilities can be exploited for water retention as in soil enhancement. For example, in [15] water released from

Table 11.1. Total shielding effectiveness of a 10 mm thick cement composite sample [11].

	Shielding efficiency (dB) at			
Sample	0.9 GHz	1.56 GHz	2.46 GHz	10 GHz
Pristine cement	1.0	1.8	2.7	5.9
0.2 wt.% HS	2.6	3.8	4.7	8.2
0.2 wt.% PS	3.7	5.1	5.7	9.5
0.5 wt.% HS	4.4	5.7	6.1	10.3
0.5 wt.% PS	4.5	5.8	6.2	10.7

biochar, used to replace 5 wt.% of cement, contributed to mechanical strength improvement by 12% compared to reference mortar at 28-day age. Other results [17, 18] suggested that biochar, from food and wood waste pyrolysis, behaved as a filler and led to an increase in strength and ductility by 15% to 20% compared to reference samples. These results were due to the local reduction of w/c ratio in biochar–concrete mixes at an early age and improved hydration due to new nucleation sites on biochar particles. The filler effect of biochar further decreased the capillary water absorption and penetration in biochar mortar composites of 30% to 40% with respect to pristine materials [18, 19].

This capacity was also observed in ultra-high-performance concretes in [20]. UHPCs are characterized by high strength (usually above 120 MPa) because of their dense microstructure [21]. This compact microstructure also confers them a higher durability performance, which leads to longer service life and reduced maintenance compared to conventional concrete. These results are due to a water to cement ratio (w/c) around 0.2, below the quantity of water needed for full hydration of cement powder [22]. In [20], biochar was obtained by pyrolysis at 500 °C for 1 h of local mixed wood sawdust. This temperature was chosen because it leads to methylene, alcohol, and phenol group removal in softwood and hardwood, as these groups are harmful for cement hydration [23]. The produced material was manually ground and sieved into three size ranges: coarse (250 to 500 μm), medium (125 to 250 μm), and fine (<125 μm). These powders were first pre-soaked into water and then used to replace cement at 2 wt.%, 5 wt.%, and 8 wt.%. The degree of hydration estimated from thermo-gravimetric analysis showed an increase from 42% in the reference sample up to 59% in samples with 5 wt.% of fine biochar. Moreover, fine biochar addition at 5 wt.% also led to comparable 28 d compressive strength (144 MPa) as the control mix (150 MPa). In contrast, the mix with 8 wt.% of coarse biochar evidenced a 30% drop in compressive strength. In general, UHPC mixes with fine biochar addition presented higher mechanical properties amongst the three size groups. These results were due to an improved hydration that balanced the weakness caused by porous biochar incorporation. Medium and coarse particles were not as effective as small ones because of the presence of more macro-pores that caused a major decrease in strength. Nonetheless, scanning electron microscopy observations of hydrated paste microstructures showed the deposition of cement hydrates on the surface and inside the pores of biochar. Dense interfacial transition zones were also evidenced, further demonstrating the efficacy of biochar in improving hydration. Finally, fine biochar particles proved to significantly accelerate the hydration rate, as shown by isothermal calorimetry tests.

Biochar was also associated with MgO expansive additive to mitigate the autogeneous shrinkage of cementitious materials [24]. Autogenous shrinkage is a macroscopic volume contraction of concrete at an early age (less than 24 h after casting) occurring without moisture evaporation. Autogenous shrinkage is attributed to water consumption due to the hydration of the cement paste and is the result of chemical shrinkage (the volume change before and after a complete hydration reactions) [25]. With the development of high-performance concretes and the use of advanced admixtures such as superplasticizers, w/c ratios lower than 0.42 are common in concrete practice. Thus, autogenous shrinkage has become an important issue for modern concretes.

Incorporation of biochar provided efficient internal curing and led to a reduction of the autogenous shrinkage by 16.3% at an age of 180 h. Of course, a synergetic effect on decreasing the autogenous shrinkage of cement paste was noticed when the biochar was added in combination with MgO. For instance, with the addition of 2 wt.% biochar and 8 wt.% MgO, an expansion rather than a shrinkage was evidenced after 180 h.

Magnesium phosphate cements (MPCs) are formed by the chemical reaction between ammonium dihydrogen phosphate (ADP) and dead burned magnesium oxide [26]. Recently, the popularity of MPCs has been increasing due to their excellent bonding strength, rapid hardening, higher early strength, lower permeability, and good durability properties [27]. Equation (11.1) illustrates the chemical reaction occurring between magnesia and ADP:

$$MgO + NH_4H_2PO_4 + 5H_2O \rightarrow MgNH_4PO_4 \cdot 6H_2O. \qquad (11.1)$$

The setting times are very short at room temperature (10 to 20 min) and the strengths of 14 MPa can be reached within 2 h. MPC is normally used for repairs in which the downtime is limited and in cold weather since the setting is fast because of the exothermic reaction [26]. In [27], biochar MPC mortars were prepared by mixing MgO, ADP, fly ash (FA, a by-product of an electricity-generating plant where coal is used as fuel), biochar, and sand in different proportions. The ratio of MgO:ADP:FA was kept at 3:1:1 for all mortar mixtures. Borax and sodium tripolyphosphate were used to retard the setting time of the mixtures. Biochar was prepared from wheat straw biomass heated to 650 °C for 4 h. The addition of biochar increased the compressive strength and flexural strength of MPC mortars. The compressive strength of MPC mortar with 0.5 wt.%, 1.0 wt.%, and 1.5 wt.% biochar with respect to binder was, respectively, 4.1%, 12.8%, and 17.3% higher than the control MPC mortar at 30 d. Flexural strength results showed a similar trend of strength increase. These results were attributed to the micro-filler effect, the reduction of local w/c, and internal curing action due to biochar addition. In fact, biochar addition decreased the total porosity and permeable porosity of MPC mortars.

Commercial biochar, in the size range of 0.841 to 1.19 mm, made from softwood was mixed (0.5 to 2.5 wt.% of the binder mass) with ground granulated blast slag (GGBS, from cast iron production process) and hydrated lime ($Ca(OH)_2$) to stabilize calcareous sands in [28]. The samples were cured at 20 °C for 3, 7, and 28 d. After 28 d, 1 wt.% biochar specimen with 15 wt.% binder had a compressive strength of 1.80 MPa, like the reference sample. The progressive release of water by biochar particles favored slag hydration, producing a tight and strong bonding between biochar, cementitious hydration products, and sand particles.

11.4 Cargo for self-healing cementitious materials

Concrete has a limited tensile strength and, consequently, is very sensitive to crack formation. Cracks limit the durability of concrete structures because aggressive liquids and gases may penetrate into the cementitious matrix and cause further damage in the reinforcement bars [29]. Small crack (<300 μm in width) occurrence in concrete is almost unavoidable and impairs its functionality, accelerates its degradation, and reduces its service life and sustainability. Thus, smart self-healing materials and preventive repair methods can be part of the answer to increase the safety and sustainability of structures [29]. Autonomic self-healing relies on the encapsulation of healing agents and their internal release at the location of the damaged zone, allowing *in situ* repair [29]. To this aim, microencapsulation (diameter $\leqslant$ 1 mm) [29, 30] and macroencapsulation [29, 31, 32] were investigated throughout the years. In addition, self-healing solutions based on certain bacteria species (*Bacillus*) involve the sealing of cracks by precipitation of calcium carbonate from direct precipitation or ureolytic decomposition of calcium compounds such as calcium nitrate/lactate. Immobilization of bacteria and of nutrients is thus mandatory. To this aim, several carriers have been studied such as lightweight clay aggregate, diatomaceous earth, polyurethane and silica gel, hydrogel, graphene nano-platelets, zeolite, polymer microcapsules, and extruded cementitious tubes [29]. In [33], wood (softwood) sawdust biochar was produced at 500 °C and used as a cargo for bacteria spores due to its porosity ranging from 2 nm to 20 μm. In addition to biochar, different combinations of superabsorbent polymer (SAP, 200–300 μm in diameter) and polypropylene (PP) micro-fiber (19 mm in length and an aspect ratio of 190) were investigated. The aim was, on one hand, to achieve a controlled cracking of the matrix and an increase of the fracture toughness through a crack bridging process during loading [34]. In addition, SAP supplied moisture to the cracked zone over the long term for bacterial metabolism. In fact, SAP can absorb high volumes of water from the surroundings and retain it within its structure without dissolving [35]. In case of cement filtrate, the absorption capacity of the chosen SAP (sodium polyacrylate based commercial product) was equal to (56.90 ± 1.01) g/g, (48.93 ± 0.42) g/g, and (50.19 ± 4.25) g/g for the first, second, and third cycle, respectively [33]. The retained moisture is then released under a humidity gradient for bacteria to precipitate calcium carbonate. Biochar was added at 2 wt.% of cement, while SAP and PP fiber were added at 0.70 wt.% of cement and 0.30 vol.%, respectively. The nutrients included yeast extract, urea, and calcium nitrate and were incorporated at 0.85 wt.%, 4 wt.%, and 8 wt.% with respect to cement. Samples with biochar immobilized spores, SAP, and PP fibers showed high recovery in terms of resistance to water penetration and a similar or slightly higher strength than the undamaged matrix.

11.5 Carbon sink

Today, 54% of the total world population lives in towns and cities and this proportion is expected to increase up to 66% by 2050 [36]. For this reason, buildings and constructions have a high significant impact and contribute up to 39% of energy-related carbon dioxide emissions when including the manufacturing of materials and products for building construction [37]. Using biochar in construction materials is an efficient mode of waste management and a strategy for storing carbon in the built environment. During pyrolysis, the carbon from biomass feedstock is sequestered in the chemical structure of the produced biochar, which would otherwise be released to the atmosphere through decay or degradation of biomass.

Concrete is the most widely used man-made material on Earth because of its high compressive strength and relatively low cost; 4.1 Gt of cement were produced in 2017, according to Cembureau estimations [38]. It is second only to water as the most-consumed resource on the planet. Emissions of CO_2 from the cement industry account for 6% of the total emissions of CO_2 from stationary sources. The current CO_2 impact of Portland cement can be divided into three steps: from 55% to 60% due to the thermal decomposition of limestone (with clay, the raw materials necessary for cement production), from 30% to 35% because of the energy needed to reach the high temperature (1450 °C) of the process, and 10% for transportation [39].

In [40], biochar made from corn stover pyrolyzed at 550 °C for 25 min was added as a filler material in cement (50 to 80 wt.%)/fly ash (20 to 50 wt.%) blocks in the range of 2 to 8 wt.% of the binder (cement + FA). The obtained blocks were subjected to 2 h mineral carbonation (pure CO_2 at 1 MPa pressure) after demolding to enhance the CO_2 uptake and sequester carbon in the specimen (figure 11.3). Accelerated carbonation curing (ACC) is one of the carbon capture and storage (CCS) processes and the fastest way of CO_2 storage in cementitious materials [41–43]. Basic carbonation can be summarized as the chemical reaction between calcium hydroxide, resulting from Portland cement hydration, with CO_2.

According to [44], one block of 15 kg made with 13 wt.% of cement has a potential capacity of storage of 0.454 kg of CO_2. In addition, around 15 million tonnes of CO_2 can be potentially sequestered in concrete by ACC [44, 45]. The results showed that there was an enhanced CO_2 uptake when biochar was added to the mix and biochar dosage was optimized at 4% and 6%, depending on the cement dosage. The high surface area and porosity of biochar particles provided the additional CO_2 sites that increased uptake. The 28 d normal curing compressive strength increased when biochar was added as well. These results were attributed to the reduction in the amount of water evaporation in the first days after casting, as well as to the filler and internal curing effects.

The 28 d compressive strength of un-carbonated blocks showed that there was an increase in the strength as biochar addition increased to a certain extent, however, thereafter it decreased with the further addition of biochar.

11.6 Patents: an updated survey

In the last few years, the increasing interest in the use of biochar addition to inorganic binders has driven new patent requests. Recently, a patent was presented by Zhang *et al* [46] to overcome the problem of concrete crack sensitivity—the CN110357658 patent is an invention for a lightweight crack-resistant adhesive mortar. In this patent, the inventors reported that 'the lightweight crack-resistant adhesive mortar, provided by the invention, has the characteristics of small self-weight, high bonding strength, good thermal insulation property and good crack resistance, and is suitable for the building exterior wall thermal insulation system'.

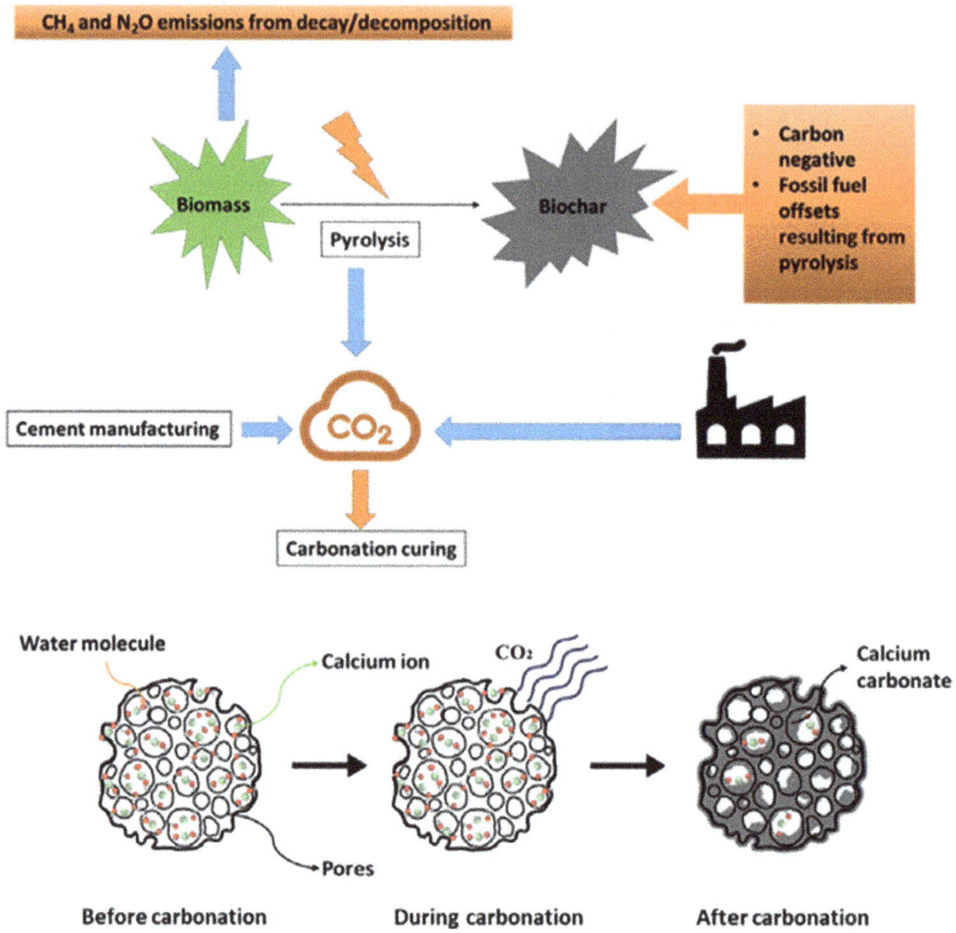

Figure 11.3. Scheme of CO_2 storage in cementitious materials due to biochar addition and CO_2 curing [40]. Reproduced with permission from [40]. Copyright 2020 Elsevier.

At the end of 2019, a patent on the shielding effectiveness properties of concrete with added biochar was presented by Suji and Kumar [47] which reported the process of preparing a bio-shield panel for absorbing electromagnetic field radiation.

Moreover, a series of international patents (WO/2018/203829, publication date: 08/11/2018, applicant: National University of Singapore) was also presented in several countries (Singapore, SG11201909610P, publication date: 28/11/2019; China, CN110621634, publication date: 27/12/2019; US, US20200062646, publication date: 27/02/2020; India, IN201937043826, publication date: 22/05/2020) with the title of *Sustainable construction material and method of preparation and use thereof* [48]. In those cases, Shuan *et al* reported that the invention relates to the use of biochar for the preparation of construction material and concludes that the wadding materials were found to be effective in sequestering carbon dioxide from an indoor environment. Moreover, biochar not only promotes the recycling of waste but also sequesters huge volumes of carbon in cementitious composites.

11.7 Conclusions

In this chapter, different applications of biochar in cementitious materials are illustrated. Being electrically conductive, biochar particles proved to be effective in increasing the electromagnetic shielding efficiency of cementitious materials. In addition, when mixed with cement or other inorganic binders, they allow carbon sequestering for a long time. This latter point is of paramount importance considering climate change and the necessity to cut anthropic CO_2 emissions in a strategic industrial sector such as construction and building. Moreover, saturated porous biochar aggregates can constitute reservoirs inside concrete and provide water for concrete internal curing as well as favor the nucleation of hydrated cement phases. This crucial feature allows partial cement replacement even in ultra-high-performance concrete. Taking sustainability aspects into account, as concrete is the most widely used construction material, the possibility to partially replace cement in its manufacturing is beneficial. Last, but not least, when entrapping bacteria in a porous biochar structure, the self-healing of cracks was demonstrated. Self-healing ability is strongly desirable for concrete because cracks limit the durability of concrete structures and cause further damage to the reinforcement bars.

References

[1] Neuhoff K *et al* 2020 Carbon control and competitiveness post 2020: the cement report *Climate Strategies* http://climatestrategies.org/wp-content/uploads/2014/02/climate-strategies-cement-report-final.pdf (Accessed: 18 July 2020)

[2] Woolf D, Amonette J E, Street-Perrott F A, Lehmann J and Joseph S 2010 Sustainable biochar to mitigate global climate change *Nat. Commun.* **1** 56

[3] Li Z 2011 *Advanced Concrete Technology* (Hoboken, NJ: Wiley) p 19

[4] Singh A P, Gupta B K, Mishra M, Govind C A, Mathur R B and Dhawan S K 2013 Multiwalled carbon nanotube/cement composites with exceptional electromagnetic interference shielding properties *Carbon.* **56** 86–96

[5] Li B Y, Duan Y P, Zhang Y F and Lin S H 2011 Electromagnetic wave absorption properties of cement-based composites filled with porous materials *Mater. Des.* **32** 3017–20

[6] Bantsis G, Mavridou S, Sikalidi C, Betsiou M, Oikonomou N and Yioultsis T 2012 Comparison of low cost shielding-absorbing cement paste building materials in X-band frequency range using a variety of wastes *Ceram. Int.* **38** 3683–92

[7] Dai Y, Sun M, Liu C and Li Z 2010 Electromagnetic wave absorbing characteristics of carbon black cement-based composites *Cem. Concrete Compos.* **32** 508–13

[8] Pretorius J C and Maharaj B T 2013 Improvement of electromagnetic wave (EMW) shielding through inclusion of electrolytic manganese dioxide in cement and tile-based composites with application for indoor wireless communication systems *Int. J. Phys. Sci.* **8** 295–301

[9] Kong L, Yin X, Yuan X, Zhang Y, Liu X, Cheng L and Zhang L 2014 Electromagnetic wave absorption properties of graphene modified with carbon nanotube/poly (dimethyl siloxane) composites *Carbon.* **73** 185–93

[10] Figarol A, Pourchez J, Boudard D, Forest V, Tulliani J M, Lecompte J P, Cottier M, Bernache-Assollant D and Grosseau P 2014 Biological response to purification and acid functionalization of carbon nanotubes *J. Nanopart. Res.* **16** 2507

[11] Khushnood R A, Ahmad S, Savi P, Tulliani J-M, Giorcelli M and Ferro G A 2015 Improvement in electromagnetic interference shielding effectiveness of cement composites using carbonaceous nano/micro inerts *Constr. Build. Mater.* **85** 208–16

[12] Khushnood R A, Ahmad S, Restuccia L, Spoto C, Jagdale P, Tulliani J-M and Ferro G A 2016 Carbonized nano/microparticles for enhanced mechanical properties and electromagnetic interference shielding of cementitious materials *Front. Struct. Civ. Eng.* **10** 209–13

[13] Nam I W, Kim H K and Lee H K 2012 Influence of silica fume additions on electromagnetic interference shielding effectiveness of multi-walled carbon nanotube/cement composites *Constr. Build. Mater.* **30** 480–87

[14] Mahmood A, Khushnood R A and Zeeshan M 2020 Pyrolytic carbonaceous reinforcements for enhanced electromagnetic and fracture response of cementitious composites *J. Cleaner Prod.* **248** 119288

[15] Choi W C, Yun H D and Lee J Y 2012 Mechanical properties of mortar containing biochar from pyrolysis *J. Korea Inst. Struct. Maint. Insp.* **16** 67–74

[16] Gupta S, Kua H W and Low C Y 2018 Use of biochar as carbon sequestering additive in cement mortar *Cement Concr. Compos.* **87** 110–29

[17] Ahmad S, Khushnood R A, Jagdale P, Tulliani J-M and Ferro G A 2015 High performance self-consolidating cementitious composites by using micro carbonized bamboo particles *Mater. Des.* **76** 223–29

[18] Gupta S, Kua H W and Pang S D 2018 Biochar-mortar composite: manufacturing, evaluation of physical properties and economic viability *Constr. Build. Mater.* **167** 874–89

[19] Gupta S, Kua H W and Koh H J 2018 Application of biochar from food and wood waste as green admixture for cement mortar *Sci. Total Environ.* **619620** 419–35

[20] Dixit A, Gupta S, Pang S D and Kua H W 2019 Waste valorisation using biochar for cement replacement and internal curing in ultra-high performance concrete *J. Cleaner Prod.* **238** 117876

[21] PCA 2019 Ultra-high performance concrete *PCA* https://cement.org/learn/concrete-technology/concrete-design-production/ultra-high-performance-concrete (Accessed: 18 July 2020)

[22] Brandt A M 2009 *Cement-Based Composites: Materials, Mechanical Properties and Performance* 2nd edn (New York: Taylor and Francis) p 442

[23] Parihar A, Sripada P, Bambery K, Garnier G and Bhattacharya S 2017 Investigation of functional group changes in biomass during slow pyrolysis using synchrotron based infra-red microspectroscopy and thermogravimetry-infra-red spectroscopy *J. Anal. Appl. Pyrolysis.* **127** 394–401

[24] Mo L, Fang J, Huang B, Wang A and Deng M 2019 Combined effects of biochar and MgO expansive additive on the autogenous shrinkage, internal relative humidity and compressive strength of cement pastes *Constr. Build. Mater.* **229** 116877

[25] Li Z 2011 *Advanced Concrete Technology.* (Hoboken, NJ: Wiley) p 198

[26] Fowler D W 2009 Repair materials for concrete structures *Failure, Distress and Repair of Concrete Structures* (Cambridge: Woodhead) pp 194–207

[27] Riaz Ahmada M, Chen B and Duan H 2020 Improvement effect of pyrolyzed agro-food biochar on the properties of magnesium phosphate cement *Sci. Total Environ.* **718** 137422

[28] Han X-L, Jiang N-J and Wang Y-J 2020 Stabilization of calcareous sand by applying the admixture of alkali-activated slag (AAS) and biochar *Proc. of Geo-Congress 2020* (Reston, VA: American Society of Civil Engineers) pp 469–75

[29] De Belie N *et al* 2018 A review of self-healing concrete for damage management of structures *Adv. Mater. Interfaces.* **5** 1800074

[30] Kanellopoulos A, Qureshi T and Al-Tabbaa A 2015 Glass encapsulated minerals for self-healing in cement based composites *Constr. Build. Mater.* **98** 780–91

[31] Maes M, van Tittelboom K and De Belie N 2014 The efficiency of self-healing cementitious materials by means of encapsulated polyurethane in chloride containing environments *Constr. Build. Mater.* **71** 528–37

[32] Formia A, Terranova S, Antonaci P, Pugno N M and Tulliani J M 2015 Setup of extruded cementitious hollow tubes as containing/releasing devices in self-healing systems *Mater.* **8** 1897–923

[33] Gupta S, Kua H W and Pang S D 2018 Healing cement mortar by immobilization of bacteria in biochar: an integrated approach of self-healing and carbon sequestration *Cem. Concr. Compos.* **86** 238–54

[34] Snoeck D, Van Tittelboom K, Steuperaert S, Dubruel P and De Belie N 2014 Self-healing cementitious materials by the combination of microfibres and superabsorbent polymers *J. Intell. Mater. Syst. Struct.* **25** 13–24

[35] Jensen O M 2013 Use of superabsorbent polymers in concrete *Concr. Int.* **35** 48–52

[36] United Nations 2014 World's population increasingly urban with more than half living in urban areas *Department of Economic and Social Affairs* https://un.org/development/desa/en/news/population/world-urbanization-prospects.html (Accessed: 18 July 2020)

[37] UN Environment 2017 Global Status Report 2017 *World Green Building Council* https://worldgbc.org/sites/default/files/UNEP%20188_GABC_en%20%28web%29.pdf (Accessed: 18 July 2020)

[38] CEMBUREAU (The European Cement Association) 2018 *2017 Activity Report* https://cembureau.eu/media/1716/activity-report-2017.pdf (Accessed: 18 July 2020)

[39] Biernacki J J *et al* 2017 Cements in the 21st century: challenges, perspectives, and opportunities *J. Am. Ceram. Soc.* **100** 2746–73

[40] Praneeth S, Guo R, Wang T, Dubey B K and Sarmah A K 2020 Accelerated carbonation of biochar reinforced cement-fly ash composites: enhancing and sequestering CO_2 in building materials *Constr. Build. Mater.* **244** 118363

[41] Ashraf W 2016 Carbonation of cement-based materials: challenges and opportunities *Constr. Build. Mater.* **120** 558–70

[42] Kashef-Haghighi S, Shao Y and Ghoshal S 2015 Mathematical modeling of CO_2 uptake by concrete during accelerated carbonation curing *Cem. Concr. Res.* **67** 1–10

[43] Kashef-Haghighi S and Ghoshal S 2010 CO_2 sequestration in concrete through accelerated carbonation curing in a flow-through reactor *Ind. Eng. Chem. Res.* **49** 1143–49

[44] El-Hassan H, Shao Y and Ghouleh Z 2014 Effect of initial curing on carbonation of lightweight concrete masonry units *ACI Mater. J.* **110** 441–50

[45] Sharma D and Goyal S 2018 Accelerated carbonation curing of cement mortars containing cement kiln dust: an effective way of CO_2 sequestration and carbon footprint reduction *J. Clean. Prod.* **192** 844–54

[46] Zhang D, Zhang K, Lu F, Mei Y and Zhang C 2019 Lightweight crack-resistant adhesive mortar *Patent* 113057658 https://patentscope.wipo.int/search/en/detail.jsf?docId=CN275773568&_cid=P12-KD0C57-27612-1 (Accessed: 8 July 2020)
[47] Suji D and Ashok Kumar L 2019 Bio shield panel for absorbing radiation *Patent* 201941040130 https://patentscope.wipo.int/search/en/detail.jsf?docId=IN275766037&_cid=P12-KD0C6J-28266-1 (Accessed: 8 July 2020)
[48] Ng M S, Kua H W and Gupta S 2018 Sustainable construction material and method of preparation and use thereof *Patent* WO/2018/203829 https://patentscope.wipo.int/search/en/detail.jsf?docId=WO2018203829 (Accessed: 24 July 2020)

IOP Publishing

Biochar
Emerging applications
Alberto Tagliaferro, Carlo Rosso and Mauro Giorcelli

Chapter 12

Insight into the mechanical performance of biochar containing reinforced plastics

Carlo Rosso, Oisik Das and Mattia Bartoli

The worldwide production of reinforced plastics represents one of the largest economic markets with a value of over 77 billion dollars in 2017. Among the reinforced polymeric materials, carbon-based materials represent more than half of the composite market with plenty of applications. Carbon-based reinforced plastic is a very large market with an annual production of up to 150 kton year^{-1} in 2018 with around 62% of the total carbon-based reinforced plastic represented by thermoplastic polymers and the remaining by thermoset resin.

The increased demand for environmental accountability has focused attention on carbonaceous materials with improved sustainability compared to traditional filler. Biochar is the perfect case-study due to its properties, sustainability, and low cost. To date, biochar has found large-scale applications in soil health improvement and as a solid fuel. However, these applications are limited and unable to exploit the full potential of biochar to become a ground-breaking material. The future of biochar will not be in the soil applications but in other industrial sectors such as the production of carbon-based reinforced plastics. In this chapter, we report an overview of biochar-reinforced plastics with a focus on mechanical properties. We evaluate the bulk properties together with the surface properties, proposing deep insight into biochar and related composite characteristics.

12.1 Towards non-conventional carbonaceous fillers: biochar as a potential resource for the production of reinforced plastics

The worldwide production of reinforced plastic represents one of the largest economic markets, with a value of over 77 billion of dollars in 2017 [1]. Among the reinforced polymeric materials, carbon-based materials represent more than half of the composite market with a plethora of applications [2, 3]. Carbon-based reinforced plastic is also a very large market with an annual production of up to

doi:10.1088/978-0-7503-2660-5ch12

150 kton year^{-1} in 2018 [4] with around the 62% of the total carbon-based reinforced plastic represented by thermoplastic polymers and the remaining by thermoset resin.

The worldwide carbon-based reinforced plastics market is currently dominated by carbon black [5] and carbon fibres [6]. These two carbonaceous fillers are the most employed in the production of reinforced polymeric commodities ranging from the manufacture of tyres [7] and high performance materials for the aerospace [8] and automotive [9] sectors. However, materials science has led the way for the use of new remarkable nanostructured carbon materials, such as carbon nanotubes and graphene. These materials and their related plastic composites showed superior properties [10, 11] in both the mechanical and electrical aspects. Despite their admirable performance, nanosized allotropic carbon composites have not gained a real market due the enormous cost of 100 k€ kg^{-1} for carbon nanotubes and up to 200 € cm^{-2} for graphene [12]. Furthermore, the increased accountability of companies with regards to environmental policies has focused great attention on the development of sustainable platforms for the production of more environmentally innocuous materials [13, 14]. While the industry awaits the large-scale commercialization of high-quality affordable nanostructured carbon, new materials have risen as affordable solutions for the promotion of innovation in the field of carbon-based reinforced plastics without compromising on the performance. In the past few years, an unconventional material has emerged as one of the most promising routes for the integration of carbon production with waste management [15]. This material is biochar, the solid residue from the thermo-chemical conversion of biomass in a limited oxygen atmosphere.

As discussed by several authors [16, 17], the future of biochar will not be in soil applications but in other industrial sectors such as the production of carbon-based reinforced plastics. In this chapter we report an overview of biochar-based reinforced plastics with a focus on the mechanical properties. We evaluate the bulk and surface properties thereby providing deep insight into biochar and related composite characteristics.

12.2 Bulk properties of biochar containing reinforced plastics

Epoxy resins represent one of the most studied thermoset polymeric hosts due the numerous applications in relevant industrial sectors [18], generally combined with high performance fillers such as carbon or glass fibres [19]. The bulk properties of epoxy resin are variable but generally are far inferior to the properties reached by carbon fibres, but can be positively improved using carbonaceous fillers such carbon nanotubes [20]. Therefore, biochar must be compared at least with this high-tech material, such as carbon nanotubes, to justify not only its sustainable production but also its properties. A very comprehensive study was reported in 2017 by Khan *et al* [21] comparing the mechanical performance of epoxy resin filled with different loadings of multiwalled carbon nanotubes, and neat and thermally annealed biochar.

As showed in figure 12.1, carbon nanotubes improved the ultimate tensile strength, elongation, and Young's modulus of neat epoxy resin for low

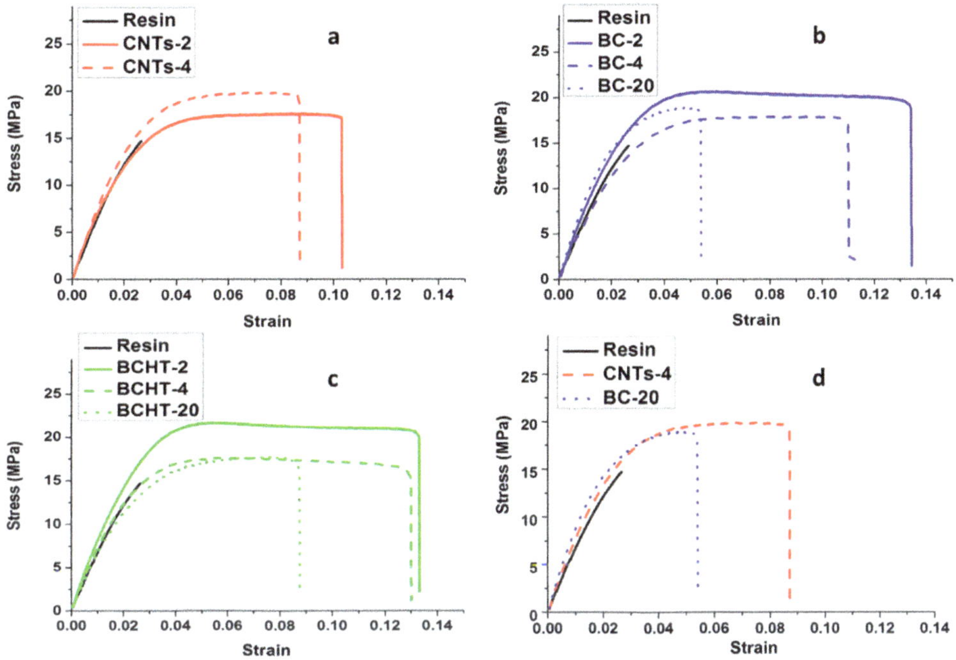

Figure 12.1. Stress versus strain behaviour of epoxy resin compared to (a) multiwalled carbon nanotubes 2 and 4 wt.%; (b) neat biochar 2, 4, and 20 wt.%; (c), neat and thermal annealed biochar 2, 4, and 20 wt.%. Additionally, (d) the comparison between blank epoxy, biochar 20 wt.%, and multiwalled carbon nanotubes 4 wt.%. Reprinted from Khan *et al* [21] with permission from MDPI.

concentrations as biochar did. A biochar concentration of 2 wt.% was proved as the best for the simultaneous improvement of all bulk mechanical properties. Higher filler content induces a shift from ductile to brittle materials, as reported by Giorcelli *et al* [22] using an epoxy filled with a 15 wt.% of coffee derived biochar. This was also attested to by Giorcelli *et al* [23] who also proposed a simple model to predict the biochar equivalent properties as follows:

$$F = \frac{|C - M(1 - f)|}{f}, \tag{12.1}$$

where F is the predicted biochar properties (i.e. ultimate tensile strength, elongation, Young's modulus), C is the experimental composite properties, M is the neat epoxy resin properties, and f is the weight fraction.

However, a biochar loading of 2 wt.% could lead also to a variation in mechanical outputs as reported by Bartoli *et al* [24] and summarized in figure 12.2.

The authors described that the divergent mechanical behaviour was related to the use of biochar produced from different feedstocks in the same pyrolytic conditions. The influence of feedstock was recognized but it was quite complex to discern. The authors suggested the simultaneous influence of particle size and particle roughness.

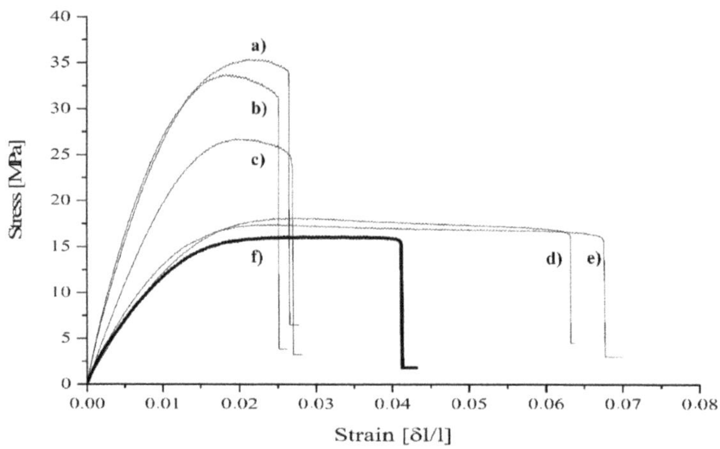

Figure 12.2. Stress versus strain curves of composites filled with 2 wt.% of biochar derived from (a) wheat, (b) oil seed rape, (c) RH, (d) mixed softwoods, and (e) *Mischantus*, and (f) neat resin as reported by Bartoli *et al* [24] reprinted with permission from MDPI.

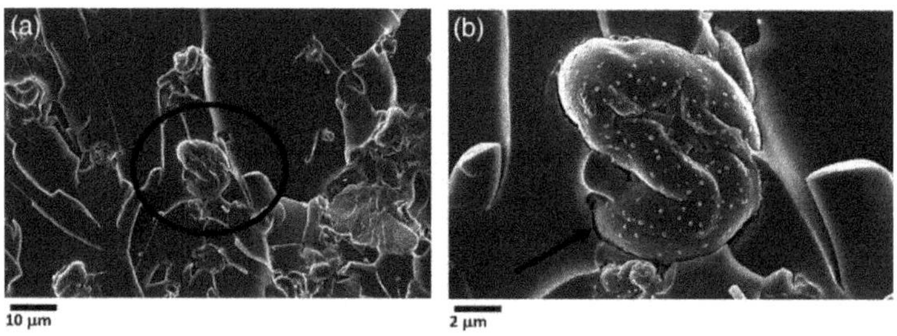

Figure 12.3. Field emission scanning microscope captures of composites containing 10 wt.% of carbonized cellulose nanocrystals with different magnifications after stress/strain fracture as reported by Bartoli *et al* [25] copyright John Wiley & Sons. The particles and interphases between the biochar particles and resin are highlighted with black circles.

Smaller and rougher particles induced brittleness while smooth surface particles promote an increment of elongation.

This hypothesis was tested more accurately by using controlled shape particles [25] produced though carbonization of cellulose nanocrystals. The spherical biochar particles showed an average radius around 2 μm and a very good dispersibility. These particles doubled the elongation of neat resin with a loading of 2 wt.% [26]. As shown in figure 12.3, the authors proved that biochar crack patterns have behaviour similar to those observed in fibre reinforced composites [27].

The properties of biochar-based composites are also related to the temperature used for pyrolization as reported by Bartoli *et al* [28]. Authors analysed an epoxy matrix filled with 2 wt.% of biochar produced using temperature ranging from 400 °C up to 1000 °C. The results showed that a high production temperature induced an increment of elongation together with a reduction of Young's modulus and ultimate tensile strength. This was due to the improved interaction between the aromatic domains of high temperature biochar and epoxy resin that granted an increased mobility of particles in the polymeric host.

Biochar was also used as an additive in glass fibre reinforced epoxy composites as reported by Dahal *et al* [29]. Three-point bending tests revealed a lower increment of up to 300 MPa and an elevated storage modulus of up to 4000 MPa together with a higher stiffness for 10% biochar composite as compared to 5% biochar and the composite without biochar. This improvement was due to the changes induced by biochar in the interfacial properties between glass fibres and polymer. Sundarakannan *et al* [30] reported the use of a different thermoset host represented by unsaturated polyester resin. This polymer was mixed with cashew nutshell biochar by a solution dispersion method at different weight percentages of 5, 10, and 15. Compared to the neat composite the 10 wt.% biochar-based composite showed an increment of tensile, impact strength, and hardness by 21%, 41%, and 37%, respectively. Interestingly, the maximum flexural strength was achieved using a biochar loading of 15 wt.% with a 40% improvement, probably due to the alteration induced by the high filler loading into the polymer matrix.

Moving from a thermoset to thermoplastic polymeric host, biochar was combined with bio-derived polyesters for the development of fully sustainable reinforced plastics. Aup-Ngoen and Noipitak [31] converted agricultural biomass residues from harvesting and processing of agricultural crops and used it as a filler for poly(lactic acid) matrix. The authors used a very low amount of biochar of up to 0.25 wt.%. As the authors reported, the mechanical performance of biochar-based composites was generally inferior to the neat matrix with the exception of the Young's modulus. The relation between biochar and poly(lactic acid) matrix was described more in detailed by Arrigo *et al* [32] using two mixing processes, namely melt mixing and solvent casting, and coffee derived biochar. This study showed an increment of tensile modulus and a decrement of tensile strength due to the presence of voids in the porous structure of biochar particles that are in any case well-dispersed, as shown in figure 12.4.

Usually, the polymer flows into the pores of biochar during processing creating a mechanical interlocking but the unique structure of the coffee limited this on the external surface leaving a massive presence of voids.

Moustafa *et al* [33] used poly(butylene adipate-co-terephthalate) and coffee derived biochar for the production of a biodegradable composite using poly(ethylene glycol) as a plasticizer. The authors reported a drastic decrement of elongation together with an increase of both Young's modulus and ultimate tensile strength. The interaction between poly(lactic acid) and fillers could be improved by biochar addition as reported by Zhang *et al* [34].

The authors studied the thermal and mechanical properties of microcrystalline cellulose /poly(lactic acid) using an activated biochar for modulating the interphase interactions between filler and polymeric host. Related reinforced plastic produced by injection moulding showed a better tensile strength of up to 16 MPa, a tensile modulus of up to around 4GPa, a flexural strength of 54 MPa, a flexural modulus of 6 GPa, and an impact strength of 4 kJ m^{-2}. This was due to the improvement of the

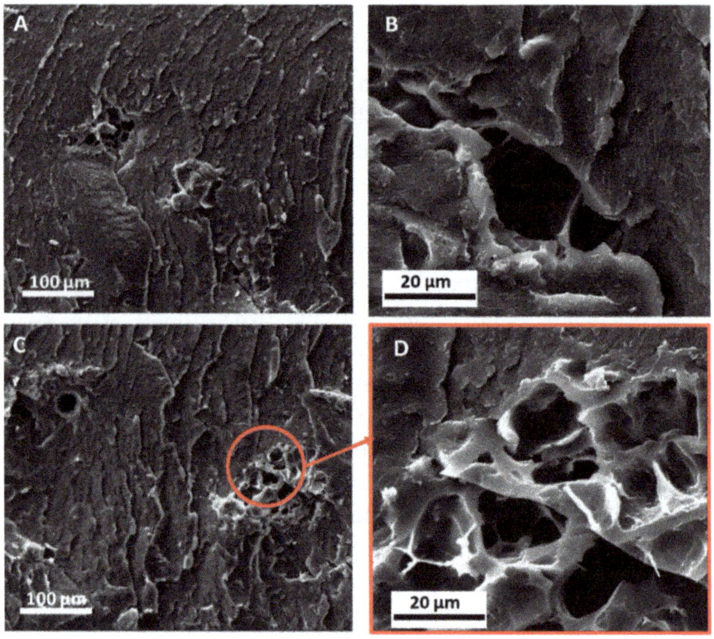

Figure 12.4. Scanning microscope captures of poly(lactic acid) composites containing 7.5 wt. % of carbonized coffee with different magnifications produced using the (a) and (b) melting mix and (c) and (d) solvent casting procedure as reported by Arrigo *et al* [32] reprinted with permission from MDPI.

creep resistance and anti-stress relaxation ability induced by the presence of activated biochar, as proved by dynamic and mechanical analysis.

The other class of polymeric hosts studied is represented by polyolefins. A first approach is represented by the study of Jiang *et al* [35] which used nanobiochar produced from lignin as a filler for an elastomeric rubber made with poly(styrene)-poly(butadiene). The authors also compared biochar to commercial carbon black N330 showing a similar reinforcing effect from both. However, the behaviour observed suggested two distinct reinforcing mechanisms of biochar and carbon black due to the difference in the surface properties and structural characteristic of the fillers.

This variation among different biochars could be lowered with the reduction in size of the biochar particles. Peterson *et al* [36] developed a methodology for reducing biochar to the nanoscale using nanosilica milling. This procedure facilitated the production of nanoscale biochar particles that were able to improve the elongation and toughness of rubber composites by over 31% and 24%, respectively.

Poly(ethylene) was also used as a polymeric host for the production of biochar-based composites. Poly(ethylene) production represent over the 34% of the overall polymer production [37] and the research for its optimization has attracted great interest [38]. Arrigo *et al* [39] reported stress relaxation measurements on coffee derived biochar-based poly(ethylene) composites revealing a 'pseudo-solid-like' behaviour for the composites containing high amounts of biochar, due to the formation of a polymer–particle network that restricts the dynamics of polymer chains. This hypothesis was confirmed by Zhang *et al* [40] who clearly showed a strong interaction between rice husk biochar and poly(ethylene), as shown in figure 12.5.

The strong interaction between biochar and poly(ethylene) were also reported in another study of Zhang *et al* [41] that achieved an improvement of flexural strength together with a decrement of impact strength.

Further studies have focused on the biochar-based poly(propylene) composite. Das *et al* [42] studied the system using biochar loading ranging from 0 to 35 wt.% produced by both compounding and injection moulding. The high surface area of biochar allowed poly(propylene) to flow in creating a mechanical interlocking and improving the mechanical properties, as shown in figure 12.6.

Furthermore, Das *et al* [43] tested the used of maleic anhydride compatibilizer in order to reduce production costs while conserving the mechanical properties of biochar-based poly(propylene) composite. The authors showed that the amount of compatibilizer could be reduced to 1 wt.% without compromising the tensile

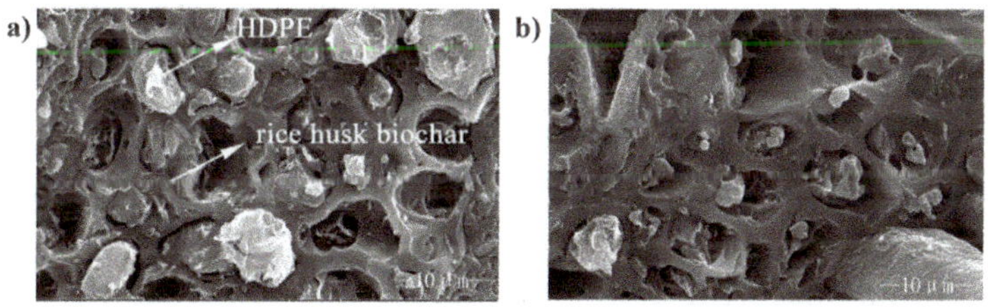

Figure 12.5. Scanning electron microscope images of poly(ethylene) composites containing rice husk biochar loaded up to (a) 30 wt.% and (b) 60 wt.%, as reported by Zhang *et al* [40] reprinted with permission from MDPI.

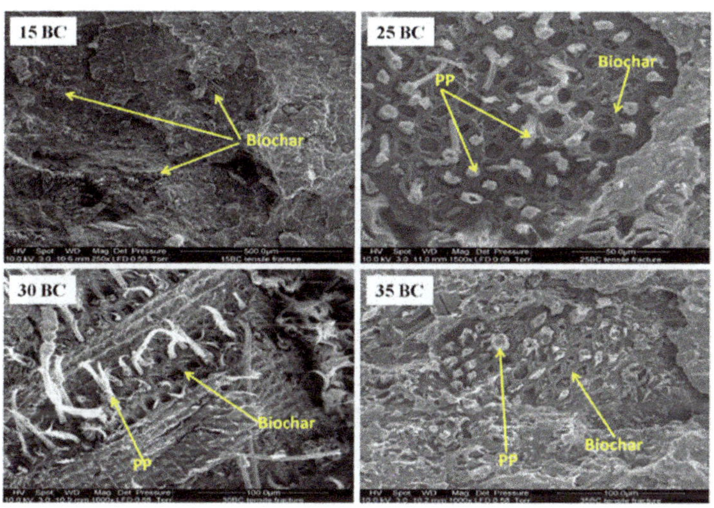

Figure 12.6. Scanning electron microscope captures of poly(propylene) composites containing biochar load from 15 wt.% up to 35 wt.% as reported by Das *et al* [42] reprinted with permission from Elsevier.

strength/modulus, impact strength, and micro-hardness. Additionally, Das *et al* [44] proved the suitability of biochar as a potent fire resistant filler when used in conjunction with ammonium polyphosphate and magnesium hydroxide. This study proved that the addition of biochar to neat poly(propylene) along with conventional fire retardants significantly impedes its flammability while enhancing flexural strength and tensile/flexural moduli and preserving the tensile strength.

Mechanical behaviour was explained by Poulose *et al* [45], proving that the incorporation of biochar into a poly(propylene) matrix significantly reduced the total crystallinity of the composites. Nonetheless, biochar addition had minimal effect on the storage modulus (G') compared to the neat polymer host up to a concentration of 15 wt.%

Noori *et al* [46] extended this work using tea derived biochar for the production of poly(propylene) composites using a twin-screw extruder. The authors compared the dynamic mechanical analysis of poly(propylene) highly loaded with up to 40 wt.% of both biochar and carbon black, as shown in figure 12.7.

The authors reported that the storage modulus of carbon black and biochar-based composites with a loading of up to 40 wt.% were identical. This proves that biochar could potentially replace carbon black, when the mechanical properties of composites are concerned. Nonetheless, the superior dispersibility of biochar was an advantage compared to carbon black that avoided the formation of particle agglomeration, as observed using carbon black [47].

Generally, the addition of biochar even in a low percentage leads to a general improvement of the mechanical properties of composites as shown in table 12.1.

Considering the data reported in table 12.1, the addition of biochar could improve the brittleness of composites or the elongation. The biochar's ability of tailoring composites' properties is one of the key advantages of its use in the production of new materials.

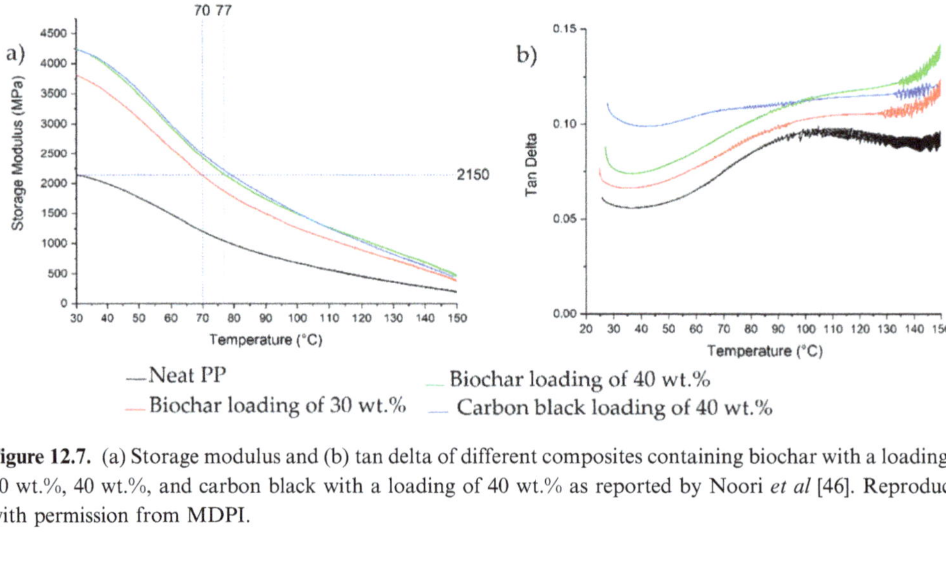

Figure 12.7. (a) Storage modulus and (b) tan delta of different composites containing biochar with a loading of 30 wt.%, 40 wt.%, and carbon black with a loading of 40 wt.% as reported by Noori *et al* [46]. Reproduced with permission from MDPI.

Table 12.1. The main mechanical properties of biochar-based epoxy composites with biochar loading of 2 wt.%. The values reported herein are calculated as an increment/decrement on neat resin properties.

Feedstock	Pyrolysis temperature	Young's modulus (%)	Toughness (%)	Ultimate tensile strength (%)	Maximum elongation (%)	Reference
Mischantus	550	112.4	3.5	90.6	−39.5	[24]
Oil seed rape	550	68.4	−8.8	50.6	−33.8	
Rice husk	550	139.0	−19.3	95.3	−53.4	
Wheat straw	550	13.2	42.1	2.4	42.4	
Mixed softwoods	550	6.0	156.5	28.0	100.2	
Cellulose nanocrystals	400	−11.0	115.6	17.8	91.1	[26]
Cotton fibres	400	112.4	3.5	90.6	−39.5	

12.3 Surface properties of biochar containing reinforced plastics

Bulk properties are not the only ones that could be used to measure the impact of biochar loading into polymers. Composite surface properties could also be positively affected by the addition of biochar to a polymeric matrix. Khan *et al* [48] showed an appreciable decrement of the friction coefficient of epoxy composites containing pyrolytic carbon. A more detailed explanation of this behaviour was reported by Bartoli *et al* [26] that showed a simultaneous decrement of the friction coefficient together with an increment of hardness for loading of 10 wt.% biochar (figure 12.8).

The authors proposed a mechanism that was previously speculated by several authors [49, 50]. The proposed mechanism suggests that the exposure of carbon fillers during the course of friction at the interface may serve as a solid-like lubricant between the steel ball face of the tribometer and composite surface.

Surfaces are interesting not only for the friction properties but also for the possibility to perform a very comprehensive analysis such as by nanoindentation. Das *et al* [51] use this technique for a direct estimation of biochar properties. This study was particularly interesting because the literature lacks biochar's individual mechanical properties. The authors reported a high hardness and elastic modulus, 4.29 and 25.01 GPa, respectively, for pine wood biochar, proving how a combination of higher heat treatment temperature and longer residence time increased those values. Nanoindentation is not the only interesting surface analysis that could be employed on biochar-based composites. George *et al* [52] described the nanomechanical behaviour of rice husk biochar-based starch polymer composite using the amplitude modulation–frequency modulation mode of an atomic force microscope. The authors studied the interaction of biochar with the starch matrix in the interfacial region. The surface morphology and viscoelastic properties of the composite were studied through atomic force microscope mapping. This methodology allowed the estimation of the Young's modulus of the biochar–starch polymer

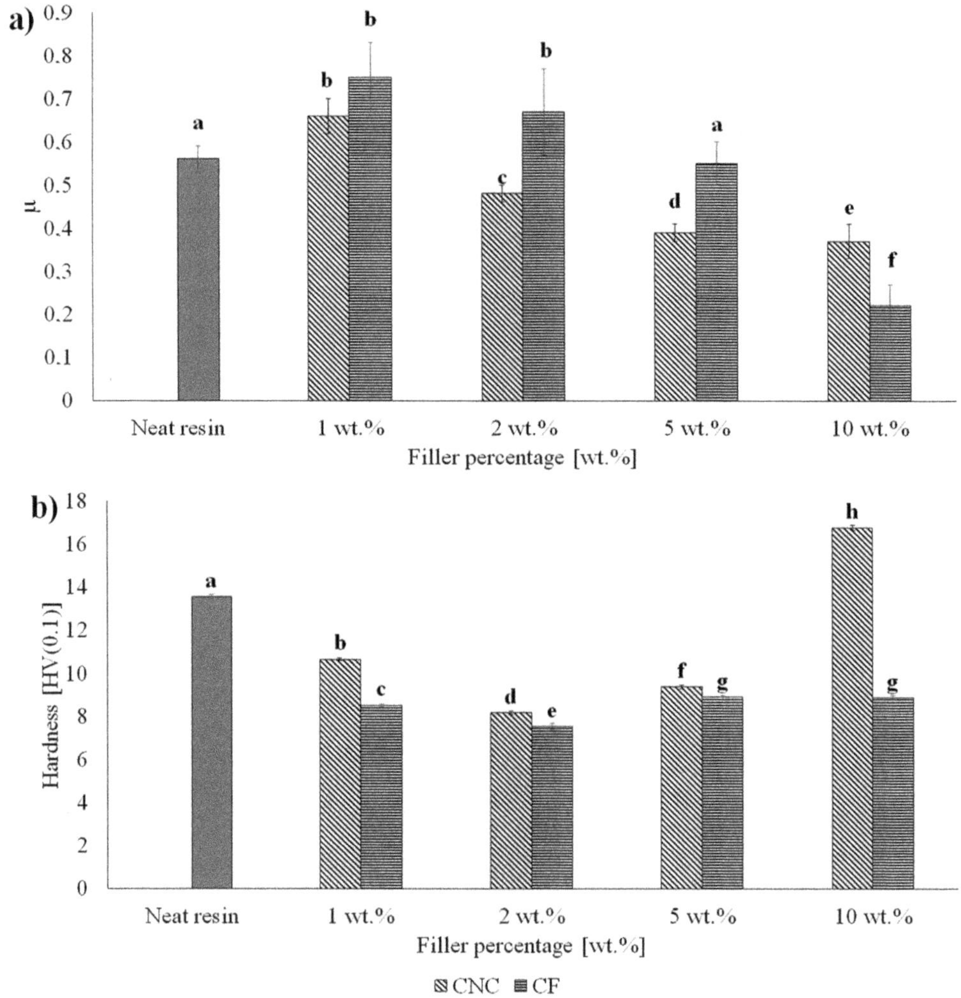

Figure 12.8. (a) Friction coefficients and (b) hardness values of neat resin and biochar-based epoxy composites containing carbonized CNC or cotton fibres with different loading, as reported by Bartololi *et al* [26], copyright John Wiley & Sons. The data are reported as average values of at least three runs and marked with different letters if they are significantly different ($p < 0.05$).

composites and its correlation with biochar loading. George *et al* attributed the decrement of Young's modulus in the tested composites to weak interactions between the biochar and matrix due to the absence of hydroxyl functionalities on the biochar surface.

12.4 Future challenges: a perspective on the uses of biochar for advanced mechanical applications

The history of biochar has seen many applications during the years. Biochar raised hope for a farming revolution with zero or negative carbon emission, but it never reached a cost that could promote a real game-changing event. If the use of biochar in soil applications does not reach its zenith then materials science could be the ground-breaking field where its real potential could be reached. If the cost of biochar is prohibitive for the agricultural sector, it will be highly competitive compared to carbon black and it outperforms carbon nanotubes and graphene for cost effectiveness. The highly tunability of biochar represents the starting point of its dissipation into multiple sectors where it can play a major role as a sustainable carbonaceous filler for the development of cutting-edge materials. However, further steps are required to engineer its properties to match those of high performance materials such as carbon and glass fibres. However, the future of biochar is now brighter than yesterday and it could lead to a 're-thinking wave' for the creation of platforms that have a low carbon footprint and also promote sustainable development.

References

[1] Grand View Research 2019 Market Value of Composite Materials Worldwide from 2015 to 2024 (in Billion U.S. Dollars) *Report*

[2] Burchell T D 1999 *Carbon Materials for Advanced Technologies* (Amsterdam: Elsevier)

[3] Chung D D 2019 *Carbon Materials* (Singapore: World Scientific)

[4] Sauer M, Kühnel M, Witten E and Mathes V 2018 Composites market report 2018—market developments, trends, outlook and challenges *Germany: AVK-TV Industrievereinigung verstrkte Kunststoffe Carbon Composite*

[5] Donnet J-B 1993 *Carbon Black: Science and Technology* (Boca Raton, FL: CRC)

[6] Morgan P 2005 *Carbon Fibers and Their Composites* (New York: CRC)

[7] Rodgers B and Waddell W 2005 *Science and Technology of Rubber* 3rd edn ed J E Mark *et al* (Burlington: Academic) p 619

[8] Cantor B, Assender H and Grant P 2015 *Aerospace Materials* (Boca Raton, FL: CRC)

[9] Mallick P K 2012 *Advanced Materials in Automotive Engineering* ed J Rowe (Philadelphia, PA: Woodhead) pp 5–27

[10] Bakshi S R, Lahiri D and Agarwal A 2010 Carbon nanotube reinforced metal matrix composites—a review *Int. Mater. Rev.* **55** 41–64

[11] Saravanan N, Rajasekar R, Mahalakshmi S, Sathishkumar T, Sasikumar K and Sahoo S 2014 Graphene and modified graphene-based polymer nanocomposites—a review *J. Reinf. Plast. Compos.* **33** 1158–70

[12] Cheap Tubes 2020 Welcome To Cheap Tubes https://www.cheaptubes.com/ (Accessed 19 October 2020)

[13] Brandt P, Ernst A, Gralla F, Luederitz C, Lang D J, Newig J, Reinert F, Abson D J and Von Wehrden H 2013 A review of transdisciplinary research in sustainability science *Ecol. Econ.* **92** 1–15

[14] Klewitz J and Hansen E G 2014 Sustainability-oriented innovation of SMEs: a systematic review *J. Clean. Prod.* **65** 57–75

[15] Lee J, Sarmah A K and Kwon E E 2019 *Biochar from Biomass and Waste* ed Y S Ok *et al* (Amsterdam: Elsevier) pp 3–18

[16] Maroušek J, Vochozka M, Plachý J and Žák J 2017 Glory and misery of biochar *Clean Technol. Environ. Policy* **19** 311–7

[17] Das O, Sarmah A K and Bhattacharyya D 2015 A sustainable and resilient approach through biochar addition in wood polymer composites *Sci. Total Environ.* **512–513** 326–36
[18] May C 2018 *Epoxy Resins: Chemistry and Technology* (New York: Routledge)
[19] Mallick P K 2007 *Fiber-Reinforced Composites: Materials, Manufacturing, and Design* (Boca Raton, FL: CRC)
[20] Montazeri A, Javadpour J, Khavandi A, Tcharkhtchi A and Mohajeri A 2010 Mechanical properties of multi-walled carbon nanotube/epoxy composites *Mater. Des.* **31** 4202–8
[21] Khan A, Savi P, Quaranta S, Rovere M, Giorcelli M, Tagliaferro A, Rosso C and Jia C Q 2017 Low-cost carbon fillers to improve mechanical properties and conductivity of epoxy composites *Polymers* **9** 642
[22] Giorcelli M and Bartoli M 2019 Development of coffee biochar filler for the production of electrical conductive reinforced plastic *Polymers* **11** 17
[23] Giorcelli M, Khan A, Pugno N M, Rosso C and Tagliaferro A 2019 Biochar as a cheap and environmental friendly filler able to improve polymer mechanical properties *Biomass Bioenergy* **120** 219–23
[24] Bartoli M, Giorcelli M, Rosso C, Rovere M, Jagdale P and Tagliaferro A 2019 Influence of commercial biochar fillers on brittleness/ductility of epoxy resin composites *Appl. Sci.* **9** 13
[25] Bartoli M, Giorcelli M, Jagdale P, Rovere M, Tagliaferro A, Chae M and Bressler D C 2019 Shape tunability of carbonized cellulose nanocrystals *SN Appl. Sci.* **1** 1661
[26] Bartoli M, Rosso C, Giorcelli M, Rovere M, Jagdale P, Tagliaferro A, Chae M and Bressler D C 2020 Effect of incorporation of microstructured carbonized cellulose on surface and mechanical properties of epoxy composites *J. Appl. Polym. Sci.* **137** 48896
[27] Schulte-Fischedick J, Zern A, Mayer J, Rühle M and Voggenreiter H 2007 The crack evolution on the atomistic scale during the pyrolysis of carbon fibre reinforced plastics to carbon/carbon composites *Composites* A **38** 2237–44
[28] Bartoli M, Nasir M A, Passaglia E, Spiniello R, Jagdale P, Rosso C, Giorcelli M, Rovere M and Tagliaferro A 2020 Influence of pyrolytic thermal history on olive pruning biochar and related epoxy composites mechanical properties *J. Composites Mater.* **54** 1863–73
[29] Dahal R K, Acharya B, Saha G, Bissessur R, Dutta A and Farooque A 2019 Biochar as a filler in glassfiber reinforced composites: experimental study of thermal and mechanical properties *Composites* B **175** 107169
[30] Sundarakannan R, Arumugaprabu V, Manikandan V and Vigneshwaran S 2020 Mechanical property analysis of biochar derived from cashew nut shell waste reinforced polymer matrix *Mater. Res. Express* **6** 125349
[31] Aup-Ngoen K and Noipitak M 2020 Effect of carbon-rich biochar on mechanical properties of PLA-biochar composites *Sustain. Chem. Pharmacy* **15** 100204
[32] Arrigo R, Bartoli M and Malucelli G 2020 Poly (lactic acid)–biochar biocomposites: effect of processing and filler content on rheological, thermal, and mechanical properties *Polymers* **12** 892
[33] Moustafa H, Guizani C and Dufresne A 2017 Sustainable biodegradable coffee grounds filler and its effect on the hydrophobicity, mechanical and thermal properties of biodegradable PBAT composites *J. Appl. Polym. Sci.* **134** doi:10.1002/app.44498
[34] Zhang Q, Lei H, Cai H, Han X, Lin X, Qian M, Zhao Y, Huo E, Villota E M and Mateo W 2020 Improvement on the properties of microcrystalline cellulose/polylactic acid composites by using activated biochar *J. Clean. Prod.* **252** 119898

[35] Jiang C, Bo J, Xiao X, Zhang S, Wang Z, Yan G, Wu Y, Wong C and He H 2020 Converting waste lignin into nano-biochar as a renewable substitute of carbon black for reinforcing styrene-butadiene rubber *Waste Manage. (Oxford)* **102** 732–42

[36] Peterson S C and Kim S 2020 Reducing biochar particle size with nanosilica and its effect on rubber composite reinforcement *J. Polym. Environ.* **28** 317–22

[37] Geyer R, Jambeck J R and Law K L 2017 Production, use, and fate of all plastics ever made *Sci. Adv.* **3** e1700782

[38] Spalding M A and Chatterjee A 2017 *Handbook of Industrial Polyethylene and Technology: Definitive Guide to Manufacturing, Properties, Processing, Applications and Markets Set* (New York: Wiley)

[39] Arrigo R, Jagdale P, Bartoli M, Tagliaferro A and Malucelli G 2019 Structure–property relationships in polyethylene-based composites filled with biochar derived from waste coffee grounds *Polymers* **11** 13

[40] Zhang Q, Yi W, Li Z, Wang L and Cai H 2018 Mechanical properties of rice husk biochar reinforced high density polyethylene composites *Polymers* **10** 286

[41] Zhang Q, Khan M U, Lin X, Cai H and Lei H 2019 Temperature varied biochar as a reinforcing filler for high-density polyethylene composites *Composites* B **175** 107151

[42] Das O, Bhattacharyya D, Hui D and Lau K-T 2016 Mechanical and flammability characterisations of biochar/polypropylene biocomposites *Composites* B **106** 120–28

[43] Das O, Bhattacharyya D and Sarmah A K 2016 Sustainable eco-composites obtained from waste derived biochar: a consideration in performance properties, production costs, and environmental impact *J. Clean. Prod.* **129** 159–68

[44] Das O, Kim N K, Kalamkarov A L, Sarmah A K and Bhattacharyya D 2017 Biochar to the rescue: balancing the fire performance and mechanical properties of polypropylene composites *Polym. Degrad. Stab.* **144** 485–96

[45] Poulose A M *et al* 2018 Date palm biochar-polymer composites: an investigation of electrical, mechanical, thermal and rheological characteristics *Sci. Total Environ.* **619–620** 311–18

[46] Noori A, Bartoli M, Frache A, Piatti E, Giorcelli M and Tagliaferro A 2020 Development of pressure-responsive polypropylene and biochar-based materials *Micromachines* **11** 339

[47] Böhm G G and Nguyen M N 1995 Flocculation of carbon black in filled rubber compounds. I. Flocculation occurring in unvulcanized compounds during annealing at elevated temperatures *J. Appl. Polym. Sci.* **55** 1041–50

[48] Khan A, Jagdale P, Rovere M, Nogués M, Rosso C and Tagliaferro A 2018 Carbon from waste source: an eco-friendly way for strengthening polymer composites *Composites* B **132** 87–96

[49] Yan L, Wang H, Wang C, Sun L, Liu D and Zhu Y 2013 Friction and wear properties of aligned carbon nanotubes reinforced epoxy composites under water lubricated condition *Wear* **308** 105–12

[50] Dong B, Yang Z, Huang Y and Li H-L 2005 Study on tribological properties of multi-walled carbon nanotubes/epoxy resin nanocomposites *Tribol. Lett.* **20** 251–4

[51] Das O, Sarmah A K and Bhattacharyya D 2015 Structure–mechanics property relationship of waste derived biochars *Sci. Total Environ.* **538** 611–20

[52] George J, Azad L B, Poulose A M, An Y and Sarmah A K 2019 Nano-mechanical behaviour of biochar-starch polymer composite: investigation through advanced dynamic atomic force microscopy *Composites* A **124** 105486

Part III

Biochar: other emerging applications

Chapter 13

Sensing properties of biochar

Daniele Ziegler, Elisabetta Di Francia and Jean-Marc Tulliani

Biochar (BC) is mainly composed of amorphous carbon with a highly functionalized surface, a high cation exchange capacity, and a large surface, which make it reactive to various compounds. For this reason, it has been proposed for the removal and/or the immobilization of contaminants such as metallic ions in water and soils. In addition, due to these features, the use of carbon paste electrodes modified with biochar has been reported for the voltammetric determination of Pb(II), Cd(II), Cu(II), Ni(II), and Zn(II) ions in beverages, pharmacologic products, biologic samples, and others. BC functionalized with gold nanoparticles was shown to be sensitive to hydroquinone and catechol determination in the nanomolar range. Sensors were also developed for the detection of organic compounds such as insecticides, herbicides, antibiotics, glucose, hormones, and phenols.

Biochar (BC) has also been proposed as a capacitive and resistive humidity sensor working at room temperature, alone or in combination with a polymer that acts as binder to improve the adhesion onto the substrate.

Therefore, this short review aims to provide a survey of recent studies dealing with BC produced from different sources used for voltammetric applications as well as for capacitive and resistive humidity sensors.

Furthermore, a discussion of the detection of organic molecules that are harmful for human health will be described in this chapter.

13.1 Introduction

Biochar is very rich in carbon and can be produced from crop residues, wood scraps, municipal and industrial solid waste, and animal manure pyrolysis at moderate temperatures, usually <700 °C, and with a low amount or in the absence of oxygen [1]. Thus, biochar is cheap and can be easily obtained from abundant raw materials. However, the type of biomass has a significant impact on the elemental composition and the properties of the as-prepared biochar. Plants typically present high H/O ratios that could benefit the development of high porosities and graphitization degrees

doi:10.1088/978-0-7503-2660-5ch13

during pyrolysis at high temperatures [2]. Moreover, plants are characterized by complex microstructures and intricate networks of interconnected struts and channels. Finally, biochar produced from plants is lightweight, nontoxic, and resistant to high temperatures [3]. This raw material consists of an inert internal structure and a highly functionalized surface based on condensed or residual aliphatic compounds and condensed aromatics, with the ability to interact with different chemical species [4]. This surface, highly rich in functional groups, can stabilize heavy metals cations by electrostatic interactions, ionic exchange, sorptive interaction, and/or specific binding of metal ions by surface ligands (e.g. carboxyl, phenol, hydroxyl) abundant in the volatile matter component of biochar [5]. Considering these properties, interesting fields of biochar's use include the development of electrochemical devices, considering its excellent electrical conductivity [6–9]. Indeed, the application of biochar in the manufacture of chemically modified electrodes (CME) or sensors has increased sharply in the last years. The addition of other species on the biochar surface can favor greater functionalization of the surface itself, creating an activated material with unique properties [10]. Then, the biochar adsorption performances can be enhanced by physical or chemical activation treatments. Among them, chemical activation exhibits a higher adsorptive capacity of organic and inorganic compounds in comparison to a material without chemical treatment. This is a consequence of the increase of functional groups on the surface after treatment [11]. In addition, promising results have been observed for the adsorption of metallic ions, such as nickel, on activated biochar [12]. The biochar surface indeed has negative charges that combine with positively charged cations through electrostatic interactions [13].

In this review, the latest researches in the field of sensors for the detection of humidity, heavy metals, and organic compounds harmful to human health are presented.

13.2 Sensor optimization

Currenty, to obtain highly sensitive electrochemical analysis, several nanomaterials such as gold nanoparticles (AuNps), platinum nanoparticles (Pt NPs), and carbon nanomaterials such as carbon nanotubes, fullerene, and graphene are used for sensor fabrication [14, 15]. They all can display great performance in improving sensitivity as a consequence of their large surface area and high conductivity. However, the investigation of cheaper, ubiquitous, and easy to synthesize nanomaterials to replace these relatively expensive and difficult to prepare materials is a challenging task. For the successful utilization of biochar for electrochemical sensing, the improvement of the electrochemical characteristics such as the conductivity of the pristine biochar is very important. By using a highly conductive biochar, combined with its great adsorption characteristic, a highly sensitive analyte detection method would be expected. However, biochars resulting from the pyrolysis of biomass are usually in the micron or millimeter size, displaying irregular shapes [16], and such a carbonaceous material may have issues in use as stable and reproducible sensors for the detection of analytes in ppm and ppb concentration range. In contrast, nanosized biochar should display better performance for low concentration targets.

In particular, the nanosized biochar has a higher surface area than the microsized biochar, which would facilitate fast electron transfer during the sensing process, furnishing more adsorption sites for analyte binding and enrichment into the biochar surface [17, 18].

13.3 Biochar as a humidity sensor

Recent advances in the sensor field have increased the demand for humidity sensors that need to be extremely accurate, precise, cheap, stable over time, and reproducible. These devices find several applications, from environmental monitoring, soil and water content determination in agriculture, air conditioning systems, food quality evaluation, medical equipment, and other areas [19]. Humidity sensors are typically grouped as resistive, capacitive, or thermal conductive [20], and display a sensitive material that may be based on metal oxide semiconductors, conducting polymers, carbon nanotubes, graphene oxides, carbon fibers, carbon films, carbon black, and biochar [21]. At room temperature, the adsorption of water molecules changes the impedance, resistance, and/or the capacitance of the sensing film [22–24]. Humidity sensors fabricated with carbon-based materials have received great attention as a consequence of their large sensitive area and superior chemical inertness [25]. As described above, biochar applications involve several fields [26], since different kind of biochar are becoming available from pilot plants generating energy and biogas.

The first paper related to biochar-based humidity sensors was published by Afify *et al* [27] in 2017. Pyrolyzed bamboo was used as the sensitive material and screen-printed in the form of thick films onto α-alumina substrates. Those bamboo components were subsequently pyrolyzed at 800 °C for 1 h under Ar. The resulting carbonized bamboo (CB) pieces were manually ground and then attrition milled for 1 h. After that, the powder was dispersed in an organic medium furnishing the appropriate rheological characteristics to the resulting ink. After screen-printing onto α-alumina substrates equipped with Pt interdigitated electrodes, the sensors were heat-treated at 300 °C in air for 1 h in order to remove organic residues. The CB thick films exhibited a significant response towards relative humidity (RH) at room temperature in terms of resistance variation with a cut-off RH at about 10%. The sensors' resistance variation began from about 937 kΩ in dry conditions decreasing to 89 kΩ upon 95% RH, with a maximum hysteresis of 28.9% under 33% RH. The response time (i.e. the time required by a sensor to achieve 90% of the total resistance variation change in the humidity adsorption) as well as the recovery times (i.e. the time necessary for a sensor to reach 90% of the total resistance change in the case of humidity desorption) were reasonably quick (around 2 min). The quick kinetics of adsorption may be due to the porosity of the film which provides great accessibility for the adsorption of water vapor. In addition, the fast recovery times denote that physisorption was presumably the main phenomenon for binding water vapor to the CB film. However, the adhesion onto ceramic substrates of this material was very poor and their use in commercial products was precluded.

For this reason, in Ziegler *et al* [21] biochar-based humidity sensors with improved adhesion onto the ceramic substrate were realized. In order to achieve this goal, two types of biochar commercially available from the UK Biochar Research Center were selected. Pyrolyzed mixed softwood pellets (SWP700) and oil seed rape (OSR700) were used for sensor fabrication by screen-printing techniques onto the same substrates used in [27]. During tests under humidity, the sensors displayed a p-type behavior at low humidity levels (in the range of 5% RH to 25% RH for SWP700 and from 5% RH to 40% RH for OSR). In more detail, for the OSR700 sensor, the impedance of the film in dry air was 163.9 kΩ, while the impedance under 99% RH dropped to 9 kΩ, with a maximum hysteresis of 52% under 59% RH. Considering the SWP sensor that displayed the highest surface area and the highest porosity compared to the other biochar, the impedance decreased from 9.5 MΩ in dry air to 222 kΩ under 99% RH. The SWP sensor displayed a lower hysteresis, with a maximum value of 23% under 55% RH. Diverse concentrations of polyvinylpyrrolidone (PVP), a n-type conductive and hydrophilic polymer [28], were added to this biochar. Sensors realized with 10 wt.% of this binder seemed to represent a satisfactory compromise between sensitivity with respect to water vapor and the adhesion of the sensitive film onto the ceramic substrate. Here, the impedance began to change under 8% RH and p-type behavior was evidenced up to 40% RH. Then, more water molecules were adsorbed onto the sensor surface and clustered to form a liquid-like multilayer of hydrogen-bonded water molecules increasing the conductivity [29]. In addition, these physisorbed water molecules were able to condense into pores with a size in the range of 1 to 250 nm. Since the generation of clusters of H_2O and the hydration of H^+ into H_3O^+ are energetically favored in liquid water, H^+ ions are the main charge carriers in the water adsorbed in the mesopores. Since the amount of H^+ increases at higher RH amounts, protons can move freely in liquid water, decreasing the grain surface resistance and impedance [29]. Conversely, at low RH levels, H_2O molecules can adsorb on the biochar surface due to a weak binding of the hydrogen atoms of water to the biochar surface carbon atoms [26]. Oxygen derivatives on the carbon surface make it hydrophilic, increasing the adsorption of water molecules [26], leading to electron exchange with a p-type semiconductor and their transfer to the valence band, increasing both the resistance and the impedance. The maximum hysteresis was equal to 42% under 60 RH% for 10 wt.% PVP composites.

The third paper in which biochar was successfully used as a humidity sensor at room temperature was in Jagdale *et al* [30], where the sensitive material was derived from waste brewed coffee powder (WBCP). The pyrolysis of the coffee waste was performed at 700 °C for 1 h under N_2 with 5 °C min^{-1} of heating ramp. After the pyrolysis, the material was manually ground before being screen-printed onto the same electrodes as used in [21] and [27]. In this case, the sensitive film displayed an n-type semiconductor trend, since the conductivity increased when the humidity concentration was higher: the impedance of the film began to change at around 20% RH. The impedance dropped from 25.2 MΩ in dry conditions to 12.3 MΩ in humid conditions. The maximum hysteresis was equal to 15% under 74% RH, confirming that the kinetics of adsorption and desorption were similar.

13.4 Biochar in electrochemical sensing applications—electrode modifier

In this part, the main applications of biochar as an electrode modifier are discussed. Figure 13.1 shows a representative scheme of modifications performed on carbon paste and solid electrodes as well as the film in a glassy carbon electrode (GCE) [31].

In Cancelliere *et al* [32], biochar from brewers' spent grain, consisting of grain husks and other residual compounds not converted into fermentable sugars in the mashing process, was used for the first time to fabricate electrodes. In this work, different electroactive species, such as ferricyanide, benzoquinone, epinephrine, ascorbic, and uric acids were tested in order to understand the electrochemical potentiality and performance of the as-prepared biochar/sensor tool. The achieved results were put in comparison with those of the same electrodes modified with commercial graphene. Moreover, a tyrosinase biosensor, an enzyme selective for catecholamines [33], was realized by direct immobilization of the enzyme on the biochar/screen-printed electrode in order to assess the potential of biochar for the

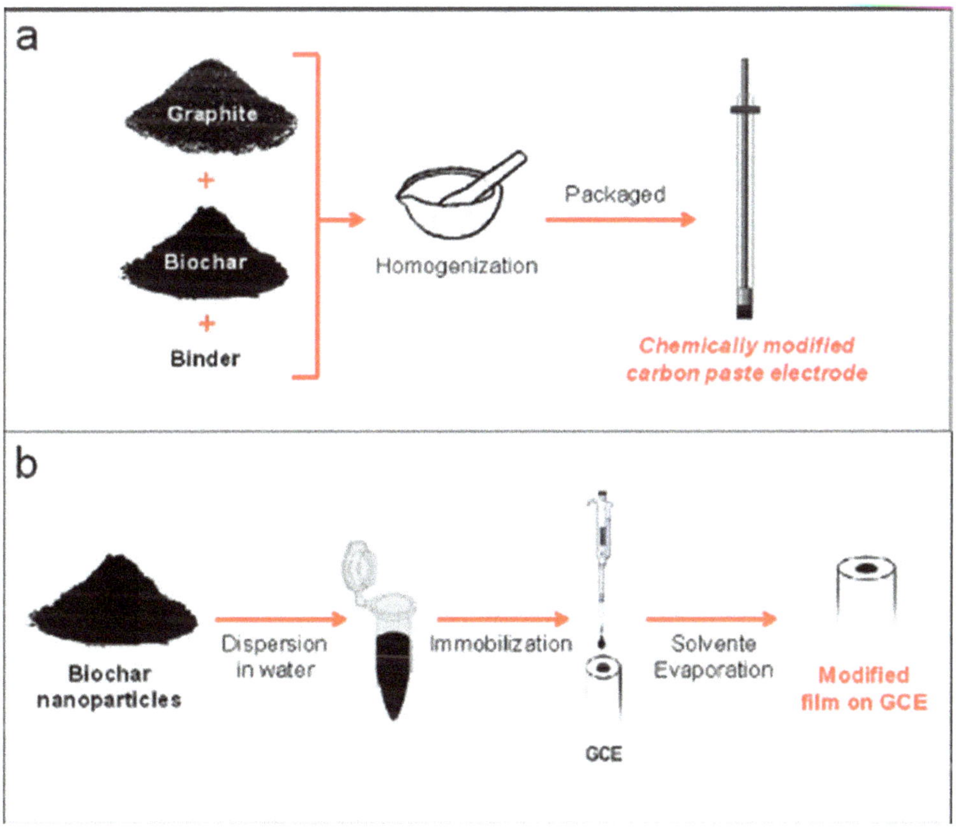

Figure 13.1. Scheme of modifications performed on carbon paste and solid electrodes (figure 13.2(a)) and on the film in a glassy carbon electrode (figure 13.2(b)) [31]. Reproduced with permission from [31]. Copyright 2020 Elsevier.

development of efficient biosensors. The detection was realized by measuring the current from the reduction of the corresponding epinephrine quinone at low potential, whic corresponds to −0.310 V for epinephrine. A linear working range of epinephrine was obtained from 0.05 mM up to 0.5 mM and the detection limit was equal to 2×10^{-4} mM for the sensor.

In the literature, many applications of the biochar carbon paste electrode (BCPE), modified with commercial biochar, determine both inorganic and organic species in the environment [34–36], because of the interaction among pollutants and biochar. An example of the application of biochar from tea waste was reported by Bal Altuntas *et al* [37] for the development of a glucose biosensor. In that study, the biochar was mixed with graphite and mineral oil to develop the BCPE and to assess the possibility of using this waste material for the production of novel composite electrodes. The authors of [32] demonstrated that the use of biochar, instead of graphene, for the fabrication of screen-printed electrodes is extremely promising because it is possible in this way to reduce the cost of these devices. Pellets from brewers' spent grain were also used as a feedstock for biochar production in [38]. The biochar samples were manually ground and used without further modifications for sensor manufacturing [39]. All the electrochemical characterizations were realized by cyclic voltammetry (CV) and amperometry (CA). The measurements of epinephrine showed a linear current response in a concentration range of 0.02 to 0.5 mM. The detection limit resulted in 2.4×10^{-4} mM. This value was within the range of the detection limit values found for tyrosinase biosensors in the literature.

13.5 Biochar in electrochemical sensing applications—heavy metals detection

In this section, a discussion of the application of different biochars as sensitive materials or as precursors for the development of electrochemical sensors able to detect heavy metals (M^{n+}) is presented.

Heavy metal ions pose serious threats to the environment and human health. Anthropogenic activities have enhanced the risk to people of heavy metal contamination as well as the occurrence of environmental problems, generating a strong need for on-site, quick, and accurate sensing techniques for the detection of heavy metal ions in aqueous systems [40]. Therefore, innovative sensing methods must be able to measure heavy metal ions below and above the regulation limits (i.e. Cd^{2+} 0.005 mg L^{-1}, Cu^{2+} 1.3 mg L^{-1}, Pb^{2+} 0.015 mg L^{-1}, and Ni^{2+} 0.1 mg L^{-1}). The use of a carbon paste electrode (CPE) modified with biochar can be a promising approach for the fabrication of effective analytical sensors. Typically, the presented voltammetric methods are combined with the utilization of a composite electrode which is easy to prepare, as an alternative probe to evaluate the capability of biochar in the preconcentration of several electroactive ions such as Pb, Ni, Cu, Zn, Cd, Ni, and Hg. In [41], microalgae biochar derived carbon dots were applied for the detection of Pb^{2+}, Cu^{2+}, and Ni^{2+}. In fact, carbon dots are 0D materials with advantageous physical properties for the detection of heavy metals through fluorescence quenching, where the metal ions (M^{n+}) interact with the carbon dot fluorophore groups reducing their fluorescence. Here, the biochar was depolymerized with a 10% solution of $KMnO_4$ that was mixed with microalgae biochar (5%). The depolymerization was realized at 120 °C for 1 h at 15 psi in an autoclave. The quenching level and the metal concentration needed to achieve quenching changed between the metal ions. The different amount of quenching measured for each heavy metal ion is related to its chemical, electronic, and vibrational properties [42]. Cu^{2+} achieved the highest level of quenching, generating a reduction in the fluorescence of >60% at 20 mM. The Cu^{2+} amount and the carbon dot fluorescence reduction were linearly correlated by a logarithmic linearization in the range of 1 μM to 10 mM. After Cu^{2+}, the Ni^{2+} quenching displayed the second largest quenching, producing a reduction in fluorescence of around 50% at 20 mM. Ni^{2+} and carbon dot fluorescence reduction were linearly correlated from 3 μM to 10 mM. Pb^{2+} was able to quench the fluorescence of carbon dots up to 30% using around 2 mM of Pb^{2+}. The Pb^{2+} concentration and fluorescence reduction were linearly correlated from 0.012 to 5 μM. Cd^{2+} produced the lowest fluorescence quenching with a maximum fluorescence reduction of <10%. The Cd^{2+} and fluorescence reduction were linearly correlated in the range of 0.01 to 1 μM. From FT-IR spectra, it was evident that the carbon dot structure was full of C–O, C═O, and C–OH bonds, denoting similarities to graphene oxide (GO) dots and their interactions with M^{n+}. The prevalent groups in the interaction with heavy metals are carboxylic acids and phenolic OH groups, which are hard donors with electrostatic forces as well as entropy gains driving their interactions with M^{n+}. These interactions result in the measured fluorescence quenching of the biochar derived carbon dot.

In [43], the same authors implemented, using multivariate analysis, the previous study to investigate biochar derived carbonaceous nanomaterials (BCNs) produced from three different feedstocks for the monitoring and discrimination of Pb^{2+}, Ni^{2+}, Cu^{2+}, and Hg^{2+} in aqueous systems. The BCNs were obtained from dairy manure (DMB-CN), rice straw (RSB-CN), and sorghum straw (SSB-CN) biochar thanks to the depolymerization/chemical oxidation and purification of nanomaterials with solvent extraction [40]. Principal component analysis (PCA) and discriminant analysis were used as statistical methods in order to discriminate among M^{n+} in water samples. DMB-CN displays the highest quenching for Ni^{2+} and Hg^{2+} whereas RSB-CN and SSB-CN are more sensitive, respectively, towards Pb^{2+} and Cu^{2+}. An important achievement of this study was that the discriminant analysis showed a 100% accuracy for the detection of Pb^{2+}, 66% for Cu^{2+} and Ni^{2+}, but only 33% for the measurement of Hg^{2+}.

In [44] a new route was presented for the determination of nickel by redox couple Ni(II)/Ni(III) using a carbon paste modified electrode (CPME) with activated biochar. The carbonaceous material exhibits a significant Ni preconcentration capacity when compared with the precursor biochar. The improvement was a result of the increase of oxygenated functional groups after chemical activation of biochar using HNO_3. This analytical method was used for the detection of nickel in ethanol fuel. Nickel is a moderate toxic pollutant that cause skin diseases, allergic reactions, and can also be carcinogenic in high concentrations [45]. Moreover, it is also a contaminant that can be present in fuels, such as ethanol, which may occur due to the absorption of metals from the soil by the plant utilized as feedstock [46]. Generally, inorganic pollutants in fuels can affect the automotive engine performance through the formation of salts that are soluble and able to induce corrosion, increasing the pollutant emission amounts [47]. In [44] activated biochar with HNO_3 resulted in a spontaneous preconcentration of Ni(II) ions in solution, both in bioethanol fuel (recovery 103%) and discharge water samples (recovery 109%). There was no need for a complexing agent and the spontaneous preconcentration with no potential application of Ni occurred in 300 s, and a LOD of 0.25 μM L^{-1} was achieved for the analyte. The main advantages of these findings are that they do not require electrodes with a wide potential range, such as Hg or similar. The reactions of Ni(II)/Ni(III) happened in a milder potential and therefore could be easily controlled with a carbon-based electrode. The activation of biochar was very easy through grafting functional groups on its surface, increasing its adsorptive performances. Moreover, the activated biochar increased the capacity of the preconcentration and the response signal for the determination of Ni. Such a device could be used as a passive sensor, aiming at the remote preconcentration species and posterior voltammetric tests.

In [48] biochar from castor oil cake was used as an electrode modifier in combination with differential pulse adsorptive stripping voltammetric (DPAdSV) methods for preconcentration and determination of Cu(II) ions in spirit drink samples (cachaça, vodka, gin, and tequila). CPE was fabricated by mixing the graphite powder (45% to 75% (w/w)) with the biochar sample at varying ratios (0% to 30% (w/w)) and then adding it to 0.25 g of mineral oil (25% (w/w)). The components were mixed manually in a mortar with a pestle to obtain an appropriate homogenization. The composite was packed into the support of an electrode with a plastic cylindrical tube and a copper rod inside used as the external electric contact. After the optimization of the experimental conditions, a linear response for copper ions in the range of concentration from 1.5×10^{-6} to 3.1×10^{-5} mol L^{-1} and an LOD equal to 4.0×10^{-7} mol L^{-1} were observed. The effect of foreign species such as Fe, Zn, As, and Pb on the electrochemical response was also studied. No significant influence was measured in the presence of foreign species (Fe, Zn, As, and Pb), demonstrating good selectivity for the proposed device when applied in these kinds of samples. The determination of Cu(II) ion content in different spirit drink samples was also achieved by means of inductively coupled plasma optical emission spectroscopy (ICP-OES) and the results were in accordance at a 95% of confidence level.

The same source of biochar was used in [49], where it was adopted as an electrode modifier for the preconcentration and determination of lead(II) and cadmium(II) under DPAdSV conditions. Samples of biochars were obtained using the same method as used in [48]. The proposed method was applied for the determination of Pb(II) and Cd(II) in spiked simulated industrial effluents and LODs equal to, respectively, 9.8×10^{-9} mol L^{-1} and of 6.9×10^{-8} mol L^{-1} were determined. For all the analyzed samples, the recoveries ranged from 95% to 104%. No interference in the presence of Hg(II) ions was measured as a result of their low concentration when compared to the Pb(II) and Cd(II) ions. However, a significant decrease in the anodic peak currents was detected for Sn(II), Fe(III), Cu(II), Zn(II), and Ni(II), demonstrating that biochar can generate complexes with other metallic species in addition to cadmium(II) and lead(II) ions. However, the interference effect is constant and it could be minimized by using standard addition procedures.

In [39] nanomolar concentrations of Pb(II) were determined using a voltammetric stripping technique at a carbon paste electrode modified with biochar and bismuth nanostructures. After the interaction of the biochar and Pb(II) ions, the reduction of these ions happened at the Bi nanorods (in the range of 20 to 70 nm) which improved the anodic stripping current. This sensor displayed a linear range from 5 nmol L^{-1} to 1000 nmol L^{-1} and a LOD of 1.41 nmol L^{-1} for Pb(II). Here, the application of the proposed sensor was the determination of lead released from overglaze-decorated ceramic dishes. The results were in accordance at a 99% confidence level with the ICP-OES reference technique. Recently, electrode modification with Bi nanostructures was proposed in order to combine the electrochemical features of Bi together with the advantages of nanomaterials, which are an increase in the mass transport, a high surface area, and an enhanced signal-to-noise ratio [50, 51]. In these studies, biochar was used at the same time as a platform for the growth of Bi nanostructures and for spontaneous preconcentration and determination of Pb(II) ions. Under these conditions, lead(II) ions could be preconcentrated simultaneously with different electrodes that permitted rapid and consecutive measurements. The proposed sensor displayed a synergic effect based on the high adsorption ability of biochar for Bi(III) and Pb(II) ions and the enhancement of the electrochemical output due to the presence of Bi nanostructures on the electrode surface. The same raw material—biochar from castor oil cake—as well as the procedure of [51] were used to fabricate the electrochemical sensors. Several cations and anions were tested in cross-sensitivity tests, and only Cu(II), Fe(III), and Sn(II) resulted in a significant decrease of the peak current, since those cations compete with Pb(II) for the same adsorption sites on the biochar.

In [52] a modified CPE was fabricated using mercury ions (Hg^{2+}), supported at the biochar surface, for the preconcentration and determination of Zn(II) ions in commercial samples (collyrium and ointment) using the DPASV technique. Biochar was obtained from the same castor oil cake source as in [48] and used for electrode preparation. Here, spontaneous interactions between the biochar surface full of functional groups and Zn(II) ions took place, followed by a reduction of ions into mercury droplets. Hg droplets promoted a high and localized concentration of Zn. Using the best operative conditions, a linear response for Zn(II) ions was obtained for concentration ranges of 5.0×10^{-7} to 3.0×10^{-5} mol L^{-1} with a LOD of 1.7×10^{-7} mol L^{-1}. Considering interference, Ni, Cu, Pb, Cd, and Fe in ten times lower concentrations than Zn presented a variation smaller than 7.0% but interfered when present at the same concentrations. These results were compared with those obtained by ICP-OES and complexometric titration and were in agreement at a 95% confidence level, demonstrating high stability, sensitivity, and analytical accuracy.

13.6 Biochar in electrochemical sensing applications—organic compounds

In this last section, different electrochemical sensors are illustrated where biochar plays an important role in the detection of organic compounds, such as insecticides, herbicides, and antibiotics.

In [35] carbon paste electrodes, modified with biochar and Sb microparticles were applied for voltammetric stripping analysis determination of paraquat through a very sensitive procedure. Here, the synergic effect between biochar and antimony was fulfilled using square wave adsorptive stripping voltammetry. Paraquat (1,1′-dimethyl-4,4′-bipyridinium chloride, PQ^{2+}) is a bipyridinium contact herbicide commonly used for weed combat in many crops. The utilization of this chemical agent is allowed when it is used according to specifications of legislation, since its waste is extremely persistent and toxic both for the environment and human health [53].

Biochar from castor oil cake was obtained in accordance with Suguihiro *et al* [49]. At the optimized conditions, an analytical curve was obtained for paraquat determination that displayed a linear response in the range of 0.2 to 2.9 μmol L^{-1}, with a LOD of 34 nmol L^{-1}, after paraquat preconcentration. The proposed sensor was applied for the determination of PQ^{2+} in tap water and citric fruit juice spiked samples and great recovery results were obtained without any sample pre-treatment. Citric acid, ascorbic acid, and saccharose, as well as Cl^- and NO_3^-, did not cause significant interference. In figure 13.2, a scheme of the analytical voltammetric procedure is depicted.

Electrodes with the addition of biochar in the carbon paste composition provided better results in comparison to the other electrodes. The principal mechanisms are based on electrostatic and polar interactions between the polar groups present in coal and in the herbicide, since in biochar phenols and carboxyl groups are abundant on the surface due to the low temperature pyrolysis.

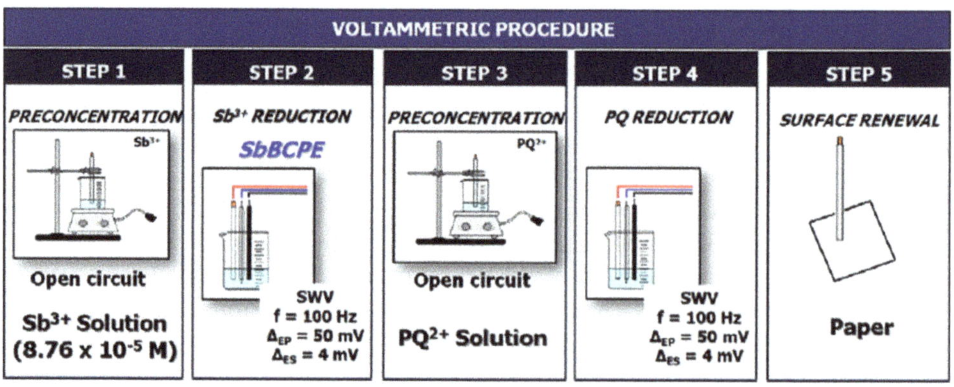

Figure 13.2. Voltammetric procedure of [35] for the determination of paraquat. Reproduced with permission from [35]. Copyright 2016 Elsevier.

In [54] the authors proposed a sensitive technique for simultaneous determination of hydroquinone (HQ) and catechol (CA) using gold nanoparticles (AuNPs) uniformly distributed in porous biochar, using a facile pyrolysis method. HQ and CA are two isomers of dihydroxybenzene that often coexist in environmental samples. These two isomers are environmental pollutants that are toxic for human health and difficult to degrade into the environment [55]. For this reason, a highly sensitive analytical technique is required for the simultaneous determination of HQ and CA. *Dracaena sanderiana* was the raw material used for the production of the biochar. It was first soaked in an aqueous solution of $HAuCl_4$ and then pyrolyzed under N_2 atmosphere at high temperatures (850 °C). In the process of decoration of biochar with AuNPs, it is essential that the Au precursor fully permeates the biochar precursor before the pyrolysis step begins. Moreover, several H^+ of $HAuCl_4$ aqueous solution dissolved some inorganic salts during soaking, resulting in an impurity decrease. The exit of the volatiles through the carbon structure led to an increase of porosity [56]. In order to fabricate electrodes, these materials were crushed into a fine powder and sieved. Then 2.0 mg of the decorated biochar was dispersed in H_2O and sonicated and 5 mL of the dispersion was dropped onto a glassy carbon electrode. In figure 13.3, a schematic diagram of the process used for the fabrication of AuNPs-decorated biochar for the determination of HQ and CA is illustrated. The recovery of HQ and CA from tap water, lake water, and industrial wastewater were in the ranges of 97.6% to 103.0%, of 97.5% to 102.7%, and of 94.8%

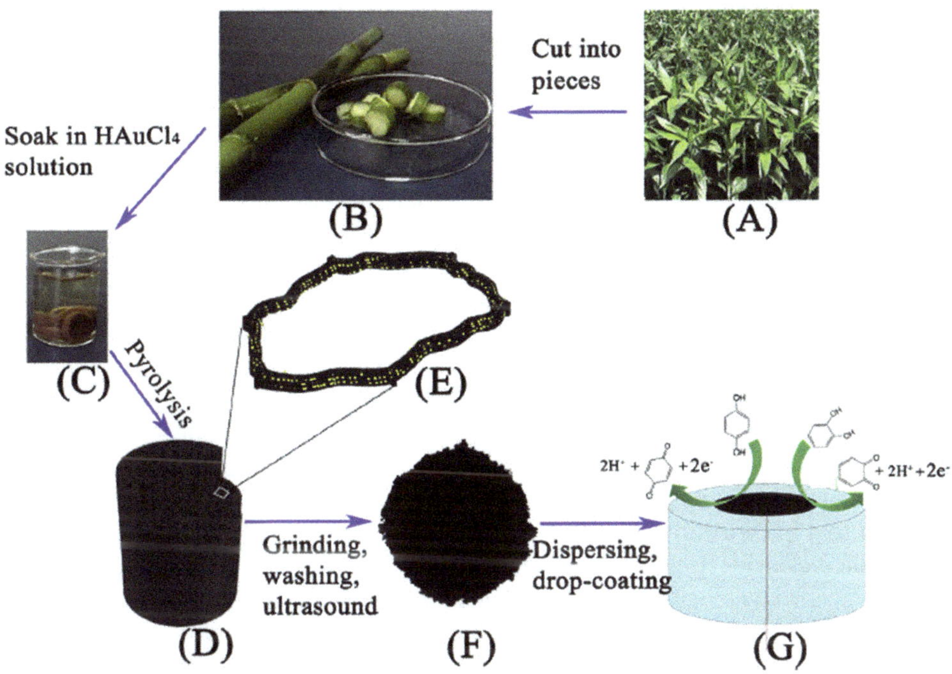

Figure 13.3. Diagram of the fabrication of electrochemical sensors [54] for the determination of HQ and CA. Reproduced with permission from [54]. Copyright 2018 Elsevier.

to 103.4%, respectively, while the LODs were, respectively, equal to 3.4 and 9.0 nM. These results confirmed that those sensors can be applied for the simultaneous detection of HQ and CA in real samples. This electrochemical activity towards the oxidation of HQ and CA is a result of the following three aspects. First, the catalytic activity of gold is enhanced by the presence of electronegative atoms such as N and O in the composite, improving its wettability and increasing the number of active sites. Second, the hierarchical porous nanostructure of the nanocomposite guarantees fast liquid diffusion by shortening the diffusion pathways. The mesoporous walls and microporous texture result in high adsorption rates of HQ and CA from the solution. Third, the high specific surface area (SSA) of the nanocomposite ensures an adequate interface for the implantation of AuNPs, resulting in a synergistic effect for their electrochemical activity towards HQ and CA.

In [18] the authors developed a cheap and sensitive electrochemical sensor using biochar nanoparticles with high conductivity and adsorptivity for the detection of 17β-estradiol. The presence of endocrine disrupting chemicals (EDCs) in the aquatic environment may results in adverse effects on both human health and wildlife [57]. 17β-estradiol is a natural estrogen excreted by humans and domestic animals with superior estrogenic activity and it is diffuse in many water samples. It is dangerous for human health even at trace levels [58]. The procedure to produce the sensors was as follows. First, bagasse from the juicing of sugarcane stalk was briefly washed with H_2O to remove residual sugar and then it was pyrolyzed under N_2 with a heating rate of 7 °C min^{-1}, and the final temperatures were set to 300 °C, 400 °C, 500 °C, 600 °C, 700 °C, and 800 °C. The obtained biochars were then ground. From Raman investigation, there were two characteristic bands of G (1583 cm^{-1}) and D (1387 cm^{-1}): the G band corresponds to the in-plane vibration of sp^2 carbon atoms in a 2D hexagonal lattice of graphene, whereas the D bands are generated by the vibration of sp^3 carbon atoms of disordered carbon or defective graphitic structures [59, 60]. The intensity ratio of D/G increased from 0.66 at 300 °C to 0.93 at 800 °C, suggesting that there are more defects or vacancies in biochar that was pyrolyzed at a higher temperature [61]. The proposed biochar-based sensor exhibited an increase in amperometric responses and a decrease of the impedance, and a lowered anodic potential together with an increase in the current signal toward 17β-estradiol in comparison to a pristine glassy carbon electrode. Moreover, the adsorption and accumulation of 17β-estradiol resulted in a superior sensing signal amplification. Thanks to the synergetic effect of high conductivity and adsorption of the proposed sensor, a LOD of 11.30 nM was obtained. Taking into account interference, K, Ca, Mg, Cu, Ni, SO_4, Cl, phenol, and progesterone (a 100-fold concentration) had no influence but 17α-ethynyl -estradiol, -estradiol, and -bisphenol presented very close oxidation to 17β-estradiol oxidation, since the peak potential separation cannot be distinguished clearly. Both the accuracy and reliability of the sensor were confirmed to be in accordance with the liquid chromatography analysis and it could be successfully used for environmental water analyses in real samples.

In [62] an easy and low cost thick-walled Moso biochar (BC) was chosen and combined with dolomite to fabricate a voltammetric sensing device for the detection of thiamethoxam in agricultural and environmental samples. Thiamethoxam is a systemic insecticide belonging to the class of neonicotinoids and it has a broad spectrum of activity against several types of insects. A sensitive material based on thick-walled Moso was realized by pyrolyzing biomass waste at 600 °C for 2 h, and the nanocomposite was obtained through milling and sieving, and then liquid-phase ultrasonic dispersion and grinding into ultra-pure water. Under the best operative conditions, the dolomite-biochar modified electrode exhibited high sensitivity as well as great stripping voltammetric response towards thiamethoxam in a linear range of 0.5 to 35 $\mu g\,mL^{-1}$ with a LOD equal to 0.22 $\mu g\ mL^{-1}$. The proposed nanocomposite sensor was used for the determination of thiamethoxam in different real samples, providing a novel route for potential applications in environmental monitoring.

In [63] an array of carbon paste electrodes composed of different forms of carbon (graphite, carbon nanotubes, and activated biochar) was studied for the development of an electronic tongue that was able to discriminate catechol (CAT), 4-ethylcatechol (4-EC), and 4-ethylguaiacol (4-EG) phenolic compounds by a stripping voltammetric technique. These molecules are natural constituents of wine and food, furnishing some of their intrinsic characteristics [64], however, if some phenolic compounds in foods and drinks are present, they can result in undesired flavors and aromas [65]. At high concentrations, phenols can cause organoleptic trouble, deteriorating the quality of wines [66]. Considering the sensor areas, CPEs exhibit intriguing characteristics such as low cost, low background currents, versatility, and simplicity of construction and modification [67]. The composition (w/w) of a CPE unmodified electrode was equal to 25% mineral oil and 75% graphite powder, while the modified electrodes were realized with a proportion (w/w) of 25% mineral oil, 60% graphite powder, and 15% modifier (biochar or carbon nanotubes). Among all the sensors, the highest values of sensitivity were achieved using biochar. The LODs for the biochar-based electrode were equal to 1.85 $\mu mol\ L^{-1}$, 0.0863 $\mu mol\ L^{-1}$, and 0.0937 $\mu mol\ L^{-1}$ towards CAT, 4-EC, and 4-EG, respectively. These findings are due to the fact that biochar presents a highly porous and functionalized surface, with mainly oxygenate functional groups. Here, the sorption occurs through π–π interactions between the carbonaceous surface and an aromatic ring in target molecules. This effect enhances the physical adsorption of phenolic compounds at the surface of the modified electrode [68]. This sensor array exhibits analytical potentiality, permitting the discrimination of CAT, 4-EC, and 4-EG by means of chemometric processing. The proposed method is very sensitive and reproducible. A spontaneous preconcentration of the three analytes was possible using all three sensors, permitting their application as passive samplers for remote determination of phenolic molecules in both wine and food samples.

In [68] the authors used a CPE modified with biochar activated with HNO_3 for spontaneous preconcentration of methyl parathion (MP) and for its quantitative determination in drinking water. MP (O,O-dimethyl O-4-nitrophenyl phosphorothioate) is an organophosphate insecticide which is highly toxic through inhalation and ingestion, and moderately toxic through dermal absorption. First, precursor biochar samples were produced from castor oil cake biomass by pyrolysis at 400 °C, with a heating rate of 5 °C min^{-1} and a residence time of 60 min. The sample was then surface treated using nitric acid 50% (v/v). The dispersion was placed in a reflux system under constant stirring at 60 °C for 3 h. After this stage, the dispersion was filtered, washed, dried in an oven at 100 °C for 1 day, and stored, for later use, as a modifier agent of the electrode. The electrode showed good sensitivity and accuracy, and a LOD equal to 39.0 nmol L^{-1}. Cu, Zn, and glucose did not present any influence but Fe and ascorbic acid presented significant interference above 10 μmol L^{-1}. This procedure was applied to determine MP in tap water samples with recovery values varying between 102% and 110% for three different concentration levels. In addition, it should be emphasized that the proposed method has a superior simplicity and very low cost in the construction of the electrodes compared to the traditional methods.

In [69] nickel ions supported on an activated biochar carbon paste electrode were used, coupled in a microfluidic thread-based electroanalytical device for non-enzymatic glucose determination. The research on novel robust and stable techniques for non-enzymatic determination of glucose is a consequence of the large number of people who have issues due to an excess of glucose in the blood, causing diseases such as diabetes. The microfluidic device permits a reduction of both the reagent use, and waste as well as the consumption of samples, providing a quick response for chemical analysis without requiring a mechanical pumping system. Biochar obtained from castor oil cake at 400 °C and chemically activated by HNO_3 exhibited an excellent sensing performance towards spontaneous preconcentration of Ni(II) ions. In an alkaline environment, a stable voltammetric output is associated with a $Ni(OH)_2$/NiOOH redox pair and a significant catalytic effect was observed in the presence of glucose. This amperometric glucose sensor presented good repeatability and a linear dynamic range of 5.0–100.0 μmol L^{-1} and a LOD of 0.137 μmol L^{-1}. For the cross-sensitivity test, the presence of 65 mmol L^{-1} of sodium, 0.65 mmol L^{-1} of chloride, 22 mmol L^{-1} of lactic acid, 10 mmol L^{-1} of uric acid, and 0.13 mmol L^{-1} of ascorbic acid had no influence in the 1.0 mmol/glucose response. This amperometric sensor was used for determining glucose in real biological samples of human saliva and blood serum.

Lastly, in [70] the authors describe the fabrication of a CPE modified with biochar decorated with nanostructured copper hexacyanoferrate (nCuHCF) and its application as an amperometric sensor for isoniazid (INZ) detection, an active molecule found in medicines for tuberculosis treatment [71]. However, this drug causes hepatotoxicity in not less than 20% of patients, resulting in severe levels in about 2% [72]. This drug has a well-known voltammetric behavior [73] that permits the evaluation of a modified electrode for INZ electrochemical determination. Here, the decoration of biochar was performed with copper ions previously adsorbed at the activated biochar surface, followed by electrochemical synthesis of nCuHCF. For the fabrication of the INZ sensor, a carbon paste electrode modified with biochar was realized (25 wt.% biochar, 50 wt.% graphite, and 25 wt. % mineral oil). Biochar was obtained from the biomass of castor oil cake after pyrolysis at 400 °C, heating rate of 5 °C min^{-1} and residence time of 60 min. The surface activation was performed also in this case by refluxing the biochar dispersion in 50% (v/v) HNO_3 solution for 3 h at 60 °C. After optimization of all the significant parameters, the nCuHCF-BM-CPE showed a linear response in the range of 1.0×10^{-6} mol L^{-1} to

1.0×10^{-5} mol L^{-1} of INZ and an LOD of 6.3×10^{-8} mol L^{-1} were obtained. Glucose, dopamine, Na, Ca, Cl, and SO_4^{2-} do not promote significant interference but high concentrations of ascorbic acid and rifampicin interfere mildly. Recovery values changed from 98.0% to 99.4%, showing an excellent analytical performance for INZ quantification. Thus biochar was an excellent support for preparation of active nCuHCF nanostructures furnishing a high electroactive area as well as great stability to the modifier. Moreover, INZ determination was realized using a facile, quick, accurate, and selective technique, in addition to the low cost of the sensor by usage of biochar as the redox mediator support matrix.

13.7 Conclusions

To conclude, in this chapter different novel applications of biochar as electrochemical and chemical sensors were presented. This carbonaceous material has a great potential in the development of efficient sensors for the detection of humidity, heavy metals, and organic molecules, both in air and in aqueous environments, as a result of its physical, chemical, and electrical characteristics. Currently, the market requires a rapid and inexpensive analysis. In this frame, electroanalytical methods, based on the use of pure biochar or using biochar as an electrode modifier, fit perfectly. Indeed, they show sensitivity and selectivity even using small volumes of reagents and samples. Moreover, biochar-based sensors can be fabricated from waste and low cost materials enabling portability for analyses in remote locations.

References

[1] Lehmann J and Joseph S 2012 *Biochar for Environmental Management: Science and Technology* (London: Routledge)

[2] Liu D, Song J, Anderson D P, Chang P R and Hua Y 2012 Bamboo fiber and its reinforced composites: structure and properties *Cellulose* **19** 1449–80

[3] Qambrani N A, Rahman M M, Won S, Shim S and Ra C 2017 Biochar properties and eco-friendly applications for climate change mitigation, waste management, and wastewater treatment: a review *Renew. Sustain. Energy Rev.* **79** 255–73

[4] Shu R, Wang Y and Zhong H 2016 Biochar amendment reduced methylmercury accumulation in rice plants *J. Hazard. Mater.* **313** 1–8

[5] Uchimiya M, Bannon D I, Wartelle L H, Lima I M and Klasson K T 2012 Lead retention by broiler litter biochars in small arms range soil: impact of pyrolysis temperature *J. Agric. Food Chem.* **20** 5035–44

[6] Cheng B H, Zeng R J and Jiang H 2017 Recent developments of post-modification of biochar for electrochemical energy storage *Bioresour. Technol.* **246** 224–33

[7] Thomas P, Lai C W and Johan M R B 2019 Recent developments in biomass-derived carbon as a potential sustainable material for super-capacitor-based energy storage and environmental applications *J. Anal. Appl. Pyrolysis* **140** 54–85

[8] Goglio A, Tucci M, Rizzi B, Colombo A, Cristiani P and Schievano A 2019 Microbial recycling cells (MRCs): a new platform of microbial electrochemical technologies based on biocompatible materials, aimed at cycling carbon and nutrients in agro-food systems *Sci. Total Environ.* **648** 1349–61

[9] Mian M M, Liu G and Fu B 2019 Conversion of sewage sludge into environmental catalyst and microbial fuel cell electrode material: a review *Sci. Total Environ.* **666** 525–39

[10] Bazuła P A, Lu A-H, Nitz J-J and Schüth F 2008 Surface and pore structure modification of ordered mesoporous carbons via a chemical oxidation approach *Microporous Mesoporous Mater.* **108** 266–75

[11] Vithanage M, Rajapaksha A U, Zhang M, Thiele-Bruhn S, Lee S S and Ok Y S 2015 Acid activated biochar increased sulfamethazine retention in soils *Environ. Sci. Pollut. Res.* **22** 2175–86

[12] Li J-h, Lv G-h, Bai W-b, Liu Q, Zhang Y-c and Song J-q 2016 Modification and use of biochar from wheat straw (*Triticum aestivum* L.) for nitrate and phosphate removal from water *Desalination Water Treat.* **57** 4681–93

[13] Shi K, Xie Y and Qiu Y 2015 Natural oxidation of a temperature series of biochars: opposite effect on the sorption of aromatic cationic herbicides *Ecotoxicol. Environ. Saf.* **114** 102–8

[14] Wen T, Xue C, Li Y, Wang Y, Wang R, Hong J, Zhou X and Jiang H 2012 Reduced graphene oxide-platinum nanoparticles composites based imprinting sensor for sensitively electrochemical analysis of 17β-estradiol *J. Electroanal. Chem.* **682** 121–7

[15] Yuan L, Zhang J, Zhou P, Chen J, Wang R, Wen T, Li Y, Zhou X and Jiang H 2011 Electrochemical sensor based on molecularly imprinted membranes at platinum nanoparticles modified electrode for determination of 17β-estradiol *Biosens. Bioelectron.* **29** 29–33

[16] Liu G B, Li L J, Zhang K, Wang X Q, Chang J, Sheng Y Y, Bai L and Wen Y P 2016 Facile preparation of water-processable biochar based on pitch pine and its electrochemical application for cadmium ion sensing *Int. J. Electrochem. Sci.* **11** 1041–54

[17] Mayo J T, Yavuz C, Yean S, Cong L, Shipley H, Yu W, Falkner J, Kan A, Tomson M and Colvin V L 2007 The effect of nanocrystalline magnetite size on arsenic removal *Sci. Technol. Adv. Mater.* **8** 71–5

[18] Dong X, He L, Liu Y and Piao Y 2018 Preparation of highly conductive biochar nanoparticles for rapid and sensitive detection of 17b-estradiol in water *Electrochim. Acta* **292** 55–62

[19] Farahani H, Wagiran R and Hamidon M N 2014 Humidity sensors principle, mechanism, and fabrication technologies: a comprehensive review *Sensors* **14** 7881–939

[20] Duraia E M and Beall G W 2015 Humidity sensing properties of reduced humic acid *Sens. Actuators* B **220** 22–6

[21] Ziegler D, Palmero P, Giorcelli M, Tagliaferro A and Tulliani J M 2017 Biochars as innovative humidity sensing materials *Chemosensors* **5** 35–51

[22] Tulliani J M and Bonville P 2005 Influence of the dopants on the electrical resistance of hematite-based humidity sensors *Ceram. Int.* **31** 507–14

[23] Esteban-Cubillo A, Tulliani J M, Pecharromán C and Moya J S 2007 Iron-oxide nanoparticles supported on sepiolite as a novel humidity sensor *J. Eur. Ceram. Soc.* **27** 1983–9

[24] Cavalieri A, Caronna T, Natali Sora I and Tulliani J M 2012 Electrical characterization of room temperature humidity sensors in $La_{0.8}Sr_{0.2}Fe_{1-x}Cu_xO_3$ ($x = 0, 0.05, 0.10$) *Ceram. Int.* **38** 2865–72

[25] Chu J, Peng X, Feng P, Sheng Y and Zhang J 2013 Study of humidity sensors based on nanostructured carbon films produced by physical vapor deposition *Sens. Actuators* B **178** 508–13

[26] Hu B, Wang K, Wu L, Yu S H, Antonietti M and Titirici M M 2010 Engineering carbon materials from the hydrothermal carbonization process of biomass *Adv. Mater.* **22** 813–28

[27] Afify A S, Ahmad S, Khushnood R A, Jagdale P and Tulliani J M 2017 Elaboration and characterization of novel humidity sensor based on micro-carbonized bamboo particles *Sens. Actuators* B **239** 1251–6

[28] Azmer M I, Zafar Q, Ahmad Z and Sulaiman K 2016 Humidity sensor based on electro spun MEH-PPV: PVP micro structured composite *RSC Adv.* **42** 35387–93

[29] Traversa E 1995 Ceramic sensors for humidity detection *Sens. Actuators* B **23** 135–56

[30] Jagdale P, Ziegler D, Rovere M, Tulliani J M and Tagliaferro A 2019 Waste coffee ground biochar: a material for humidity sensors *Sensors* **19** 801–14

[31] de Almeida L S, Oreste E Q, Maciel J V, Heinemann M G and Dias D 2020 Electrochemical devices obtained from biochar: advances in renewable and environmentally-friendly technologies applied to analytical chemistry *Tr. Environ. Anal. Chem.* **26** 1–8

[32] Cancelliere R, Carbone K, Pagano M, Cacciotti I and Micheli L 2019 Biochar from brewers' spent grain: a green and low-cost smart material to modify screen-printed electrodes *Biosens.* **9** 139–44

[33] Apetrei I M, Popa C V, Apetrei C and Tutunaru D 2014 Biosensors based on graphene modified screen-printed electrodes for the detection of catecholamines *Rom. Biotechnol. Lett.* **19** 9801–9

[34] Xue X, Wei Q, Wu D, Li H, Zhang Y, Feng R and Du B 2014 Determination of methyl parathion by a molecularly imprinted sensor based on nitrogen doped graphene sheets *Electrochim. Acta* **116** 366–71

[35] Gevaerd A, De Oliveira P R, Mangrich A S, Bergamini M F and Marcolino-Junior L H 2016 Evaluation of antimony microparticles supported on biochar for application in the voltammetric determination of paraquat *Mater. Sci. Eng.* C **62** 123–9

[36] Agustini D, Mangrich A S, Bergamini M F and Marcolino-Junior L H 2015 Sensitive voltammetric determination of lead released from ceramic dishes by using of bismuth nanostructures anchored on biochar *Talanta* **142** 221–7

[37] Bal Altuntas D, Akgül G, Yanik J and Anik U 2017 A biochar-modified carbon paste electrode *Turk. J. Chem.* **41** 455–65

[38] Sperandio G, Amoriello T, Carbone K, Fedrizzi M, Monteleone A, Tarangioli S and Pagano M 2017 Increasing the value of spent grain from craft microbreweries for energy purposes *Chem. Eng. Trans.* **58** 487–92

[39] Abenavoli L M, Longo L, Proto A R, Gallucci F, Ghignoli A, Zimbalatti G and Colantoni A 2016 Characterization of biochar obtained from olive and hazelnut prunings and comparison with the standards of European Biochar Certificate (EBC) *Procedia Behav. Sci.* **223** 698–705

[40] Gogoi N, Barooah M, Majumdar G and Chowdhury D 2015 Carbon dots rooted agarose hydrogel hybrid platform for optical detection and separation of heavy metal ions *ACS Appl. Mater. Interfaces* **7** 3058–67

[41] Plácido J, Bustamante-López S, Meissner K E, Kelly D E and Kelly S L 2019 Microalgae biochar-derived carbon dots and their application in heavy metal sensing in aqueous systems *Sci. Total Environ.* **656** 531–9

[42] Shtepliuk I, Caffrey N M, Iakimov T, Khranovskyy V, Abrikosov I A and Yakimova R 2017 On the interaction of toxic heavy metals (Cd, Hg, Pb) with graphene quantum dots and infinite graphene *Sci. Rep.* **7** 3934–50

[43] Plácido J, Bustamante López S, Meissner K E, Kelly D E and Kelly S L 2019 Multivariate analysis of biochar-derived carbonaceous nanomaterials for detection of heavy metal ions in aqueous systems *Sci. Total Environ.* **688** 751–61
[44] Kalinke C, Oliveira P R, Oliveira G A, Mangrich A o S, Marcolino L H Jr. and Bergamini M F 2017 Activated biochar: preparation, characterization and electroanalytical application in an alternative strategy of nickel determination *Anal. Chim. Acta* **983** 103–11
[45] Das K, Das S and Dhundasi S 2008 Nickel, its adverse health effects and oxidative stress *Indian J. Med. Res.* **128** 412–25
[46] Nascimento D S, Insausti M, Band B S and Lemos S G 2014 Simultaneous determination of Cu, Pb, Cd, Ni, Co and Zn in bioethanol fuel by adsorptive stripping voltammetry and multivariate linear regression *Fuel* **137** 172–8
[47] Tormin T F, Narciso L C D, Richter E M and Munoz R A A 2014 Batch-injection stripping voltammetry of metals in fuel bioethanol *Fuel* **117** 952–6
[48] Oliveira P R, Lamy-Mendes A C, Pissinati Rezende E I, Mangrich S, Marcolino-Junior L H and Bergamini M F 2015 Electrochemical determination of copper ions in spirit drinks using carbon paste electrode modified with biochar *Food Chem.* **171** 426–31
[49] Suguihiro T M, Roberto de Oliveira P, de Rezende E I P, angrich A S M, Marcolino-Junior L H and Bergamini M F 2013 An electroanalytical approach for evaluation of biochar adsorption characteristics and its application for lead and cadmium determination *Bioresour. Technol.* **143** 40–5
[50] Lee G J, Kim C K, Lee M K and Rhee C K 2010 Effect of phase stability degradation of bismuth on sensor characteristics of nano-bismuth fixed electrode *Talanta* **83** 682–5
[51] Campbell F W and Compton R G 2010 The use of nanoparticles in electroanalysis: an updated review *Anal. Bioanal. Chem.* **396** 241–59
[52] de Oliveira P R, Lamy-Mendes A C, Gogola J L, Mangrich A S, Marcolino-Junior L H and Bergamini M F 2015 Mercury nanodroplets supported at biochar for electrochemical determination of zinc ions using a carbon paste electrode *Electrochim. Acta* **151** 525–30
[53] Bromilow R H 2004 Paraquat and sustainable agriculture *Pest Manag. Sci.* **60** 340–9
[54] Xiang Y, Liu H, Yang J, Shi Z, Tan Y, Jin J, Wang R, Zhang S and Wang J 2018 Biochar decorated with gold nanoparticles for electrochemical sensing application *Electrochim. Acta* **261** 464–73
[55] Kumar A A, Swamy B E K, Ganesh P S, Rani T S and Reddy G V 2017 Voltammetric determination of catechol and hydroquinone at poly(-niacinamide) modified glassy carbon electrode *J. Electroanal. Chem.* **799** 505–11
[56] Singh G, Kim I Y, Lakhi K S, Srivastava P, Naidu R and Vinu A 2017 Single step synthesis of activated bio-carbons with a high surface area and their excellent CO_2 adsorption capacity *Carbon* **116** 448–55
[57] Moraes F C, Rossi B, Donatoni M C, de Oliveira K T and Pereira E C 2015 Sensitive determination of 17 beta-estradiol in river water using a graphene based electrochemical sensor *Anal. Chim. Acta* **881** 37–43
[58] Han Q, Shen X, Zhu W Y, Zhu C H, Zhou X M and Jiang H J 2016 Magnetic sensing film based on Fe_3O_4@Au-GSH molecularly imprinted polymers for the electrochemical detection of estradiol *Biosens. Bioelectron.* **79** 180–6
[59] Chen C M, Huang J Q, Zhang Q, Gong W Z, Yang Q H, Wang M Z and Yang Y G 2012 Annealing a graphene oxide film to produce a free standing high conductive graphene film *Carbon* **50** 659–67

[60] Smith M W *et al* 2016 Structural analysis of char by Raman spectroscopy: improving band assignments through computational calculations from first principles *Carbon* **100** 678–92
[61] Liu J Q, Lin Z, Liu T, Yin Z, Zhou X, Chen S, Xie L, Boey F, Zhang H and Huang W 2010 Multilayer stacked low temperature-reduced graphene oxide films: preparation, characterization, and application in polymer memory devices *Small* **6** 1536–42
[62] Chen S, Li L, Wen Y, Yang G, Liu G, Yi Y, Shang Q, Yang X and Cai S 2019 Voltammetric analysis of thiamethoxam based on inexpensive thick-walled Moso bamboo biochar nanocomposites *Int. J. Electrochem.* **14** 10848–61
[63] Kalinke C, de Oliveira P R, San Emeterio M B, Calabuig A G, del Valle M, Mangrich A S, Marcolino-Junior L H and Bergamini. M F 2019 Voltammetric electronic tongue based on carbon paste electrodes modified with biochar for phenolic compounds stripping detection *Electroanalysis* **31** 2238–45
[64] Sanchez-Gomez R, Zalacain A, Pardo F, Alonso G L and Salinas M R 2016 An innovative use of vine-shoots residues and their 'feedback' effect on wine quality *Innovative Food Sci. Emerging Technol.* **37** 18–26
[65] Gonzalez-Calabuig A and Del Valle M 2018 Electronic tongues for food and security applications *Talanta* **179** 70–4
[66] Buron N, Coton M, Legendre P, Ledauphin J, Kientz-Bouchart V, Guichard H, Barillier D and Coton E 2012 Implications of *Lactobacillus collinoides* and *Brettanomyces/Dekkera anomala* in phenolic off-flavour defects of ciders *Int. J. Food Microbiol.* **153** 159–65
[67] Švancara I, Vytřas K, Barek J and Zima J 2001 Carbon paste electrodes in modern electroanalysis *Crit. Rev. Anal. Chem.* **31** 311–45
[68] de Oliveira P R, Kalinke C, Gogola J L, Mangrich A S, Marcolino L H and Bergamini J M F 2017 The use of activated biochar for development of a sensitive electrochemical sensor for determination of methyl parathion *J. Electroanal. Chem.* **799** 602–8
[69] Kalinke C, Wosgrau V, Oliveira P R, Oliveira G A, Martins G, Mangrich A S, Bergamini M F and Marcolino-Junior L H 2019 Green method for glucose determination using microfluidic device with a non-enzymatic sensor based on nickel oxyhydroxide supported at activated biochar *Talanta* **200** 518–25
[70] Oliveira P R, Kalinke C, Mangrich A S, Marcolino-Junior L H and Bergamini M F 2018 Copper hexacyanoferrate nanoparticles supported on biochar for amperometric determination of isoniazid *Electrochim. Acta* **285** 373–80
[71] Caon T, Maduro Campos C E, Oliveira Simoes C M and Segatto Silva M A 2015 Novel perspectives in the tuberculosis treatment: administration of isoniazid through the skin *Int. J. Pharm.* **494** 463–70
[72] Huang Y S 2014 Recent progress in genetic variation and risk of antituberculosis drug-induced liver injury *J. Chin. Med. Assoc.* **77** 169–73
[73] Yan X, Bo X J and Guo L P 2011 Electrochemical behaviors and determination of isoniazid at ordered mesoporous carbon modified electrode *Sens. Actuators* B **155** 837–42

Chapter 14

Monolithic wood biochar properties and supercapacitor performance relationships

Aldrich Ngan, Johnathon N Caguiat, Li Tao, Donald W Kirk and Charles Q Jia

Wood-derived carbon, or wood biochar, was investigated as low-cost monolithic electrodes in electrical double-layer capacitors (EDLCs) or supercapacitors. This study used a high-level approach to reveal relationships between the biochar properties and supercapacitor performance in a statistically relevant sample size of over 20 samples. The samples were characterized using CO_2 physisorption and density functional theory (DFT), and electrochemical testing was undertaken using cyclic voltammetry (CV) and galvanostatic charge and discharge cycles (GC). This work revealed that biochar samples were highly microporous. However, their specific surface area (SSA in $m^2\ g^{-1}$) values were independent of the bulk density, suggesting a negligible contribution of micropores to the overall porosity. The data showed that the specific capacitance (in F g^{-1}) was not a function of SSA, probably due to the limited ionic accessibility of micropores. Moreover, the device resistance was found to relate to the dry resistance measured across the electrodes, highlighting the importance of biochar conductivity to supercapacitor performance. Finally, there was an inverse proportionality between device capacitance and resistance, underlining the importance of developing low-resistance wood biochar to achieving energy and power densities at the same time.

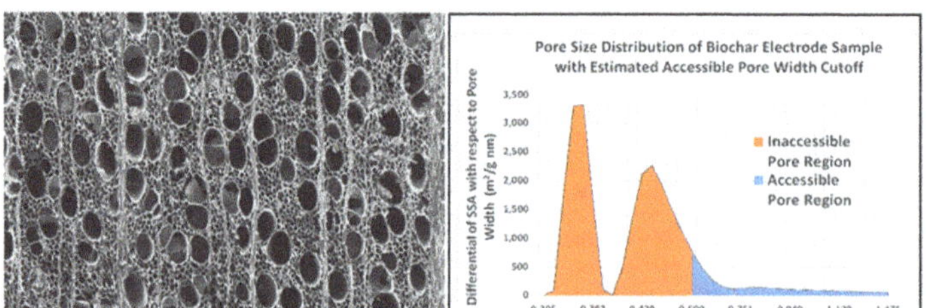

Left: SEM image of biochar macrostructure   Right: Measured biochar microstructure distribution

doi:10.1088/978-0-7503-2660-5ch14

14.1 Introduction

Traditional forms of energy storage, namely batteries and capacitors, have been reported to be unable to meet the energy storage requirements of intermittent renewable energy sources such as wind and solar energy [1]. Supercapacitors form one of the most promising developing technologies in electrical energy storage, which would serve to meet the demands of energy buffering, enabling renewable energy systems as well as providing the efficient use of electrical power in the transportation, commercial, and residential sectors. Also known as electrical double-layer capacitors (EDLCs), supercapacitors store charge electrostatically rather than faradaically [2]. They are typically constructed from nanoporous carbon materials with high surface areas, such as activated carbon, graphene, carbon nanotubes, carbon aerogel, and templated carbons [3–5]. These materials are often difficult to produce and therefore expensive, thus inhibiting the adoption of supercapacitors.

Biomass-derived carbon, or biochar, is a potential source of nanoporous carbon material suitable for EDLC electrodes since it is inexpensive and readily available. Biochar, produced through an oxygen-deficient heat treatment process known as pyrolysis or carbonization, can be subsequently activated to enhance its porosity further [6].

Biochar has been explored extensively for application as an active electrode material in supercapacitors. Various types of biomass, including cotton stalk, palm tree waste, sunflower seed shells, bamboo, soybean roots, almond shells, seaweed, and different types of wood have been used for producing biochar electrodes via pyrolysis [7–10]. After pyrolysis, biochar can retain the microporous, mesoporous, and microporous structures of the biomass, which provide a large specific surface area and short ion transport distance [11]. For example, Qiu *et al* tested KOH-activated corn straw biochar and reported an electrode capacitance of 327 F g^{-1}, with a remarkable cycle stability of 12 000 cycles at 5 A g^{-1} [12]. They attributed the performance to the hierarchical porous structure of biochar. To further improve the electrochemical performance, conducting materials and pseudocapacitive materials have often been introduced into biochar. Kouchachvili and Entchev added Ag to infested ash tree biochar and improved the electrode capacitance due to the enhanced conductivity of Ag [13]. Genovese and Lian investigated a polyoxometalate (POM) modified pinecone biochar electrode for a supercapacitor and reported a higher areal capacitance of 1.19 F cm^{-2} [14]. The high capacitance was attributed to the pseudocapacitive POM and the pore structure of biochar that enabled a high POM loading. Other hetero-elements such as Mn, Ni, Co, and N have been explored for their capacity in inducing pseudocapacitance [15].

Although powder biochar has been widely used for supercapacitor electrodes in the past, the use of monolithic biochar is new. Some recent studies demonstrated the advantages of monolithic wood biochar over biochar powder for electrode applications, citing improved conductivity and power density [16, 17]. A monolithic biochar electrode (1 mm thickness) from cotton rosewood achieved a specific capacitance of 133 F g^{-1} at 10 mA cm^{-2} and an energy density of 9.4 Wh kg^{-1} [18]. The hierarchical porous structure of the monolithic biochar electrode was found to be responsible for the high performance. However, the relationships among the physical properties of biochar monoliths have not been explored systematically. Furthermore, the relationships between biochar properties and supercapacitor performance also remain unexploited. This research objective was to fill this knowledge gap, focusing on biochar density, specific surface area, resistance, and supercapacitor performance parameters such as capacitance and device resistance. A key goal is to better understand the biochar properties that control supercapacitor performance. This work specifically tested the following four hypotheses:

1. The specific surface area (SSA) would decrease with the increase in density due to the porosity decrease.
2. The biochar density would be inversely proportional to electrode dry resistance. A higher porosity would mean a less solid carbon, which serves as the current conductor in a biochar electrode.
3. The specific capacitance would be proportional to the specific surface area, as a larger surface area would allow the formation of a more extensive electrical double layer that produces capacitance.
4. The device resistance would depend on electrode dry resistance. The device resistance consists of the resistive components from the electrodes and the electrolyte in electrodes, the contact resistances between current collectors and electrodes, and the resistance through the separator, assembled in series.

14.2 Experimental details

14.2.1 Biochar electrode preparation and characterization

Commercial biochar originating from *Acer saccharum* (sugar maple) wood was used as the material for study. Biochar pieces were first sorted based on the approximate resistance across each piece measured by the voltage drop with respect to the current passed through the sample. Biochar pieces of less than 40 Ω measured resistance were used to create the biochar electrodes. Electrode samples were created in pairs, each pair originating from the same biochar piece and were made into cubes ($5 \times 5 \times 5\ mm^3$).

For orientation, the direction parallel to the vessels and tracheids in the wood was dubbed the z-direction, with the other two axes labeled x and y, respectively. Due to the vessels and tracheids' openings, the faces perpendicular to the z-direction were darker, rougher, and less reflective. The orientation is depicted in the schematic shown in figure 14.1. The dry resistance was measured along the z-orientation using a voltmeter for a voltage drop at a given current. At the same time, all side lengths and cubes masses were determined using callipers and a Sartorius microbalance, respectively.

After the measurement of the biochar electrodes' dimensions, masses, and resistances, each pair of cubes was characterized by CO_2 gas physisorption using a Quantachrome Autosorb surface area and pore size analyzer. The resultant isotherm was then analysed with the ASiQwin software provided by Quantachrome, by applying density functional theory (DFT) isotherm reduction methods to produce a pore size distribution and specific surface area value.

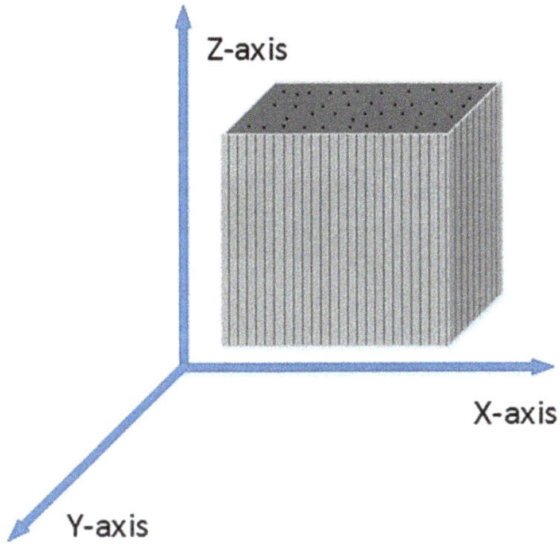

Figure 14.1. The electrode orientations, vessels, and tracheids along the *z*-axis.

Although the Brunauer–Emmet–Teller (BET) method remains widely used in gas physisorption, it does not measure microporosity accurately [19] and the DFT method has gained increasing acceptance. The BET's inadequacy is primarily due to its limitations in distinguishing different pore structure morphologies, accounting for the effects of microporosity, and predicting the pore sizes in microporous media [19, 20]. CO_2 was selected as the adsorbate, since the biochar was predominantly microporous [21]. The Quantichrome non-local DFT (NLDFT) kernel was chosen as it is the most suitable for CO_2 data for the reasons outlined in [22].

14.2.2 Electrode performance in a supercapacitor

Electrodes were submerged in 4M KOH for a minimum of 1 week before testing to ensure electrolyte penetration throughout the electrode. Supercapacitor cells were assembled, as shown in figure 14.2, after the soaking period.

Supercapacitor cells were connected to a Metrohm Autolab potentiostat/galvanostat for testing. The supercapacitor cell was made up of two 5 mm cubic electrodes separated by a porous sheet to prevent the direct contact of electrodes. The separator was made of polyphenylene sulfide or Ryton®. Current collectors were nickel mesh (nickel gauze, 40 mesh woven from 0.13 mm wire) purchased from Alfa Aesar. Nickel mesh with spot-welded wires was used as a current collector and the entire cell was held together by $4 \times 3 \times 0.5$ cm^3 acrylic sheets, with stainless steel bolts and nuts tightened to finger tightness.

Cyclic voltammetry (CV) was first performed on the supercapacitors using the voltage range of –0.8 to 0.8 V at a sweep rate of 10 mV s^{-1}. The CV scanning continued until the cell stabilized. The extended period of scanning ensured stability by removing the surface species that contribute to faradaic reactions and pseudocapacitance. All of the samples were held under cyclic voltammetry for a minimum of 100 cycles. However, in some cases, additional cycles, up to 400 cycles, were

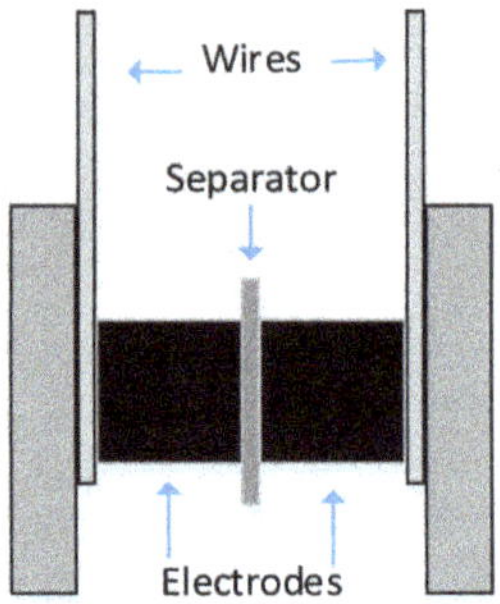

Figure 14.2. Schematic of a supercapacitor cell (not to scale) with two symmetrical monolithic biochar electrodes ($5 \times 5 \times 5$ mm^3)

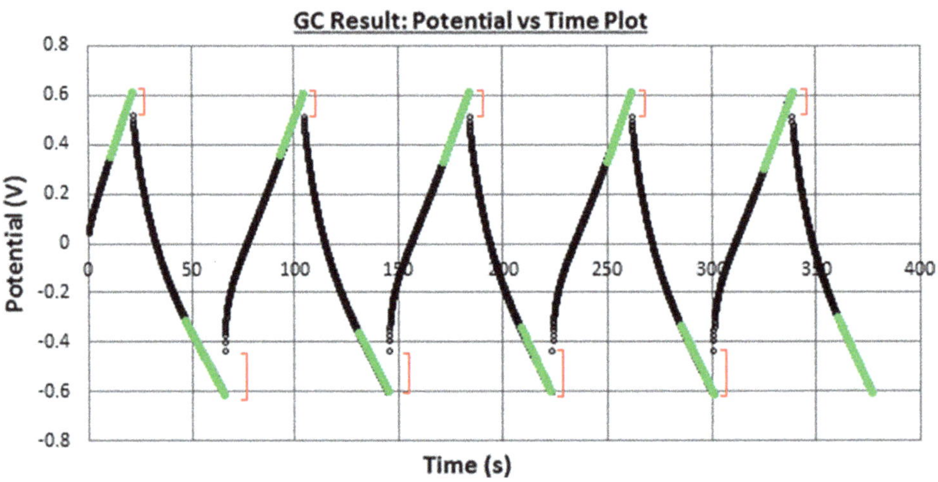

Figure 14.3. Typical galvanostatic cycling (gc), potential versus time plot, featuring linear portions preceding peaks (green), and voltage drops immediately following peaks (red).

required to achieve the desired stability. Stability was judged by observing the changes in the potential versus current plots of cyclic voltammetry cycles. A difference of less than 5% between five consecutive cycles was considered as stable.

After the electrodes stabilized, galvanostatic charge and discharge (GC) cycles were conducted at five separate current densities. The selected charge and discharge current densities were 100, 50, 25, 10, and 5 mA g^{-1}. At each current density, five charge and discharge cycles were performed within the voltage window of -0.6 and 0.6 V. From the GC plots depicting voltage versus time, such as the one depicted in figure 14.3, the capacitance values for the supercapacitor cells and the device resistances were calculated from the following equations.

Capacitance (C in F) was calculated from the slope of linear portions of the GC spectra using

$$C = \frac{I}{\mathrm{d}V/\mathrm{d}t}, \tag{14.1}$$

where V is voltage in V, t is time in seconds, and I is current in mA.

As each charge–discharge cycle has one linear region for the charging process and another linear region for the the discharging process, ten values for capacitance were calculated for every current density. These values could then be normalized to mass or surface area. Then, the average values were reported. Resistance across the entire device was calculated from the voltage drops at every peak using

$$R = \frac{\Delta V}{2I}, \tag{14.2}$$

where R is resistance in ohms and I is the current density in mA g^{-1}. Nine voltage drops and rises were recorded at every current density, allowing nine resistance values to be calculated. Average values were reported. The capacitance and device resistance values calculated from the lowest current density, namely 5 mA g^{-1}, were deemed the closest to the intrinsic capacitance and resistance values of the biochar electrodes. Interestingly, within the nine cycles, the capacitance value was less variable than the device resistance. Of the 21 samples measured, 18 capacitance values had a relative standard deviation (RSD) less than 5%, while the remaining three samples were between 5% and 6.7 %. On the other hand, device resistance values showed a more significant variability within the nine cycles. Ten samples had an RSD less than 5%, eight had an RSD between 5% and 10%, and three between 50% and 100%. The high RSD values (>50%) were likely caused by asymmetry in the electrodes, as deviations occurred primarily between the resistances calculated for charge versus discharge and not for the replication of cycles. Thus, the higher variation in device resistance averages are indicative of asymmetry rather than issues of repeatability and uncertainty.

14.3 Results and discussion

14.3.1 Relationships between biochar physical properties

The experiments were carried out with 21 electrode pairs. The results are summarized in table 14.1 below. The electrode dry resistance refers to the resistance of the $5 \times 5 \times 5$ mm^3 cubes along the z-orientation measured in the dry state. Normalized capacitance refers to the capacitance of a single electrode in each sample measured at the 5 mA g^{-1} current density, normalized to a specific surface area (SSA). Device resistance refers to the total resistance in the supercapacitor device, including the contact resistance between current collectors and electrodes, the electrolyte-filled electrode resistances, and the separator resistance, measured at the 5 mA g^{-1} current density.

It was hypothesized that the NLDFT-measured surface area would be inversely proportional to the electrodes' bulk density. An increase in porosity would lower the relative volume of solid material, thereby reducing the electrode density. The NLDFT-measured surface area is plotted against the electrode densities in figure 14.4.

As shown in figure 14.4, the specific surface area remained relatively uniform throughout all measurements, with a relative standard deviation of less than 7%.

Table 14.1. Overview of results of biochar properties (dry resistance, density, and SSA) and supercapacitor performance parameters (21 pairs of monolithic biochar electrodes).

	Electrode dry resistance (Ω)	Density (g cm^{-3})	NLDFT SSA (m^2 g^{-1})	Normalized capacitance (μF cm^{-2})	Device resistance (Ω)
Mean	10.84	0.425	549	2.749	14.92
Std. Dev.	14.55	0.068	35	0.645	4.31
Median	3.43	0.430	544	2.800	16.00
Min. value	0.73	0.279	454	1.358	4.98
Max. value	55.00	0.545	616	3.964	21.85

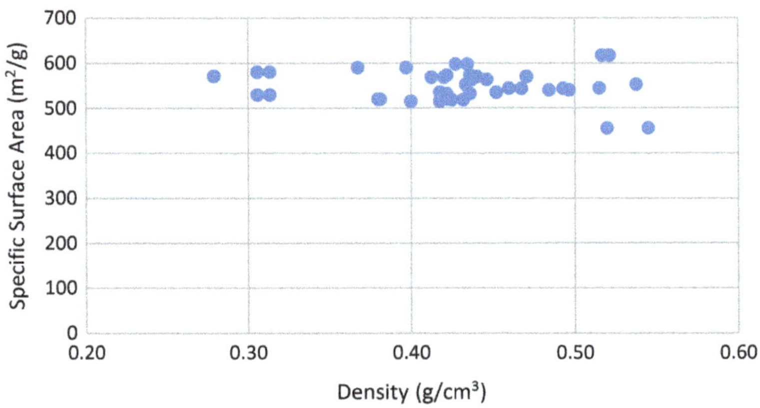

Figure 14.4. NLDFT-measured specific surface areas (SSAs) plotted against biochar bulk density.

The nature of the porosity can explain the lack of correlation between the density and the SSA in biochar monoliths. The SSA value is mostly controlled by the presence of micropores that are in nanometer scales. The small variation in SSA suggests relatively high homogeneity at the microscale level among the samples. On the other hand, macroporous features, such as tracheids and vessels, determine the porosity and the density of biochar monoliths. Figure 14.4 indicates that the distributions of tracheids and vessels in biochar monoliths were less uniform, probably due to natural variation. The vast majority of pores found in the samples were in the sub-nanometer pore width range. These small micropores have a substantial effect on the specific surface area relative to their contributions to pore volume. Figure 14.5 shows the differential change in pore volume and specific surface area as a function of pore width.

In figure 14.5 most of the micropores present in the electrodes are found near the 0.35 nm pore width and 0.55 nm pore width regions, with the majority of the SSA originating from pores of those sizes. It is important to note that the cumulative pore volume is integral to the differential change in volume function, dV(d). The pore volume from micropores was approximately 0.14 cm^3 g^{-1} in this sample.

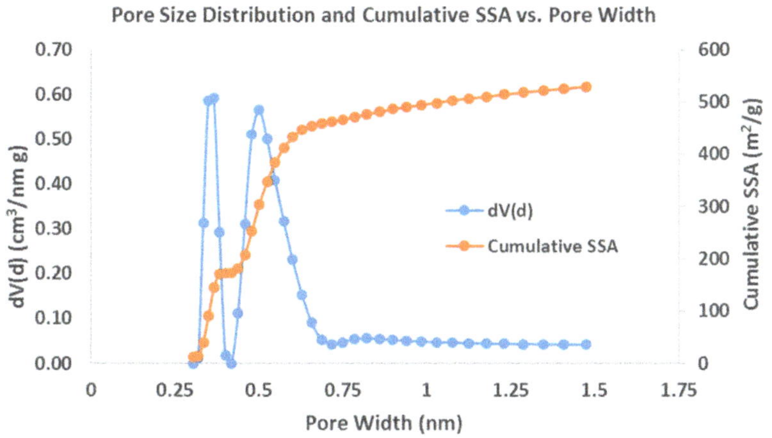

Figure 14.5. Typical pore size distribution (blue) and cumulative SSA versus pore width (orange).

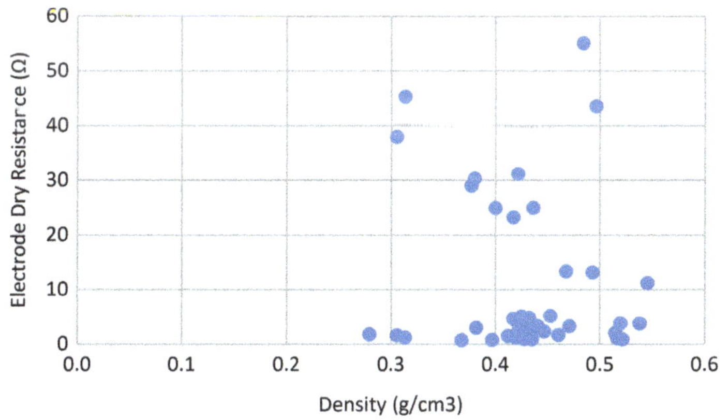

Figure 14.6. Biochar electrode dry resistance plotted against electrode density.

This finding effectively proves the first hypothesis negative. A potentially important implication of this finding is that high density and high SSA could be achieved simultaneously. High density and high SSA biochar materials could be advantageous for electrode workability, durability, and device size.

Electrode resistance was expected to be inversely proportional to electrode density, as a decrease in density, or increase in porosity, would reduce the amount of conductive material—carbon in biochar—and hence increase the resistance across the electrode. Figure 14.6 shows the measured dry resistance values of the cubic biochar electrodes.

As shown in figure 14.6, a relationship between the dry resistance and the density was lacking, proving the second hypothesis negative, which also suggests that other factors affect the dry resistance. The factors may include the solid-phase carbon's intrinsic properties, such as the bonding and crystalline structures. All these structural characteristics are expected to determine the electrical conductivity of biochar and should be studied in greater detail.

14.3.2 Dependence of supercapacitor performance on biochar properties

14.3.2.1 Capacitance and surface area

The development of high SSA carbons is an approach to increasing capacitance [23]. As such, it was expected that for the electrodes used in this study, capacitance would be proportional to the surface area available for EDL formation. Figure 14.7 shows the plot of specific capacitance versus DFT-measured specific surface area.

Figure 14.7 does not show a strong correlation between the SSA and specific capacitance, which may be explained by the pore size distribution. The vast majority of the pores present in the electrodes were in the pore width range of 0.35 nm and 0.6 nm. Studying the carbide-derived carbon (CDC), Chmiola *et al* [24] observed an abnormal increase in capacitance in pore sizes below the size of the solvation shell. The pore sizes present in the sugar maple electrodes were smaller than the smallest pores in the CDC. A follow-up study by Largeot *et al* [25] sought to determine the relationship between ion size and pore size in EDLCs. They found that in the case of an ionic liquid electrolyte, which is solvent-free and has no solvation shells, the pore size leading to the maximum double-layer capacitance is close to the size of the ions themselves.

If the conclusion from the study by Largeot *et al* applies to the sugar maple biochar electrodes, most of the micropores present in the biochar electrodes would be accessible to the hydrated ions in aqueous potassium hydroxide (KOH) electrolyte. However, as the study by Largeot *et al* arrives at their conclusion for ionic liquid electrolytes, the accessibility of the micropores remains a function of the extent to which the solvation shells for each ion can be distorted or discarded, which in turn is a function of ion size and charge density. Higher charge density should result in a tighter solvation sphere, which causes a higher resistance for the removal of any solvent molecule within the solvation shell, in addition to physically limiting

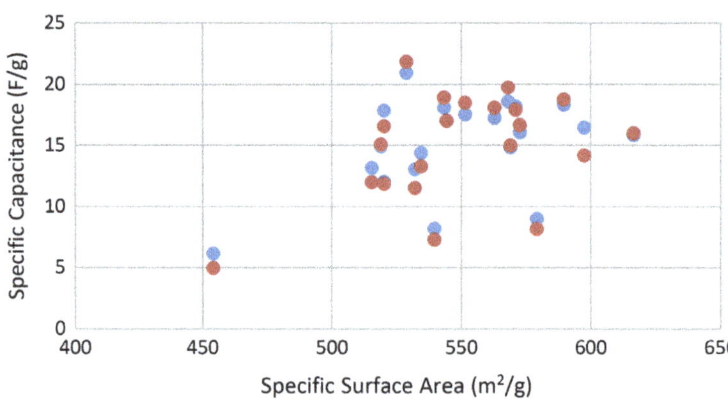

Figure 14.7. Specific capacitance measured at 5 mA g^{-1} (red) and 10 mA g^{-1} (blue) plotted against the measured specific surface area.

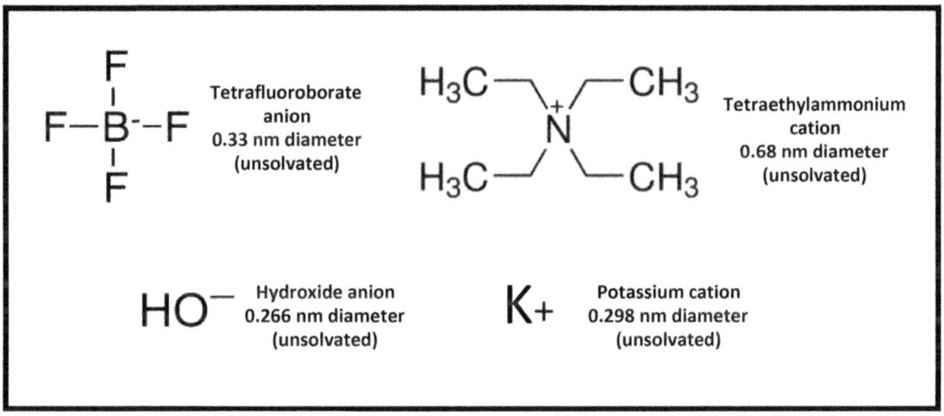

Figure 14.8. Unsolvated electrolyte anion and cation diameters for electrolytes used in work by Chmiola *et al* [24] and this study [26, 27].

Table 14.2. Unsolvated ion sizes and charge densities for anions and cations from electrolytes used in work by Chmiola *et al* and in this study.

Ion	Molecular formula	Ion diameter (nm)	Charge density (C/nm^3)
Tetrafluoroborate anion	BF_4^-	0.33	8.51×10^{-18}
Tetraethylammonium cation	$C_8H_{20}N^+$	0.68	9.73×10^{-19}
Hydroxide anion	OH^-	0.266	1.63×10^{-17}
Potassium cation	K^+	0.298	1.16×10^{-17}

the amount of possible compression and distortion of the solvation shell. Figure 14.8 shows the ions from the electrolyte used in the study by Chmiola *et al*, as well as the ones used in this study, and the unsolvated ion diameters for each ion [24, 26, 27].

Consider the charge densities of ions shown in table 14.2 below where charge density was calculated by dividing the electric charge of a single electron or proton, $\pm$ 1.602×10^{-9} C, by the volume of the ions (assumed as spheres).

As seen in table 14.2, there is a significantly higher charge density in KOH electrolyte ions than the organic electrolyte ions, particularly for cations, which differ by nearly 12-fold. The higher charge density would lead to the stronger retention of the solvation shell of potassium and hydroxide ions. The diameters of the hydrated potassium and hydroxide ions were estimated to be 0.662 and 0.600 nm, respectively [26, 27]. Hence, a large portion of pores in the biochar electrodes would remain inaccessible due to the size of hydrated ions in the KOH electrolyte. The inaccessibility of pores is also supported by the relatively low normalized capacitance values across the electrode samples, ranging from 1.358 to 3.964 μF cm^{-2} normalized to total DFT-measured surface area. The theoretical area-normalized capacitances are about 20 μF cm^{-2}, depending on the electrolyte used [28, 29]. This finding suggests a need to quantify the accessible surface area based on the size of hydrated ion sizes.

14.3.2.2 Electrode dry resistance and device resistance

Device resistance is one of the most crucial device performance measures, as it affects the power and the charging efficiency of a supercapacitor. Figure 14.9 shows the dependence of measured device resistance on electrode dry resistance, which confirms the importance of the resistance of biochar material to the device performance. Recent studies [30, 31] show that the electrical conductivity of monolithic biochar is determined by carbonization and graphitization and can be significantly enhanced by changing pyrolysis conditions such as temperature.

As shown in figure 14.9, there is a positive offset for device resistance when the electrode dry resistance is zero, attributed to the contact resistance and the resistance across the separator.

14.3.2.3 Capacitance and device resistance

Specific capacitance and resistance are the two most important parameters that measure a device's performance, as they determine the device's energy and power density. In figure 14.10, the specific capacitance is plotted against the device resistance, which shows a decreasing trend of capacitance with resistance. The standard equivalent circuit model treats a supercapacitor as a resistor and a capacitor in series. In the model, the capacitance and the resistance are independent from of each other [2].

The data suggest an inversely proportional relationship that is not expected from the supercapacitor equivalent circuit. It is well known that the measured capacitance is dependent on the current density used in a GC measurement. Specifically, the measured capacitance increases with a decrease in current density. There is an intrinsic capacitance determined by the structure of a device, not by the measurement condition. The measured capacitance is expected to approach the intrinsic value as the current density decreases. A smaller current density allows a slower and longer charging process. Fundamentally, the efficiency of the charging process is controlled by the rate of ion migration in the electrolyte within the porous electrode

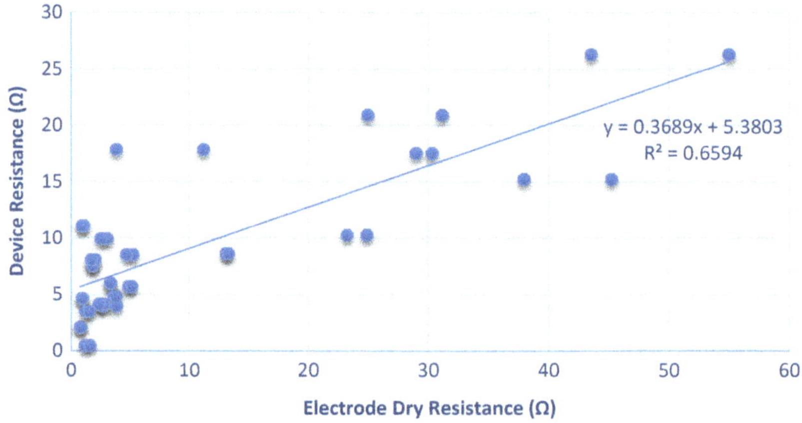

Figure 14.9. Dependence of measured device resistance on electrode dry resistance.

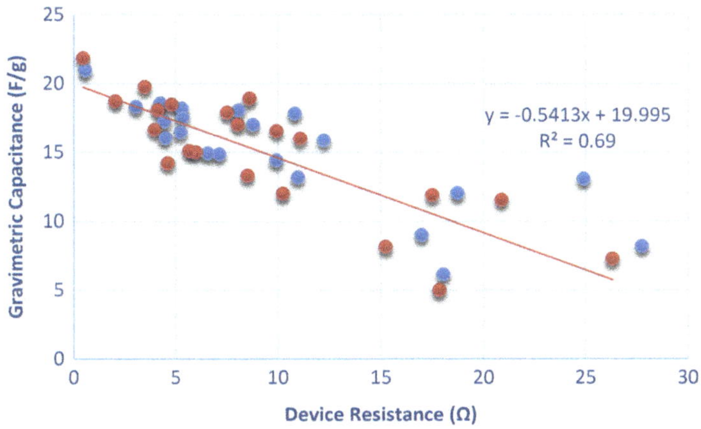

Figure 14.10. Dependence of gravimetric capacitance on device resistance measured at 5 mA g^{-1} (red) and 10 mA g^{-1} (blue).

under an electric field. A high device resistance could be related to a greater resistance to ion migration. An inefficient ion migration process would directly slow down the formation of a double-layer and result in a greater difference between the measured capacitance and the intrinsic capacitance. Consequently, there would be a dependence of the measured capacitance on the device resistance. More systematic work is needed to determine the source of dependency of measured capacitance on device resistance and to re-evaluate the equivalent circuit model's validity.

14.4 Conclusions

Twenty-one monolithic biochar electrode pairs were tested for supercapacitor applications in the KOH electrolyte. The nanoporous electrodes were characterized using CO_2 physisorption and NLDFT data reduction, and the performance metrics for supercapacitor applications were studied using cyclic voltammetry (CV) and galvanostatic charge and discharge (GC) cycles. The measured specific surface area was found to be independent of biochar bulk density, due to the massive amounts of sub-nanometer micropores present in the electrodes. A high specific surface area is achievable without lowering the density. Hence, structurally stable and dense biochar electrodes with high surface area are obtainable. The dependence of device capacitance on the total surface area was weak, due to the limited accessibility of hydrated ions to the sub-nanometer pores. The dependence of device resistance on the dry electrode resistance was found, highlighting the importance of biochar's electrical conductivity to the supercapacitor performance. Finally, an inverse proportionality between device capacitance and device resistance was observed, suggesting ion migration in the electrolyte within the electrode plays essential roles in determining both the energy and power densities of supercapacitors. This observation also calls for the development of new equivalent circuit models that accurately illustrate the dependence of measured capacitance on device resistance.

References

[1] Goodenough J B, Abruna H and Buchanan M 2007 Basic research needs for electrical energy storage *Report of the Basic Energy Sciences Workshop on Electrical Energy Storage, April 2–4, 2007* (U.S. Department of Energy) p 186
[2] Conway B E 2013 *Electrochemical Supercapacitors: Scientific Fundamentals and Technological Applications* (Berlin: Springer)
[3] Zhu Y *et al* 2011 Carbon-based supercapacitors produced by activation of graphene *Science* **332** 1537–41
[4] Davies A and Yu A 2011 Material advancements in supercapacitors: from activated carbon to carbon nanotube and graphene *Can. J. Chem. Eng.* **89** 1342–57
[5] Iyyamperumal E, Wang S and Dai L 2012 Vertically aligned BCN nanotubes with high capacitance *ACS Nano* **6** 5259–65
[6] Sohi S, Krull E, Lopez-Capel E and Bol R 2010 A review of biochar and its use and function in soil *Adv. Agron.* **105** 47–82
[7] Chen M *et al* 2013 Preparation of activated carbon from cotton stalk and its application in supercapacitor *J. Solid State Electrochem.* **17** 1005–12
[8] Farma R *et al* 2013 Preparation of highly porous binderless activated carbon electrodes from fibres of oil palm empty fruit bunches for application in supercapacitors *Bioresour. Technol.* **132** 254–61
[9] Li X *et al* 2011 Preparation of capacitor's electrode from sunflower seed shell *Bioresour. Technol.* **102** 1118–23
[10] Liu W J, Jiang H and Yu H Q 2019 Emerging applications of biochar-based materials for energy storage and conversion *Energy Environ. Sci.* **12** 1751–79
[11] Tan X F *et al* 2017 Biochar as potential sustainable precursors for activated carbon production: multiple applications in environmental protection and energy storage *Bioresour. Technol.* **227** 359–72
[12] Qiu Z P *et al* 2018 Biochar-based carbons with hierarchical micro-meso–macro porosity for high rate and long cycle life supercapacitors *J. Power Sources* **376** 82–90
[13] Kouchachvili L and Entchev E 2017 Ag/biochar composite for supercapacitor electrodes *Mater. Today Energy* **6** 136–45
[14] Genovese M and Lian K 2018 Polyoxometalate modified pine cone biochar carbon for supercapacitor electrodes *J. Mater. Chem.* A **5** 3939–47
[15] Cheng B H, Zeng R J and Jiang H 2017 Recent developments of post-modification of biochar for electrochemical energy storage *Bioresour. Technol.* **246** 224–33
[16] Caguiat J N, Arpino G, Krigstin S G, Kirk D W and Jia C Q 2018 Dependence of supercapacitor performance on macro-structure of monolithic biochar electrodes *Biomass Bioenergy* **118** 126–32
[17] Yanchus D A, Kirk D W and Jia C Q 2018 Investigating the effects of biochar electrode macrostructure and dimension on electrical double-layer capacitor performance *J. Electrochem. Soc.* **165** A305
[18] Ma Y *et al* 2020 Ultra-thick wood biochar monoliths with hierarchically porous structure from cotton rose for electrochemical capacitor electrodes *Electrochim. Acta* **352** 136452
[19] Sing K S 1995 Physisorption of nitrogen by porous materials *J. Porous Mater.* **2** 5–8
[20] Landers J, Gor G Y and Neimark A V 2013 Density functional theory methods for characterization of porous materials *Colloids Surf Physicochem. Eng. Aspects* **437** 3–32

[21] Thommes M *et al* 2015 Physisorption of gases, with special reference to the evaluation of surface area and pore size distribution (IUPAC Technical Report) *Pure Appl. Chem.* **87** 1051–69

[22] Caguiat J N, Kirk D W and Jia C Q 2014 Uncertainties in characterization of nanoporous carbons using density functional theory-based gas physisorption *Carbon* **72** 47–56

[23] Simon P and Gogotsi Y 2012 Capacitive energy storage in nanostructured carbon–electrolyte systems *Acc. Chem. Res.* **46** 1094–103

[24] Chmiola J, Yushin G, Gogotsi Y, Portet C, Simon P and Taberna P L 2006 Anomalous increase in carbon capacitance at pore sizes less than 1 nanometer *Science* **313** 1760–63

[25] Largeot C, Portet C, Chmiola J, Taberna P, Gogotsi Y and Simon P 2008 Relation between the ion size and pore size for an electric double-layer capacitor *J. Am. Chem. Soc.* **130** 2730–1

[26] Volkov A, Paula S and Deamer D 1997 Two mechanisms of permeation of small neutral molecules and hydrated ions across phospholipid bilayers *Bioelectrochem Bioenerget* **42** 153–60

[27] Conway B E 1981 *Ionic Hydration in Chemistry and Biophysics* (Amsterdam: Elsevier)

[28] Xia J, Chen F, Li J and Tao N 2009 Measurement of the quantum capacitance of graphene *Nat. Nanotechnol.* **4** 505–9

[29] Ji H *et al* 2014 Capacitance of carbon-based electrical double-layer capacitors *Nat. Commun.* **5** 3317

[30] Gabhi R S, Kirk D W and Jia C Q 2017 Preliminary investigation of electrical conductivity of monolithic biochar *Carbon* **116** 435–42

[31] Gabhi R, Basile L, Kirk D W, Giorcelli M, Tagliaferro A and Jia C Q 2020 Electrical conductivity of wood biochar monoliths and its dependence on pyrolysis temperature *Biochar* **2** 369–78

Chapter 15

Applications of biochar in gas/water purification and in contaminated soil remediation

Hanieh Bamdad, Griffin Loebsack, Naomi Klinghoffer, Ken Yeung, Kelly Hawboldt and Franco Berruti

Biochar, a natural carbonaceous framework, is a pyrogenic product of an anaerobic thermochemical conversion process, called pyrolysis. Biochar can be produced from the carbonization of various types of feedstock including lignocellulosic biomass (e.g. forestry residues, bark, etc) and biosolid wastes (e.g. manure, digestate, etc). The unique structure of biochar with developed porosity, high surface area, and diverse functional groups has resulted in numerous emerging applications. Adsorption with the use of biochar is one of the most effective technologies applied for the removal of impurities from the biogas stream. Purification of biogas, as a renewable energy source, could improve its quality for the sake of energy efficiency, health, and the environment. Global warming caused by greenhouse gas emissions can be alleviated by implementing sustainable and cost-effective adsorbents. Of the porous sorbents, biochar is a promising candidate to remove harmful gases from fuel/fuel gas streams. As well, biochar is a highly effective adsorbent for the removal of contaminants in aqueous solutions which pose a large threat to both the environment as well as human and aquatic life. Treatment methods for natural water resources are necessary to remove organic, inorganic, and biological contaminants, all of which biochar is capable of removing with significant efficacy. Another beneficial application of biochar is the restoration of contaminated soil from petrochemical effluents and heavy metals, released from a wide variety of anthropogenic sources. The scope of this chapter includes a critical overview of the biochar applications that are used for gas purification and treatment/remediation of wastewater, contaminated soil, sediments, and surface/groundwater.

doi:10.1088/978-0-7503-2660-5ch15

15.1 Biochar as an adsorbent

Biochar is a porous carbonaceous structure that is produced by thermal decomposition of a wide variety of biomass materials in an oxygen lean environment. Liquid (bio-oil) and non-condensable gas can be obtained along with biochar in the thermochemical conversion process. Many types of biomass are used as organic precursors, such as lignocellulosic materials (e.g. forestry residues, bark, and crop/plant residues) and bio-wastes (e.g. manure, food and fishery wastes, digestate, and sewage sludge). Process efficiency and product yields and characteristics depend on the operating parameters, such as feedstock selection, reaction temperature, heating rate, residence time, carrier gas, and reactor operation (e.g. batch versus continuous). Given this, biochar is potentially tunable in terms of properties and further can enhance the feasibility of the overall thermochemical conversion process. Biochar has a high porosity, large surface area (specifically after modification), and multiple surface functional groups, and therefore is a versatile material for use in different applications [1].

The application of biochar as an adsorbent in gas/water purification has recently attracted increased attention [2, 3]. Biochar-based adsorbents exhibit high potential for the removal of various environmental contaminants in gas systems, such as greenhouse gases and contaminants in biogas [4]. Biochar is also capable of removing various inorganic (e.g. heavy metals, fertilizers) and organic pollutants (e.g. pharmaceuticals, pesticides, herbicides) from aqueous solutions. The preparation and modification of biochar with respect to a targeted adsorbate is the key to the biochar's efficacy in contaminant management [5]. Herein, the impact of biochar on the adsorption of different types of pollutants, as well as the mechanisms in gaseous and aqueous phases is reviewed.

15.1.1 Biochar in gas adsorption

Agriculture and fossil fuel burning are the two major contributors to greenhouse gas (GHG) emissions and consequently, climate change. To address serious environmental issues, renewable and economically attractive resources can be utilized. Biogas is one alternative to fossil fuels and is derived from the anaerobic digestion of organic matter, however this gas can contain gases such as hydrogen sulfide. Biochar can be used as an adsorbent to capture GHGs (such as CO_2, CH_4, N_2O) and other problematic gases produced from different anthropogenic and livestock operations and from biogas production. In the following sections, the application of biochar in biogas purification and GHG adsorption is discussed.

15.1.1.1 Biogas purification

Biogas is a renewable and sustainable energy source that can be substituted for conventional fossil fuels in several applications. Biogas can be produced through anaerobic processes (i.e. digestion and thermal conversion) of different types of wet organic substrates including agricultural, municipal, livestock, and industrial wastes. The produced biogas is typically composed of 50%–70% (v/v) methane, 30%–45% (v/v) carbon dioxide, a fraction of water vapor, and trace amount of impurities such as sulfur, chlorine, aromatics, and siloxanes [6]. The biogas composition depends on the nature of the feedstock. For example, sulfurs, carbonyl compounds, carboxylic acids, and terpenes are the most abundant trace components in municipal solid waste biogas [7]. A possible application of biogas is using it as a fuel in energy generators, such as fuel cells (follow steam reforming of methane), as well as combined heat and power plants. Although biogas plays a prominent role in waste management and reduction of GHGs [8], the biogas quality must be improved prior to any energy use due to potential health, environmental, and operational concerns.

The treatment of biogas occurs in two steps: cleaning (removal of undesired materials) and upgrading (CO_2 removal) [9]. There are a number of processes used to purify biogas, including physical (adsorption), chemical (absorption), and biological. The most effective method, particularly in remote regions, is adsorption using solid porous adsorbents, such as biochar, due to its simple technology, small scale equipment, and reduced operational intensity [10]. Hydrogen sulfide is flammable, toxic, corrosive, and a catalyst poison, causing major threats to human health, environment, and facilities. Thus, the removal of H_2S from biogas is of utmost importance from a scientific point of view. Han *et al* [11] revealed the H_2S adsorption characteristics of biochars (from two typical macroalgae at 800 °C) was mainly controlled by physisorption enhanced by chemisorption. The obtained adsorption capacities (500–5800 mg kg^{-1}) were comparable to biochar sourced from sludge at 800 °C (4600 mg kg^{-1}) [12].

The trace compounds in biogas interfere with fuel cell performance for generating electricity. To raise the efficiency of a fuel cell fed with biogas, the VOCs clean-up section is highly recommended [13]. Biochar was found to be capable of removing 2-butanone (158.8 mg g^{-1}), toluene (140.1 mg g^{-1}), and limonene (64 mg g^{-1}) with greater efficiency when compared to siloxane (1.28 mg g^{-1}) compounds [14]. The low calorific value of biogas is attributed to the presence of non-CH_4 gases, indicating that upgrading is required for removing undesirable gases and specifically CO_2.

Scientific studies pointed out that thermal/chemical modification [15] and metal impregnation (e.g. Fe, Al, Ni, Mg, and Ca) [16] can enhance the CO_2 adsorption capacity of biochar compared to pristine biochar. After biogas treatment, the final product is called bio-methane with 95%–99% CH_4 and 1%–5% CO_2 [17]. The clean biogas can be used as an alternative to natural gas as the properties are almost identical.

15.1.1.2 Greenhouse gas mitigation

Biochar is receiving increasing interest from the scientific community partly because it is sustainable, nature-based, and produced from renewable agricultural or forestry sources. One of the major sources of greenhouse gases is the agricultural sector [18]. Applying biochar in soil represents a win–win strategy since biochar can simultaneously have a positive impact on both soil/plant interactions and the environment. Biochar is highly stable against biotic and abiotic factors and has the ability to sink carbon in soil, thus representing an effective means for carbon sequestration. In addition, biochar has the potential to reduce GHG emissions even when applied to soil.

Agricultural soil amendment with mineral and organic N fertilizers are a major source of nitrous oxide (N_2O), which is produced through biotic and abiotic transformations. Mandal *et al* [19] have reported the potential mechanisms in which biochar impacts N cycling and consequently N_2O emission (figure 15.1).

The soil application of biochar for N_2O mitigation can be obtained by hindering the enzyme activity for converting of NO_3^- and NO_2^- to N_2O (nitrification), inhibiting soil NH_4^+ and NO_3^- ions (denitrification) leaching, adsorbing nitrous oxide, and promoting complete denitrification process to produce nitrogen (N_2) [20, 21].

The application of biochar produced from banana peel waste to soil resulted in the reduction of greenhouse gas emissions, CO_2 (20%–24%) and N_2O (37%–55%), in comparison to soil amendment solely from the banana peel [22].

In a study based on a 117 d incubation period, CO_2 emission from slow pyrolysis biochar amended soils was found independent of the biochar source (wood and digestate) [23]. Conversely, the results obtained by Liu *et al* [24] demonstrated that biochar from rice straw (produced at 600 °C) was more efficient in the reduction of CH_4 (~91%) and CO_2 (~40%) emissions compared to the bamboo biochar (produced at 600 °C) in paddy soil during a 49 d incubation period. In another study, woody biochar (produced at 550 °C) addition to soil successfully reduced the CO_2 emissions up to 5.9% over the 60 d incubation or 0.58–1.72 Mg CO_2-eq ha^{-1} annually [25]. The potential dominant mechanisms for CO_2 mitigation using biochar were reported as enzymatic activity reduction, carbon-use efficiency improvement during photosynthesis, and adsorption on the biochar surface [26].

15.1.2 Biochar in liquid adsorption

The quality and hence availability of freshwater is decreasing due to contamination of natural water resources [23]. The major sources for water contamination are from population growth, industrialization, agricultural activities, and geological, environmental, and global changes [27]. Adsorption is considered one of the best options for universal water treatment due to its low cost, wide range of applications, and ease of operation [27]. Biochar is a promising adsorbent due to its excellent capacity to remove both soluble and insoluble organic, inorganic, and biological pollutants from aqueous solutions thus constituting an untapped technology for water

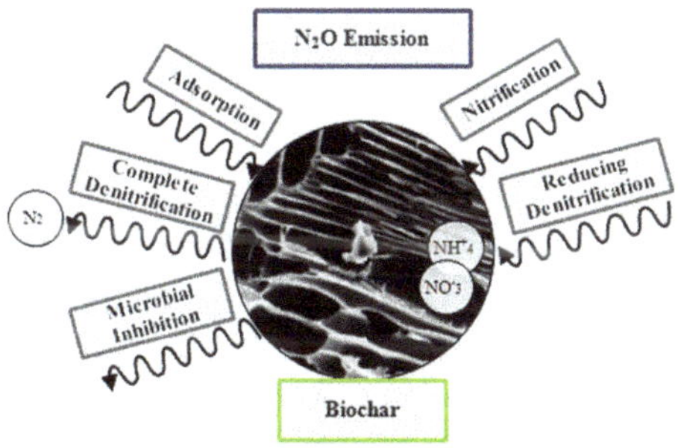

Figure 15.1. Mechanisms of reduction of nitrous oxide (N_2O) by biochar. Adapted from Mandal *et al* [20].

treatment [28]. Compared to the existing water purification methods such as sand filtration, boiling, solar disinfection, and chlorination, biochar presents several advantages as it is a low-cost and renewable adsorbent made using readily available biomaterials and skills which can be optimized for specific adsorbates [28]. Biochars also maintain the organoleptic properties of water whereas other methods such as chlorination and boiling generate carcinogenic by-products and increase the concentrations of compounds such as metals [28]. The following sections discuss the application of biochar as an adsorbent in the removal of various contaminants from water originating from agricultural, industrial, and municipal wastes.

15.1.2.1 Agricultural waste removal

The increased use of agricultural chemicals has correlated with an increase in the concentration of nitrogen and phosphorus-containing compounds in the surrounding ground and surface water [29, 30]. Even at low concentrations, nitrogen and phosphorus can create serious problems, such as eutrophication of rivers, lakes, and reservoirs, deterioration of water quality, and a potential hazard to human health when released into the environment [30, 31]. In addition, nitrate and nitrite both have the potential to form carcinogenic N-nitrous compounds in water [31]. In addition to fertilizer, natural soil, rare geologic deposits, manure, sewage, and other waste disposals, and even precipitation, all contribute to adding phosphorus and nitrogen to the soil–plant system, and consequently to groundwater [29]. However, these natural phosphorus reserves are expected to run out within 50–100 years [32] necessitating not just the removal of phosphorus from the water systems, but also its recovery and subsequent utilization.

In the past, methods for the removal of nitrates and phosphates from water have been limited to physiochemical and biological methods, whose processes include ion exchange, electrocatalytic reduction, and membrane separation methods [33]. The physiochemical ion exchange process is feasible from an economic standpoint but is nonspecific for the ions of interest and yields concentrated brines containing copious ions which must be disposed of separately [33]. As well, membrane separation and biological techniques present the issue of economic feasibility [33].

Adsorption by biochar, however, is potentially an economic and practical process for nitrogen and phosphorus removal, as well as recovery. However, it has two opposing mechanisms for ammonia and nitrate/phosphate adsorption. Positively charged ammonia has shown higher sorption for more polar biochar, which is attributed to electrostatic interactions with the negatively charged oxygen containing groups on the surface [34]. This is confirmed when looking at the effects of pH. As pH increases, the concentration of negatively charged sites increases due to the deprotonation of carboxylic acid, lactones, and phenolic functional groups on the surface [34]. As well, there is less competition for adsorption sites by hydrogen ions at more alkaline pH. Nitrate and phosphate sorption, on the other hand, increases with increasing pyrolysis temperature and decreasing pH as it is primarily the result of an anion exchange mechanism [34]. Lawrineko *et al* [35] hypothesized that biochar anion exchange capacity (AEC) resulted from basic surface groups such as oxonium and pyridinium groups on its surface, with oxonium sites being pH-independent. This implies a decrease in AEC at greater solution pH as a result of lowered adsorption sites due to deprotonation of pyridinium groups and competition from hydroxide ions in solution [35]. As well, the surface of the biochar is capable of being modified to improve its affinity towards both nitrate and phosphate anionic pollutants by increasing the number of positive adsorption sites through the introduction of basic functional groups such as pyridine [36]. The opposite polarity of ammonia and nitrate/phosphate suggests that a single type of biochar cannot be simultaneously optimized for the sorption of both nitrogen species, nor can there be an optimal pH of solution for the removal of both ions. However, this represents an excellent example of how biochar can effectively be tailored for specific applications, depending on the targeted contaminant.

15.1.2.2 Industrial waste treatment

Heavy metal pollution is a major environmental problem plaguing the world's water sources today. Of the important metals, mercury, lead, cadmium, and chromium (VI) are regarded as toxic, whereas others such as copper, nickel, cobalt, and zinc are not as toxic, but their extensive usage and increasing levels in the environment are of serious concern [37]. These metal ions are non-degradable, toxic, carcinogenic, and are persistent in the environment, making the elimination of these metals from wastewater essential to public health [37]. Industrial effluents remain the most prevalent contributor of inorganic contamination from corrosion of water pipes, waste dumping, electroplating, electrolysis, mining, surface finishing, energy, and fuel-producing, electroplating, iron and steel, photography, and leather production [37]. Heavy metal contaminated water is most commonly treated with activated carbon; however, because of the heterogeneous nature of the surface of activated carbon, there remains some uncertainty surrounding its adsorption mechanisms of heavy metals.

This limits the design of treatment systems as it is difficult to predict the adsorption capacity when the adsorption mechanisms remain unclear and based solely on empirical observations. It can, however, be concluded that the main variables for the adsorption of metals onto carbon surfaces are: (i) the surface area and pore distribution, (ii) the surface chemistry of the adsorbent, as determined by the surface functional groups, (iii) the pH of the solution, and (iv) the presence of other electrolytes that can compete with the metal for active adsorption sites on the surface of adsorbent [38].

Acidic oxygen surface groups on biochar ionize in aqueous solution to produce hydrogen ions in solution [39]. As a result, the surface of the biochar has many negative charge sites, depending on the amount of these surface groups, which can adsorb positively charged metal ions through cation exchange [39]. Bautista-Toledo *et al* noticed a positive correlation between the density of oxygenated surface groups on the biochar and the adsorption of chromium metal cations, implying the significance of cation exchange in the adsorption mechanism for metals [40]. The role of oxygenated functional groups was additionally highlighted when upon removal of carboxylic groups from activated carbon via thermal treatment a decrease of cation adsorption was observed [40]. However, this was hypothesized not to be the only mechanism through which metal ions are adsorbed onto biochar. Shim *et al* observed relatively high adsorption of nickel and copper by pitch-based activated carbon fibers even after the removal of oxygen containing groups with NaOH [41]. These results are explained by cation–π interactions, which were studied by Montes-Moran *et al* [42]. Cation–π interactions are the result of electrostatic attraction between the metal cation and the quadrupole charge distribution above the benzene rings present on the adsorbent surface [42]. This adsorption mechanism is then not dependent on the pH of the solution nor the charge of the biochar, making it a practical adsorbent for toxic metals in varying conditions.

15.1.2.3 Municipal wastewater treatment

The widespread occurrence of dissolved organic compounds in water poses is a threat to the environment as well as human and ecological health, as they mimic and antagonize natural hormones, hinder metabolic processes, occupy hormone receptors and cause reproductive developmental problems when consumed by human and aquatic species [43]. The spread of these bioactive compounds can also potentially lead to long-term adverse consequences on various ecosystems [43]. These contaminants are composed of pharmaceuticals, personal care products, surfactants, and pesticides and are most often a result of municipal wastewater discharge, as these compounds are not completely removed during treatment [44]. Pesticides, in particular, have been identified as having toxicological interactions including mutagenicity, carcinogenicity, immunomodulant effects, etc, and are released into the environment as both agricultural and municipal wastes [45]. Given how ubiquitous these potential organic contaminants are in everyday life, sufficient and effective methods for their removal from water sources are essential. Current treatment methods for pharmaceutical contaminants include oxidation, enzymatic, photocatalytic, and photo-electrocatalytic degradation, and biodegradation by activated sludge, however, remediation using biochar is a potentially more sustainable approach [43]. The mechanisms through which adsorption of pharmaceutical compounds onto biochar takes place are mainly reliant on electrostatic and π–π interactions which are greatly dependent on a variety of factors including the pH of the aqueous solution, and the surface chemistry of the biochar.

The significance of electrostatic interactions in terms of adsorption is responsible for the general trend that has been observed for increased adsorption of organic compounds at lower solution pH. The charge of the organic pollutants is dependent on the pKa or pKb of the ionizable groups of the pollutant molecules and the pH of the solution, where the interaction with negatively charged biochar is seen to be greatest when compounds are positive or neutral, and weakest when negative. Many authors have noticed the greater adsorption affinity of biochar when the reaction solution pH falls below the pKa of the adsorbent [46, 47].

Biochar adsorption mechanisms are complex, however, and not solely dependent on electrostatic interactions. π– π interactions, including π– π electron donor–acceptor (EDA) interactions, π–H bonding, and π stacking, have been found to play a large role in the adsorption of pharmaceutical compounds [46, 48]. Specifically, π–π EDA interactions have been investigated heavily for their binding interactions between pharmaceuticals and biochar [46, 48]. Electron rich (π-donor) aromatic compounds such as phenanthrene, naphthalene, and pentamethylbenzene showed better sorption under more acidic conditions. According to Zhu *et al* who this was not the trend for electron-poor (poor donor) aromatic and aliphatic compounds [48]. Given this prevalence of π–π EDA interactions in the absorbance capacity of biochar, surface aromaticity, and polarity of biochar are two very important factors when considering organic adsorption.

15.2 Biochar soil remediation

Exponential population growth and rapid industrialization inevitably lead to a huge release of hazards, which jeopardized human health and the environment. Soil contamination occurs both by natural mechanisms and anthropogenic sources, such as volcanic eruptions, agrochemicals, mining, industrial activities, municipal waste, and sewage sludge. Bioremediation, phytoremediation, soil flushing, soil washing, electrokinetic extraction, stabilization, and solidification are some of the techniques used for the treatment of pollution in soil [49]. Soil scientists have identified more sustainable and cost-effective alternatives, specifically biochar, that can enhance ecosystem performance. Over the past years, the main purpose of biochar addition to soil has been to improve soil quality and plant yield rather than restoration of contaminated sites. The role of biochar in the separation of most prevalent types of pollutants/resources from soil/sediments will be discussed in the following sections.

15.2.1 Inorganic pollutants/resources

The ideal properties of biochar, including high surface are, high porosity, the presence of a wide variety of functional groups, moderate hydrophobicity, and cation exchange capacity make it suitable for contaminant sorption. The application of co-pyrolysis biochar to contaminated agricultural soil could effectively reduce the ecological risks of Pb, Cd, and Cu due to precipitation and complexation capacities of carboxyl, hydroxyl, and phosphate functional groups [50]. Moreover, biochar can be applied as a feasible *in situ* remediation of metal(loid) contaminated sediments, e.g. As and Cd [51]. Mine-contaminated soil amended with biochar could minimize the available concentration of Cd, Pb, and Zn for a soil/plant system, which is attributed to the high surface area of biochar [52]. Biochar can be substituted for conventional fertilizers due to higher stability and lower hazards.

Livestock manure and sewage sludge may contain a number of heavy metals, antibiotics, and pathogens, which could pose serious environmental and human health issues while using as a fertilizer. Shen *et al* [53] evaluated the effect of pyrolysis temperature (300 °C–700 °C) on the environmental risk of heavy metals. They found that leaching toxicity of Cr, Mn, Cu, and Zn in pig manure biochar was profoundly decreased, in particular at an optimum pyrolysis temperature of 500 °C. One emerging application of biochar is recovering precious resources from soil such as nitrogen and phosphorus, produced by excessive use of inorganic fertilizers. The re-application of spent biochar is also plausible as a slow-release fertilizer [54].

15.2.2 Organic pollutants

Pesticides and herbicides have been widely used in agriculture to increase crop yield and control pests, weeds, and disease. Applied agrochemicals are partially taken up by the crops/plants and the remaining persists in the soil environment. Jaramillo *et al* [55] revealed that the ion-exchangeable polar groups in the biochar, in particular O– and N–, showed a significant role in the adsorption of herbicide azimsulfuron in rice paddy soils.

Organic pollutants including polycyclic aromatic hydrocarbons (PAHs), polychlorinated biphenyls (PCBs), and aromatic nitro compounds are highly toxic and persistent in soil and sediment. Biochar can immobilize PCB contaminations by inducing specific strains of bacterial community that inhabit soil [56]. Oleszczuk *et al* [57] reported that biochar reduced the bioaccessibility of PAHs after 6 months in soils vegetated with willow.

15.3 Summary and conclusions

This chapter has presented the emerging applications of biochar in gas and aqueous phases as well as a soil environment. Given that the environment and human health are increasingly threatened by many forms of pollution, biochar which can make a significant contribution to reduce pollution is required for sustainable development. Biochar provides a valuable opportunity to purify biogas in order to curtail GHG emissions and achieve energy independence from fossil fuels. As elaborated in this chapter, biochar has the multifaceted potentials of carbon sequestration, GHG mitigation, and crop yield enhancement. In terms of aqueous media, biochar is an effective adsorbent for a diverse range of pollutants, including toxic metal ions, bioactive organic compounds, and nitrates whose presence poses severe environmental and health risks. Although some techniques are established for remediation of contaminated soil, biochar is found to be a more reliable and less expensive approach. Biochar with high porosity and surface heterogeneity can interact with organic/inorganic pollutants in the soil through several mechanisms.

The conclusion is that biochar still has unknown potentials/aspects that need to be explored by the scientific community. While biochar with a range of benefits seems phenomenal, researchers have to be cautious about risks associated with deploying it in different fields. The key is engineering/tailoring the biochar structure by chemical/physical activation and changing the source and pyrolysis conditions properly. A number of biochar applications are still at the experimental stage; further expansion to actual/large scale, such as the agricultural field and industries is desired. This leads to a practical challenge for producing vast amounts of biochar to show significant efficacy.

References

[1] Nanda S, Dalai A K, Berruti F and Kozinski J A 2016 Biochar as an exceptional bioresource for energy, agronomy, carbon sequestration, activated carbon and specialty materials *Waste Biomass Valor* **7** 201–35

[2] Pelaez-Samaniego M R, Smith M W, Zhao Q, Garcia-Perez T, Frear C and Garcia-Perez M 2018 Charcoal from anaerobically digested dairy fiber for removal of hydrogen sulfide within biogas *Waste Manage. (Oxford)* **76** 374–82
[3] Bamdad H and Hawboldt K 2016 Comparative study between physicochemical characterization of biochar and metal organic frameworks (MOFs) as gas adsorbents *Can. J. Chem. Eng.* **94** 2114–20
[4] Bamdad H, Hawboldt K, MacQuarrie S and Papari S 2019 Application of biochar for acid gas removal: experimental and statistical analysis using CO_2 *Environ. Sci. Pollut. Res.* **26** 10902–15
[5] Vikrant K *et al* 2018 Engineered/designer biochar for the removal of phosphate in water and wastewater *Sci. Total Environ.* **616–617** 1242–60
[6] Rasi S 2009 Biogas Composition and Upgrading to Biomethane Saija Rasi Biogas Composition and Upgrading to Biomethane *Academic Dissertation* University of Jyväskylä, Finland
[7] Papurello D, Santarelli M and Fiorilli S 2018 Physical activation of waste-derived materials for biogas cleaning *Energies* **11** 2338
[8] Paolini V, Petracchini F, Segreto M, Tomassetti L, Naja N and Cecinato A 2018 Environmental impact of biogas: a short review of current knowledge *J. Env. Sci. Health* A **53** 899–906
[9] Kárászová M, Sedláková Z and Izák P 2015 Gas permeation processes in biogas upgrading: a short review *Chemical Papers.* **69** 1277–83
[10] Bamdad H, Hawboldt K and MacQuarrie S 2018 A review on common adsorbents for acid gases removal: focus on biochar *Renew. Sustain. Energy Rev.* **81** 1705–20
[11] Han X, Chen H, Liu Y and Pan J 2020 Study on removal of gaseous hydrogen sulfide based on macroalgae biochars *J. Nat. Gas Sci. Eng.* **73** 103068
[12] Yuan W and Bandosz T J 2007 Removal of hydrogen sulfide from biogas on sludge-derived adsorbents *Fuel* **86** 2736–46
[13] Papurello D, Silvestri S and Lanzini A 2019 Biogas cleaning: trace compounds removal with model validation *Sep. Purif. Technol.* **210** 80–92
[14] Papurello D, Boschetti A, Silvestri S, Khomenko I and Biasioli F 2018 Real-time monitoring of removal of trace compounds with PTR-MS: biochar experimental investigation *Renewable Energy* **125** 344–55
[15] Bamdad H, Hawboldt K and MacQuarrie S 2018 Nitrogen functionalized biochar as a renewable adsorbent for efficient CO_2 removal *Energy Fuels* **32** 11742–8
[16] Lahijani P, Mohammadi M and Mohamed A R 2018 Metal incorporated biochar as a potential adsorbent for high capacity CO_2 capture at ambient condition *J. CO_2 Util.* **26** 281–93
[17] Adnan O, Nomanbhay C and Show 2019 Technologies for biogas upgrading to biomethane: a review *Bioengineering* **6** 92
[18] Kavitha B, Reddy P V L, Kim B, Lee S S, Pandey S K and Kim K H 2018 Benefits and limitations of biochar amendment in agricultural soils: a review *J. Environ. Manage.* **227** 146–54
[19] Mandal S *et al* 2016 Designing advanced biochar products for maximizing greenhouse gas mitigation potential *Crit. Rev. Environ. Sci. Technol.* **46** 1367–401
[20] Zhou Y, Berruti F, Greenhalf C and Henry H A L 2017 Combined effects of biochar amendment, leguminous cover crop addition and snow removal on nitrogen leaching losses and nitrogen retention over winter and subsequent yield of a test crop (*Eruca sativa* L.) *Soil Biol. Biochem.* **114** 220–8
[21] Zhou Y, Berruti F, Greenhalf C, Tian X and Henry H A L 2017 Increased retention of soil nitrogen over winter by biochar application: implications of biochar pyrolysis temperature for plant nitrogen availability *Agri. Ecosyst. Environ.* **236** 61–8

[22] Sial T A *et al* 2019 Contrasting effects of banana peels waste and its biochar on greenhouse gas emissions and soil biochemical properties *Process Safety Environ. Protection* **122** 366–77

[23] Ameloot N, De Neve S, Jegajeevagan K, Yildiz G, Buchan D, Funkuin Y N, Prins W, Bouckaert L and Sleutel S 2013 Short-term CO_2 and N_2O emissions and microbial properties of biochar amended sandy loam soils *Soil Biol. Biochem.* **57** 401–10

[24] Liu Y, Yang M, Wu Y, Wang H, Chen Y and Wu W 2011 Reducing CH_4 and CO_2 emissions from waterlogged paddy soil with biochar *J. Soils Sediments* **11** 930–9

[25] Ramlow M and Cotrufo M F 2018 Woody biochar's greenhouse gas mitigation potential across fertilized and unfertilized agricultural soils and soil moisture regimes *GCB Bioenergy* **10** 108–22

[26] Zhang C *et al* 2019 Biochar for environmental management: mitigating greenhouse gas emissions, contaminant treatment, and potential negative impacts *Chem. Eng. J.* **373** 902–22

[27] Ali I 2014 Water treatment by adsorption columns: evaluation at ground level *Sep. Purif. Rev.* **43** 175–205

[28] Gwenzi W, Chaukura N, Noubactep C and Mukome F N D 2017 Biochar-based water treatment systems as a potential low-cost and sustainable technology for clean water provision *J. Environ. Manage.* **197** 732–49

[29] Hallberg G R 1987 The impacts of agricultural chemicals on ground water quality *GeoJournal* **15** 283–95

[30] Liu X and Zhang L 2015 Removal of phosphate anions using the modified chitosan beads: adsorption kinetic, isotherm and mechanism studies *Powder Technol.* **277** 112–9

[31] Foglar L, Briški F, Sipos L and Vuković M 2005 High nitrate removal from synthetic wastewater with the mixed bacterial culture *Bioresour. Technol.* **96** 879–88

[32] Xu K, Lin F, Dou X, Zheng M, Tan W and Wang C 2018 Recovery of ammonium and phosphate from urine as value-added fertilizer using wood waste biochar loaded with magnesium oxides *J. Clean. Prod.* **187** 205–14

[33] Peel J W, Reddy K J, Sullivan B P and Bowen J M 2003 Electrocatalytic reduction of nitrate in water *Water Res.* **37** 2512–9

[34] Fidel R B, Laird D A and Spokas K A 2018 Sorption of ammonium and nitrate to biochars is electrostatic and pH-dependent *Sci. Rep.* **8** 17627

[35] Lawrinenko M and Laird D A 2015 Anion exchange capacity of biochar *Green Chem.* **17** 4628–36

[36] Xu X, Zheng Y, Gao B and Cao X 2019 N-doped biochar synthesized by a facile ball-milling method for enhanced sorption of CO_2 and reactive red *Chem. Eng. J.* **368** 564–72

[37] Vijayaraghavan K and Yun Y-S 2008 Bacterial biosorbents and biosorption *Biotechnol. Adv.* **26** 266–91

[38] Rivera-Utrilla J and Sánchez-Polo M 2003 Adsorption of Cr(III) on ozonised activated carbon. Importance of $C\pi$–cation interactions *Water Res.* **37** 3335–40

[39] Aggarwal D, Goyal M and Bansal R C 1999 Adsorption of chromium by activated carbon from aqueous solution *Carbon* **37** 1989–97

[40] Bautista-Toledo I, Rivera-Utrilla J, Ferro-García M A and Moreno-Castilla C 1994 Influence of the oxygen surface complexes of activated carbons on the adsorption of chromium ions from aqueous solutions: effect of sodium chloride and humic acid *Carbon* **32** 93–100

[41] Shim J-W, Park S-J and Ryu S-K 2001 Effect of modification with HNO_3 and NaOH on metal adsorption by pitch-based activated carbon fibers *Carbon* **39** 1635–42

[42] Montes-Morán M A, Menéndez J A, Fuente E and Suárez D Jul. 1998 Contribution of the basal planes to carbon basicity: an *ab initio* study of the $H_3O^+-\pi$ interaction in cluster models *J. Phys. Chem.* B **102** 5595–601

[43] Kolpin D W *et al* 2002 Pharmaceuticals, hormones and other organic wastewater contaminants in US. streams, 1999–2000: a national reconnaissance *Environ. Sci. Technol.* **36** 1202–11

[44] Ternes T A, Stumpf M, Mueller J, Haberer K and Servos M 1999 Behavior and occurrence of estrogens in municipal sewage treatment plants—I. Investigations in Germany, Canada and Brazil *Sci.Total Env.* **225** 81–90

[45] Székács A, Mörtl M and Darvas B Monitoring pesticide residues in surface and ground water in Hungary: surveys in 1990–2015 *J. Chem.* **2015** 717948

[46] Jung C *et al* 2013 Adsorption of selected endocrine disrupting compounds and pharmaceuticals on activated biochars *J. Hazard. Mater.* **263** 702–10

[47] Essandoh M, Kunwar B, Pittman C U, Mohan D and Mlsna T 2015 Sorptive removal of salicylic acid and ibuprofen from aqueous solutions using pine wood fast pyrolysis biochar *Chem. Eng. J.* **265** 219–27

[48] Zhu D, Hyun S, Pignatello J J and Lee L S 2004 Evidence for $\pi-\pi$ electron donor–acceptor interactions between π-donor aromatic compounds and π-acceptor sites in soil organic matter through pH effects on sorption *Environ. Sci. Technol.* **38** 4361–8

[49] Liu L, Li W, Song W and Guo M 2018 Remediation techniques for heavy metal-contaminated soils: principles and applicability *Sci. Total Environ.* **633** 206–19

[50] Gao R, Hu H, Fu Q, Li Z, Xing Z, Ali U, Zhu J and Liu Y 2020 Remediation of Pb, Cd, and Cu contaminated soil by co-pyrolysis biochar derived from rape straw and orthophosphate: speciation transformation, risk evaluation and mechanism inquiry *Sci. Total Env.* **730** 139119

[51] Zhang W, Tan X, Gu Y, Liu S, Liu Y, Hu X, Li J, Zhou Y, Liu S and He Y 2020 Rice waste biochars produced at different pyrolysis temperatures for arsenic and cadmium abatement and detoxification in sediment *Chemosphere* **250** 126268

[52] Puga A P, Abreu C A, Melo L C A, Paz-Ferreiro J and Beesley L 2015 Cadmium, lead, and zinc mobility and plant uptake in a mine soil amended with sugarcane straw biochar *Environ. Sci. Pollution Res.* **22** 17606–14

[53] Shen X, Zeng J, Zhang D, Wang F, Li Y and Yi W 2020 Effect of pyrolysis temperature on characteristics, chemical speciation and environmental risk of Cr, Mn, Cu, and Zn in biochars derived from pig manure *Sci. Total Environ.* **704** 135283

[54] Dai Y, Wang W, Lu L, Yan L and Yu D 2020 Utilization of biochar for the removal of nitrogen and phosphorus *J. Clean. Prod.* **257** 120573

[55] García-Jaramillo M, Trippe K M, Helmus R, Knicker H E, Cox L, Hermosín M C, Parsons J R and Kalbitz K 2020 An examination of the role of biochar and biochar water-extractable substances on the sorption of ionizable herbicides in rice paddy soils *Sci. Total Environ.* **706** 135682

[56] Huang S, Bao J, Shan M, Qin H, Wang H, Yu X, Chen J and Xu Q 2018 Dynamic changes of polychlorinated biphenyls (PCBs) degradation and adsorption to biochar as affected by soil organic carbon content *Chemosphere* **211** 120–7

[57] Oleszczuk P, Godlewska P, Reible D D and Kraska P 2017 Bioaccessibility of polycyclic aromatic hydrocarbons in activated carbon or biochar amended vegetated (*Salix viminalis*) soil *Environ. Pollut.* **227** 406–13

Chapter 16

Applications of biochar catalysts

Naomi Klinghoffer

Biochar has been shown to be an effective and low cost catalyst for a variety of applications within the energy industry. Biochar catalysts can replace conventional catalysts with a renewable, low cost alternative, while increasing product yield and selectivity, or decreasing the energy inputs required. This chapter discusses the biochar properties which give rise to its catalytic activity, such as its surface area, porosity, surface functional groups, and the presence of metals in the char. Furthermore, biochar's catalytic activity can be enhanced by activation and functionalization of its surface, or via addition of catalytically active metals. This chapter discusses the use of biochar catalysts in specific applications, including tar removal, biodiesel production, NO_x reduction, electrochemical applications (microbial fuel cells, electro-oxidation of methanol), bio-oil upgrading, and biomass hydrolysis. Once the biochar catalyst is spent, it can be used in other applications such as soil remediation, creating a zero-waste process supporting a circular economy.

16.1 Introduction

Catalysts play a critical role in many important areas including fuel production, pharmaceuticals, transportation, and industrial processes. As a result, there is a large market for catalysts. Biochar has catalytic activity for a variety of these applications, many of which are within the energy industry. Conventional catalysts can be expensive and often experience deactivation, which results in process disruptions and increased cost. However, biochar is inexpensive and can often be a good substitute for conventional catalysts. As will be discussed in this chapter, the physical and chemical properties of biochar can be modified to make it a suitable catalyst for a variety of important applications.

doi:10.1088/978-0-7503-2660-5ch16

16.2 Biochar properties

In order to understand the role of biochar as a catalyst, it is important to first understand the basic principles of catalysis. Catalysts play an important role in many chemical processes, and their structure, chemistry, and bulk properties influence the catalyst performance. Typically, catalysts are highly porous materials with a high surface area in order to increase the number of active sites. Catalysts can consist of bulk materials with catalytic properties, or catalytically active materials that are dispersed onto a high surface area support.

During heterogeneous catalytic reactions, the reactants must diffuse to the surface of the catalyst, and subsequently diffuse through the pores of the catalytic material, as shown in figure 16.1. Reactants then adsorb to the active site on the catalyst surface, and the reaction takes place. Products then must desorb from the surface, diffuse through the pores, and then undergo bulk diffusion into the reaction environment. These steps are shown in figure 16.1.

Looking at this model, it is obvious that both the char morphology and its surface chemistry play an important role in determining the catalytic activity of the char. The surface chemistry determines the reactions that can be catalyzed, while the morphology influences the accessibility of these sites. For example, biochar that is rich in catalytic metals or catalytically active functional groups but contains very narrow or tortuous pore structures may have reduced catalytic activity due to limited accessibility to active sites. Alternatively, such biochar may have high catalytic activity for reactions involving small molecules, but may not be appropriate for reactions involving larger molecules. This section discusses the critical role that biochar properties, such as morphology (surface area and porosity), composition, and surface functionalities play in its catalytic activity. Other properties which can influence the catalytic activity of char are its bulk properties, such as particle size or addition of binders or supports.

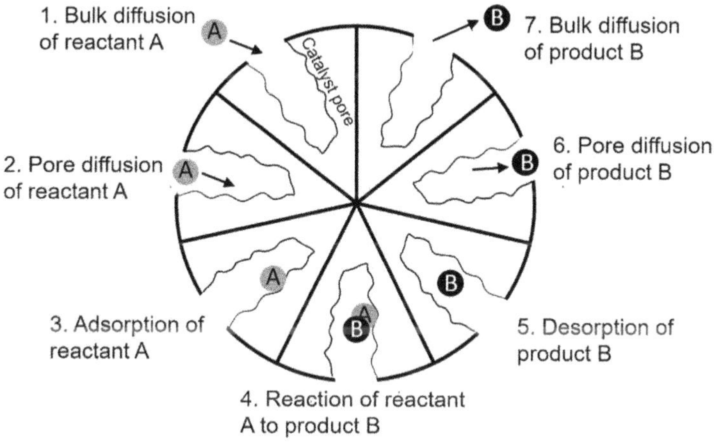

Figure 16.1. Steps involved in heterogeneous catalytic reactions.

16.2.1 The role of biochar morphology in catalytic activity

A primary reason why biochar is a good catalyst is due to its high surface area and porosity. A high surface area is important for achieving good catalytic activity, as it creates a high density of active sites per unit volume. However, the pore structure must also be considered, as the pore size can influence the accessibility of the active sites. Biochar surface area and porosity are highly dependent on the feedstock and on the conditions in which the biochar is produced. Factors such as reaction environment, temperature, residence time, and heating rate all influence the structure of biochar.

Biochar surface area is typically reported as BET surface area and can range from $<1\ m^2\ g^{-1}$ to $>1500\ m^2\ g^{-1}$. Pore volume is highly variable and in some cases may be negligible, or can be higher than $0.9\ cm^3\ g^{-1}$. In general, higher reaction temperature results in increased surface area and porosity [1]. Residence time also influences biochar structure, where increased residence time can lead to higher surface area and porosity [2]. Biochar from slow pyrolysis has been found to be more porous than from fast pyrolysis [1]. There is often a trade off between surface area and yield where biochar with a high yield has low surface area. This is because much of the pore structure develops as the biomass decomposition reactions take place.

Surface area is also influenced by reactants used during the production of biochar. For example, biochar produced under N_2 generally has a lower surface area compared to biochar produced in the presence of steam. When biochar is produced in the presence of CO_2 it can further enhance the biochar surface area. As a result, biochar produced in gasification processes, which typically contain an oxidizer such as O_2, CO_2, or H_2O, generally produces higher surface area materials compared to biochar produced during pyrolysis with N_2. While we typically observe an increase in both surface area and porosity, in some cases, co-reactants can influence these properties in different ways. The catalytic activity of biochar samples that were produced under either steam or CO_2 were compared. While the two samples had similar surface areas, the biochar produced under CO_2 had a higher micropore volume, and consequently exhibited lower catalytic activity which was attributed to diffusion limitations within the micropores [3]. As a result, when evaluating biochar materials for use in catalytic applications, both surface area and pore size must be considered while taking into account properties of the reactant molecules (such as mean free path under the desired reaction conditions). One benefit of using biochar is that even after initial production (via pyrolysis or gasification) the biochar surface properties can be modified to improve activity, as discussed in section 16.3.

16.2.2 The role of biochar composition in catalytic activity

Biochar is primarily composed of carbon, oxygen, hydrogen, and various inorganic components (minerals) that are inherently present in the biomass feedstock. The inorganic fraction of biomass is primarily composed of silica and alkali and alkaline earth metals (AAEM) such as Na, K, Ca, and Mg [4]. Biomass also contains Al, Fe, Mn, and P in smaller quantities relative to AAEM species. In many applications, the catalytic activity of biochar is influenced by both the inorganic and organic fractions of the biochar [3]. The quantity of inorganics, type of inorganics, and dispersion of inorganics on the biochar surface are factors which determine catalytic activity. While some biochar properties, such as surface area and pH are primarily dependent on reaction conditions such as temperature, the mineral content of biochar is very closely related to the feedstock that is used to produce the biochar [5]. However, overall mineral content in biochar is still affected by reaction conditions. Mineral content is inversely related to biomass conversion since ash typically has a lower volatility than organic matter and therefore remains in the biochar, whereas the carbon, oxygen, and hydrogen fraction in the biochar will decrease as conversion increases.

The catalytic activity of biochar catalysts is attributed to both the carbon structure and the AAEM species in the biochar. Catalytic activity of AAEM species during gasification reactions has been studied, and is a good indicator of the catalytic role of these species for other applications as well. The most abundant ash species that enhance catalytic activity of biochar are K and Ca, with K typically having a more significant influence on catalyzing gasification reactions. On the other hand, silica and phosphorus have been shown to decrease the catalytic activity of biochar during gasification [6]. Therefore, the presence of inorganics is sometimes expressed as $K/(Si + P)$ or $(Ca + K)/(Si + P)$ in order to reflect the overall influence of inorganic species on catalytic activity of the biochar. Fe can also have catalytic activity, but is only present in very low quantities.

When catalysts are synthesized, active sites should be well dispersed on a support in order to increase catalytic activity. Therefore in biochar catalysts it is desirable for catalytic elements to be well dispersed on the surface. This has not been widely studied, as it can be difficult to quantify dispersion due to the heterogeneous nature of the biochar. However, the conditions under which the biochar is produced, such as temperature, co-reactants, and heating rate can influence the dispersion of elements on the surface [7].

Catalytic activity of biochar has been demonstrated for biochar with and without AAEM species (i.e. with AAEM species removed), highlighting the role of both the inorganic species and the organic fraction in the catalytic activity of biochar [8]. In some cases, synergistic effects have been observed between the metals and the carbon support, further emphasizing the critical role that both play in the catalytic activity of the biochar [1].

16.2.3 Role of surface functionalities in catalytic activity

Biochar surface contains functional groups which can influence its catalytic activity for low temperature applications. Acidic functional groups, such as lactones phenols, and carboxylic groups as well as basic functional groups such as carbonyl, quinone, chromene, hydroxyl, and pyrone structures have been observed [3, 9]. The type and quantity of functional groups on the biochar surface depend on the biomass feedstock as well as the conditions under which the biochar is produced. Acidic sites are created when a surface is heated in an oxidizing environment whereas basic sites are created when an oxidized surface is heated in an inert environment [10]. During pyrolysis, higher temperatures result in lower H/C and O/C ratios in the biochar [11]. As a result, biochar produced at higher temperature typically has fewer functional groups on its surface. For example, during microwave pyrolysis, a higher reaction temperature resulted in fewer basic functional groups [11].

Biochar functional groups influence the performance of char for a variety of applications, such as in soil amendment, wastewater treatment, inorganic pollutant removal, and catalytic applications [12–14]. Hydrophilic functional groups can increase the adsorption of hydrophilic reactants, such as ethanol, which is important in biodiesel production (section 16.4.2) [1]. Surface functionalities can influence the electrical conductivity of biochar, which is an important property when biochar is used in electrochemical applications (section 16.4.4). For example, for rice straw derived biochar, the electrical conductivity was negatively correlated with aliphatic O-alkylated (HCOH) carbons, but was positively correlated with fused ring aromatic structures and aromatic C–O groups [15].

16.3 Modified biochar catalysts

16.3.1 Biochar as a catalyst support

While biochar can be used directly as a catalyst, it has also been used as a support for other catalytically active metals. This includes the addition of metals that are already present in the biochar, such as Ca and K, as well as metals that are either not present or present in trace amounts, such as Fe or Ni [16–18]. This approach leverages the benefits of biochar, such as its high surface area, complex pore structure, and tunability while expanding the applications for biochar based catalysts.

Biochar based catalysts are typically synthesized via a wet impregnation technique where metal complexes in aqueous solution are introduced onto the char surface and subsequently heated to ca. 300 °C–600 °C for calcination. For example, a Ni catalyst would be synthesized by impregnating biochar with nickel nitrate, followed by calcination to produce a surface that contains highly dispersed nickel nanoparticles. This technique is relatively straightforward and inexpensive, while producing catalysts with high dispersion of metals on the surface. Metal loading can range from ca. 4% up to 18%. Some common applications that employ supported catalysts are tar decomposition and NO_x removal, which are discussed in detail in sections 16.4.1 and 16.4.3.

16.3.2 Activation and functionalization of biochar catalysts

While biochar's catalytic activity can be enhanced by the addition of catalytically active metals, another approach is to activate biochar by enhancing either its surface area or surface functionalities. This avoids the need for the utilization of metallic species, and has less of an impact on subsequent utilization of spent biochar catalysts (for example, in soil amendment) because the composition of the biochar is not substantially changed. Furthermore, the properties of biochar arising from different processes (pyrolysis, gasification, etc) and feedstocks can be quite variable and the approaches presented here can help to create a more uniform surface from heterogeneous feedstocks.

The activation of biochar can be done via physical or chemical activation, and can significantly increase the surface area and pore volume, enabling the production of biochar with a surface area greater than 1500 $m^2 g^{-1}$ [19]. Physical activation involves heating the biochar to high temperatures (typically >700 °C) in the presence of CO_2, steam, or a combination of these. During physical activation, unstructured parts of the biochar decompose, exposing fine pores within the biochar structure [20]. These pores then expand, creating larger pores. The reaction kinetics are often influenced by the presence of inorganics in the biochar [20]. Increasing the temperature or amount of oxidant results in higher BET surface area but lower biochar yield [20, 21]. However, if the extent of reaction is too great, this can lead to lower BET surface area and pore volume due to pore collapse. For example, an increase in BET surface area has been observed up to 900 °C, at which point surface area and micropore volume started to decrease [19]. Similarly, increasing activation time resulted in higher surface area up to 5 h, after which it began to decrease.

Chemical activation is typically done with potassium hydroxide (KOH), phosphoric acid, or zinc chloride [21]. This reaction, which is typically done at room temperature or low temperature, has been shown to significantly increase surface area and porosity. For example, when biochar was activated with KOH followed by HCl, its surface area increased from 1.66 $m^2 g^{-1}$ to 990 $m^2 g^{-1}$ [22]. The porosity, which was initially not measurable, was 0.90 $cm^3 g^{-1}$ in the activated biochar.

Surface functionalization can be done by acid treatment or thermal treatment. The most common types of functional groups that are added to the biochar surface are oxygen based functional groups or sulfonic acid groups (SO_3H). Sulfonation of the biochar surface is necessary to generate biochar catalysts for biodiesel production (section 16.4.2). Sulfonation can be done by exposing biochar to SO_3 gas at elevated temperatures (25 °C–150 °C) or by immersion in sulfuric acid solution (H_2SO_4) at 90 °C–200 °C [1]. The former approach has been shown to generate higher acid density on the surface, while the latter increases the surface area and pore volume of the biochar [23].

Oxygen based functional groups can be added to the surface via acid treatment or thermal treatment. During thermal treatment, biochar is exposed to an oxygen containing stream at elevated temperatures, resulting in the addition of functional groups on the surface. For example, exposing biochar to a stream containing 8% O_2 and 92 % N_2 at 280 °C–400 °C can increase weak acid functionalities on the biochar surface without removing basic functional groups [9]. At lower temperatures (280 °C) oxygenation resulted in a decrease in surface area while higher temperatures (400 °C) resulted in an increase in surface area as well as increase in oxygen functional groups [9]. Alternatively, functional groups can be added by wet oxidation, by immersing biochar in HNO_3 and H_2SO_4 and undergoing mild heating.

Surface functionalities can also be removed from the biochar surface. Thermal treatment of biochar (for example, heating in N_2 at 800 °C) can cause functional groups to desorb from the surface, resulting in a reduction in surface functional groups [3], while increasing surface area [12]. On the other hand, in some cases the addition of functional groups to the surface can decrease surface area due to the bulky nature of the functional groups in the micropores [3].

These approaches to activation and functionalization of biochar enable the biochar surface to be tuned in order to enhance its catalytic activity and enable its application in new processes where it may otherwise not have sufficient activity.

16.4 Applications for biochar catalysts

16.4.1 Tar removal

One promising route for converting biomass or other solid feedstocks (such as plastics, mixed waste, or coal) into chemicals and fuels is through gasification. Gasification produces synthesis gas ($CO + H_2$) as a primary product which is used to synthesize other chemicals such as methanol, ethanol, synthetic natural gas, gasoline, or dimethyl ether. Synthesis gas can also be used for energy production via combustion in a gas turbine, or in a fuel cell (sometimes in combination with a water gas shift process to increase H_2 content). The gasification process also produces biochar and tar, which is a mixture of condensable hydrocarbons, composed of phenols, olefins, and polycyclic aromatic compounds [24]. Many applications for synthesis gas require tars to be removed. For example, processes that catalytically synthesize chemicals require syngas cleaning as tars can cause catalyst deactivation. If synthesis gas is used for electricity production in gas turbines, tars will also have to be removed in order to meet specifications for turbine feed composition. As a result, catalytic processes have been developed in order to decrease tar content in the outlet streams of gasifiers.

Tar removal can be done either by tar cracking or via steam reforming, and is typically done at 600 °C–900 °C [18, 25]. Commercial catalysts that are commonly used for tar removal are olivine, dolomite, nickel, or alkali metal based catalysts [24]. However, these catalysts typically experience deactivation, for example via coking (carbon deposition) which can block pores and active sites or due to agglomeration of metals. Biochar can be an effective and inexpensive alternative to conventional commercial catalysts for decomposition of tar from gasification processes. It is also beneficial to use biochar catalysts because the biochar is produced from biomass, which is already a primary feedstock in the gasification process. Furthermore, it is likely that once the biochar catalyst deactivates, it can still be used in other applications, resulting in a zero-waste process.

Biochar catalysts that have been used for tar decomposition have been generated from many different feedstocks such as switchgrass, wood, palm kernel shell, and rice husk shell [25, 26]. Tar conversion increases with temperature, and biochar surface area [7, 18]. In some cases, inorganics have been shown to improve the catalytic activity of char for this application [3]. Char derived from municipal solid waste (MSW) and biomass has been used to reform tars produced from gasification of wood chips and MSW and demonstrated 85% conversion of primary tars to permanent gases [27].

Biochar has also been used as a catalyst support for tar reduction. Rice husk biochar that was loaded with Ni and Ni–Fe catalysts showed 93% conversion of tar [16]. When iron supported catalysts were used, the activation energy for decomposition of toluene, a tar surrogate molecule, was lowered by 47% compared to using biochar alone [18]. There is a trade off between cost/complexity and performance where utilization of biochar supported catalysts is more expensive but can often generate higher conversion. Furthermore, opportunities for serial utilization of the biochar may be reduced if biochar is loaded with additional metals. Ultimately, this trade off needs to be analyzed on a case by case basis to determine the optimal solution for each application. Catalyst stability remains a challenge and future work should focus on improving the stability of biochar catalysts in tar removal applications [8]. Biochar catalysts (without the addition of metals) have lower activity compared to olivine catalysts, so their catalytic activity should be improved to be a competitive alternative [28].

16.4.2 Biodiesel production

Biodiesel is an attractive replacement for fossil derived diesel and has great potential to reduce CO_2 emissions, while enabling the use of energy-dense liquid fuels that are compatible with our energy infrastructure. Biodiesel has similar properties to fossil derived diesel and is biodegradable, non-toxic, and renewable. It is produced from vegetable oils or animal fat via transesterification of glycerides with low molecular weight alcohols such as methanol or ethanol, as shown in figure 16.2. Typically methanol is used as the alcohol in these reactions. Feedstocks containing high amounts of free fatty acids (FFAs) will undergo esterification to produce esters and water. Some acid catalysts can catalyze both esterification and transesterification reactions simultaneously [22].

A common approach to the production of biodiesel is via homogeneous alkali or acid catalysts. However, this process requires a high purity feed and additional separation steps to recover the catalyst from the process, which increases the overall cost [29]. Heterogeneous catalysts are much easier to separate, making them a desirable alternative to homogeneous catalysts. Catalysts containing Al_2O_3, TiO_2, ZrO_2, and WO_3 or zeolites can be used in heterogeneous catalytic processes to produce biodiesel [30]. However, biochar has been shown to be an effective and less expensive catalyst for biodiesel production [31]. Biochar derived from biomass feedstocks such as palm kernel shell, oat hull, coconut shell, wood, and peanut hull have been used for biodiesel production [1, 29]. Conversion rates of up to 95% have been observed [32]. When producing biodiesel, acid sites on the catalyst are important. Therefore, biochar is sulfonated with H_2SO_4 or with SO_3 in order to improve its acid strength and consequently its catalytic activity [29]. As with other catalytic applications, the catalyst should have high surface area to increase the

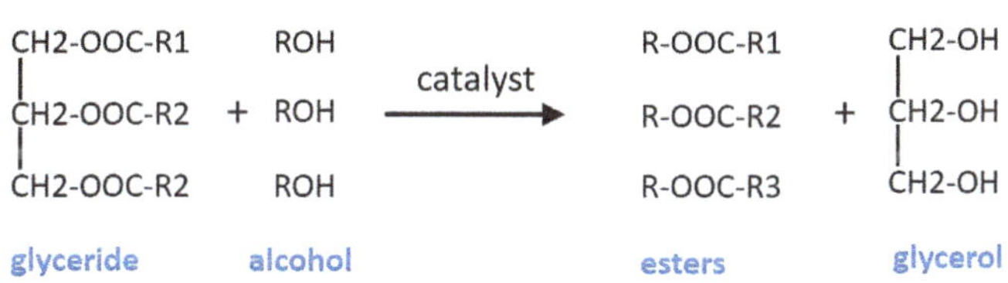

Figure 16.2. Reactions showing transesterification of glyceride using alcohol to produce esters (biodiesel) and glycerol with a sulfonated biochar catalyst.

number of available active sites. While high conversion can be achieved, catalyst stability can be a challenge. Deactivation can occur as a result of impurities in the feedstock which can poison the catalyst or via leaching of $-SO_3H$, whereby $-SO_3H$ functionalities on the biochar surface are hydrated and then leach as H_2SO_4. As a result, there is a need to generate catalysts with good stability.

16.4.3 NO_x removal

NO_x, which refers to NO and NO_2, is a pollutant that is produced through combustion of fuel and is generated via the reaction of oxygen with nitrogen at high temperatures. NO_x emissions can result in a number of harmful environmental effects, such as the production of smog or acid rain (from production of nitric acid). Furthermore, NO_x can react with volatile organic compounds in the presence of sunlight to produce ozone which can cause health issues. Some of the largest emission sources of NO_x are related to automobiles, trucks, power generation, and industrial processes such as cement production.

A common approach to NO_x removal is through selective catalytic reduction (SCR). In this process, NH_3 reacts selectively with NO_x and O_2 to produce N_2 and water, as shown in equations (16.1) and (16.2). Typically, SCR uses a V_2O_5/TiO_2 catalyst which can be modified with WO_3 or MoO_3 [33]. These catalysts operate at temperatures in the range of 300 °C–400 °C. Zeolites can be effective catalysts for removal of NO_x, but require temperatures greater than 400 °C. Platinum catalysts can be used at lower temperatures (around 300 °C), but are very expensive.

$$4NH_3 + 4NO + O_2 \rightarrow 4N_2 + 6H_2O \qquad (16.1)$$

$$4NH_3 + 2NO_2 + O_2 \rightarrow 3N_2 + 6H_2O. \qquad (16.2)$$

Biochar has the potential to be used as an SCR catalyst for NO_x removal. Furthermore, it has demonstrated good catalytic activity at temperatures below 300 °C, which is lower than the operating temperature of conventional catalysts. This is beneficial because it allows SCR to take place downstream of flue gas clean-up steps (such as desulfurization and particulate removal), without the need to reheat the flue gas to high temperatures. Biochar catalysts derived from feedstocks such as sewage sludge, municipal solid waste, refuse derived fuel, cotton stalk, and rice husk have been used for NO_x removal. In some cases, these catalysts have achieved over 80% removal efficiency [24, 34, 35]. Beyond using biochar directly as a catalyst, biochar can be activated to enhance its performance. Because NO_x reduction takes place at relatively low temperatures (below 300 °C), surface functional groups play an important role it the activity of the biochar catalyst. Therefore, surface functional groups can be added to the biochar (as discussed in section 16.3.2) in order to increase NO_x removal rates [34]. Other approaches to biochar activation, such as physical and chemical activation may increase its surface area and/or pore volume. However, for NO_x removal, the role of surface functional groups has a bigger impact on performance than biochar morphology [34].

Biochar can also be used as a catalyst support for NO_x reduction. This has the benefit of enabling lower temperature operation compared to conventional catalysts while achieving higher performance than biochar-only catalysts. For example, Mn–CeO_x can be loaded onto biochar in order to produce an SCR catalyst which has demonstrated >90% removal of NO and >65% removal of NO_2 [26, 36]. The activity of the Mn–CeO_x/biochar catalyst was attributed to its high specific surface area, porosity, surface acidity, dispersion of Mn–CeO_x on the surface, as well as surface functional groups [26].

16.4.4 Electrochemical applications

This section discusses two types of electrochemical applications in which biochar based catalysts can be used: oxygen reduction reactions (ORR) in microbial fuel cells (MFCs) and methanol electro-oxidation in direct methanol fuel cells (DMFCs).

MFCs enable the production of electricity from biomass using bacteria [37]. At the anode of an MFC, organic matter is degraded by micro-organisms, generating electrons and protons. At the cathode, oxygen is introduced and combined with electrons and protons to form water via the ORR. As a result, MFCs can simultaneously degrade organic matter and generate electricity.

Platinum (Pt) catalysts are commonly used in microbial fuel cells. However, these can be very expensive. Catalysts that are not metal based, such as biochar, can be beneficial due to lower costs, better stability, and can suffer less from crossover effects [38]. Sewage sludge biochar has been used as a catalyst for ORR in a microbial fuel cell, and demonstrated a maximum power density of 500 mW m^{-2}, which is comparable to that of a Pt cathode. The high catalytic activity was attributed to the high surface area and elevated N and Fe content [38]. This catalyst also demonstrated good stability during >50 cycles.

Biochar can also be used as a catalyst support for methanol electro-oxidation, which forms the basis of direct methanol fuel cells (DMFCs), as shown in figure 16.3. DMFCs enable the production of electricity from methanol, as shown in equations (16.3) and (16.4). DMFCs are compact, have high energy density, operate at low temperatures, and directly use methanol as a fuel, which can be produced from renewable resources and from CO_2. This is in contrast with proton exchange

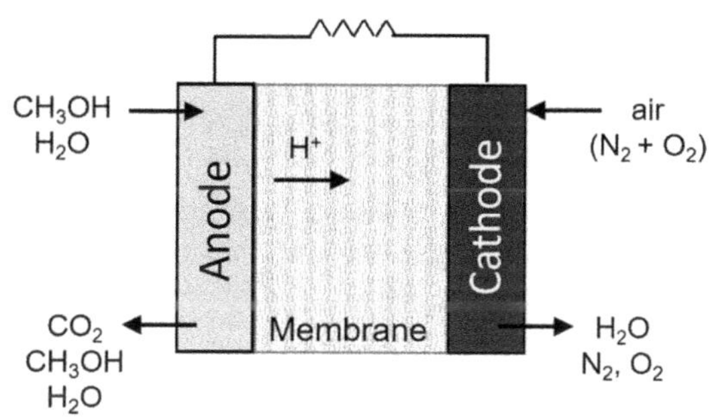

Figure 16.3. Direct methanol fuel cell operation. Biochar can be used as a catalyst for methanol electro-oxidation in a direct methanol fuel cell.

membrane (PEM) fuel cells which use H_2 as a fuel, therefore requiring either H_2 storage, or an additional steam reforming step in order to produce H_2 from methanol or other feedstocks. Typically, DMFCs use carbon supported Pt catalysts. Carbon is a good catalyst support for this application due to its stability, electronic conductivity, resistance to acidic and basic environments, and high surface area [39]. Biochar derived from pyrolysis of cellulose loaded with Cu–Ru@Pt core shell catalysts has been used as a catalyst for methanol electro-oxidation. In another case, pectin was hydrothermally carbonized, activated with KOH, and used as a support for a Pt nanowire catalyst [40]. The high performance was attributed to the high specific surface area of the biochar, as well as the high amount of surface defects on the biochar surface, which supports a smaller average particle size of the catalytic material. It is well known that a smaller catalyst particle size increases the number of active sites, thus increasing the overall product yield. Pectin supported catalysts not only demonstrated good performance compared to conventional Pt/C catalysts, but also exhibited good resistance to CO poisoning, which is a major challenge with fuel cells. The resistance to CO poisoning was attributed to surface hydroxyl and carboxyl groups.

$$\text{Anode: } CH_3OH + H_2O \rightarrow CO_2 + 6H^+ + 6e^- \tag{16.3}$$

$$\text{Cathode: } 1.5\ O_2 + 6H^+ + 6e^- \rightarrow 3H_2O. \tag{16.4}$$

16.4.5 Bio-oil upgrading and biomass hydrolysis

Hemicellulose, which is extracted from biomass in an integrated forest biorefinery, can be used as a chemical building block after it undergoes hydrolysis to monomeric carbohydrates [41]. For this application, biochar is sulfonated (as discussed in section 16.3.2) to generate an active catalyst. Biochar catalysts have demonstrated significantly higher activity than activated carbon for this application, achieving up to 85% conversion using xylan as a representative molecule [41]. Sulfonated corn stover derived biochar was used for treatment of lignocellulosic biomass to produce glucose and xylose [42]. The conversion rates were comparable to hydrolysis of model compounds over the same catalyst, indicating that sulfonated biochar catalysts can maintain high performance even in the presence of complex biomass feedstocks [24].

Ni/biochar catalysts using biochar produced from pyrolysis of microalgae have been used for bio-oil upgrading including hydrodeoxygenation and hydrodenitrogenation reactions [43]. This has the benefit of being a closed loop process, as the biochar generated from pyrolysis is used to upgrade the oil produced within the same process. The primary product generated from the process was *n*-heptadecane, which is an important component of diesel fuel.

16.5 Conclusions and outlook

This chapter has discussed the biochar properties that give rise to its catalytic activity, as well as applications in which biochar is an effective catalyst. Biochar is a highly tunable material and as a result can be modified to be an active catalyst for many different applications. A common challenge among these applications relates to catalyst stability, and this presents a focus area for future work to improve the opportunities for biochar utilization. However, biochar has the unique benefit that it is a valuable material on its own, with various applications in environmental remediation, soil amendment, and removal of inorganic, organic, and gaseous pollutants. This means that even spent biochar catalysts can have value if they can be applied to these other applications. As a result, future work should investigate approaches both for decreasing catalyst deactivation and for utilization of spent biochar catalysts. While this chapter has addressed several applications for biochar catalysts, this unique and tunable material has vast possibilities and we can expect to see the opportunities for biochar based catalysts continue to expand.

References

[1] Cheng F and Li X 2018 Preparation and application of biochar-based catalysts for biofuel production *Catalysts* **346** 1–35

[2] Dutta B, Raghavan G S V and Ngadi M 2012 Surface characterization and classification of slow and fast pyrolyzed biochar using novel methods of pycnometry and hyperspectral imaging *J. Wood Chem. Technol.* **32** 105–20

[3] Klinghoffer N B, Castaldi M J and Nzihou A 2015 Influence of char composition and inorganics on catalytic activity of char from biomass gasification *Fuel* **157** 37–47

[4] Nanda S, Dalai A K, Berruti F and Kozinski J A 2016 Biochar as an exceptional bioresource for energy, agronomy, carbon sequestration, activated carbon and specialty materials *Waste Biomass Valorization* **7** 201–35

[5] Zhao L, Cao X, Masek O and Zimmerman A 2013 Heterogeneity of biochar properties as a function of feedstock sources and production temperatures *J. Hazard. Mater.* **256–57** 1–9

[6] Dupont C, Jacob S, Marrakchy K O, Hognon C, Grateau M, Labalette F and Da Silva Perez D 2016 How inorganic elements of biomass influence char steam gasification kinetics *Energy* **109** 430–35

[7] Klinghoffer N, Castaldi M J and Nzihou A 2012 Catalyst properties and catalytic performance of char from biomass gasification *Ind. Eng. Chem. Res.* **51** 13113–22

[8] Khiari B, Jeguirim M, Limousy L and Bennici S 2019 Biomass derived chars for energy applications *Renew. Sustain. Energy Rev.* **108** 253–73

[9] Ducousso M, Weiss-Hortala E, Nzihou A and Castaldi M J 2015 Reactivity enhancement of gasification biochars for catalytic applications *Fuel* **159** 491–9

[10] Serp P and Figueiredo J L 2009 *Carbon Materials for Catalysis* (Hoboken, NJ: Wiley)

[11] Zhu L, Lei H, Wang L, Yadavalli G, Zhang X, Wei Y, Liu Y, Yan D, Chen S and Ahring B 2015 Biochar of corn stover: microwave-assisted pyrolysis condition induced changes in surface functional groups and characteristics *J. Anal. Appl. Pyrolysis* **115** 149–56

[12] de Mendonça F G, da Cunha I T, Soares R R, Tristão J C and Lago R M 2017 Tuning the surface properties of biochar by thermal treatment *Bioresour. Technol.* **246** 28–33

[13] Oliveira F R, Patela A K, Jaisib D P, Adhikaric S, Lud H and Khanala S K 2017 Environmental application of biochar: current status and perspectives *Biores. Technol.* **246** 110–22

[14] Bartoli M, Giorcelli M, Jagdale P, Rovere M and Tagliaferro A 2020 A review of non-soil biochar applications *Materials* **13** 1–35

[15] Li X, Shen Q, Zhang D, Mei X, Ran W, Xu Y and Yu G 2013 Functional groups determine biochar properties (pH and EC) as studied by two-dimensional 13C NMR correlation spectroscopy *PLoS One* **8** e65949

[16] Shen Y, Zhao P, Shao Q, Takahashi D M F and Yoshikawa K 2014 *In-situ* catalytic conversion of tar using rice husk char-supported nickel–iron catalysts for biomass pyrolysis/gasification *Appl. Catalysis* B **152–153** 140–51

[17] Feng D, Zhao Y, Zhang Y, Sun S, Meng S, Guo Y and Huang Y 2016 Effects of K and Ca on reforming of model tar compounds with pyrolysis biochars under H_2O or CO_2 *Chem. Eng. J.* **306** 422–32

[18] Kastner J R, Mani S and Juneja A 2015 Catalytic decomposition of tar using iron supported biochar *Fuel Process. Technol.* **130** 31–7

[19] Guo S, Peng J, Li W, Yang K, Zhang L, Zhang S and Xia H 2009 Effects of CO_2 activation on porous structures of coconut shell-based activated carbons *Appl. Surf. Sci.* **255** 8443–9

[20] Cha J S, Park S H, Jung S-C, Ryu C, Jeon J-K, Shin M-C and Park Y-K 2016 Production and utilization of biochar: a review *J. Ind. Eng. Chem.* **40** 1–15

[21] Azargohar R and Dalai A K 2008 Steam and KOH activation of biochar: experimental and modeling studies *Microporous Mesoporous Mater.* **110** 413–21

[22] Dehkhoda A M and Ellis N 2013 Biochar-based catalyst for simultaneous reactions of esterification and transesterification *Catal. Today* **207** 86–92

[23] Kastner J R, Miller J, Geller D P, Locklin J, Keith L H and Johnson T 2012 Catalytic esterification of fatty acids using solid acid catalysts generated from biochar and activated carbon *Catal. Today* **190** 122–32

[24] Xiong X, Yu I K M, Cao L, Tsang D C W, Zhang S and Ok Y S 2017 A review of biochar-based catalysts for chemical synthesis, biofuel production, and pollution control *Biores. Technol.* **246** 254–70

[25] Bhandari P N, Kumar A, Bellmer D D and Huhnke R L 2014 Synthesis and evaluation of biochar-derived catalysts for removal of toluene (model tar) from biomass-generated producer gas *Renewable Energy* **66** 346–53

[26] Shen B, Chen J, Yue S and Li G 2015 A comparative study of modified cotton biochar and activated carbon based catalysts in low temperature SCR *Fuel* **156** 47–53

[27] Assima G P, Paquet A and Lavoie J-M 2019 Utilization of MSW-derived char for catalytic reforming of tars and light hydrocarbons in the primary syngas produced during wood chips and MSW-RDF air gasification *Waste Biomass Valorization* **10** 1203–22

[28] Mani S, Kastner J R and Juneja A 2013 Catalytic decomposition of toluene using a biomass derived catalyst *Fuel Process. Technol.* **114** 118–25

[29] Dehkhoda A M, West A H and Ellis N 2010 Biochar based solid acid catalyst for biodiesel production *Appl. Catal.* A **382** 197–204

[30] Lee J, Kim K-H and Kwon E E 2017 Biochar as a catalyst *Renew. Sustain. Energy Rev.* **77** 70–9

[31] Ahmad J, Rashid U, Patuzzi F, Alamoodi N, Choong T S Y, Soltani S, Ngamcharussrivichai C, Nehdi I A and Baratieri M 2020 Mesoporous acidic catalysts synthesis from dual-stage and rising co-current gasification char: application for FAME production from waste cooking oil *Materials* **13** 871

[32] Shu Q, Gao J, Nawaz Z, Liao Y, Wang D and Wang J 2010 Synthesis of biodiesel from waste vegetable oil with large amounts of free fatty acids using a carbon-based solid acid catalyst *Appl. Energy* **87** 2589–96

[33] Heck R M, Farrauto R J and Gulati S T 2009 *Catalytic Air Pollution Control* (Hoboken, NJ: Wiley)

[34] Jo Y B, Cha J S, Ko J H, Shin M C, Park S H, Jeon J-K, Kim S-S and Park Y-K 2011 NH_3 selective catalytic reduction (SCR) of nitrogen oxides (NO_x) over activated sewage sludge char *Korean J. Chem. Eng.* **28** 106–13

[35] Cha J S, Choi J C, Ko J H, Park Y K, Park S H, Jeong K E and Jeon J K 2010 The low temperature SCR of NO over rice straw and sewage sludge derived char *Chem. Eng. J.* **156** 321–7

[36] Singh S, Nahil M A, Sun X, Wua C, Chen J, Shen B and Williams P T 2013 Novel application of cotton stalk as a waste derived catalyst in the low temperature SCR-deNO_x process *Fuel* **105** 585–94

[37] Logan B E 2008 *Microbial Fuel Cells* (Hoboken, NJ: Wiley)

[38] Yuan Y, Yuan T, Wang D, Tang J and Zhou S 2013 Sewage sludge biochar as an efficient catalyst for oxygen reduction reaction in an microbial fuel cell *Bioresour. Technol.* **144** 115–20

[39] Lobos M L N, Sieben J M, Comignani V, Duarte M, Volpe M A and Moyano E L 2016 Biochar from pyrolysis of cellulose: an alternative catalyst support for the electro-oxidation of methanol *Int. J. Hydrogen Energy* **41** 10695–706

[40] Fan Y, Liu P-F, Yang Z-J, Jiang T-W, Yao K-L, Han R, Huo X-X and Xiong Y-Y 2015 Bi-functional porous carbon spheres derived from pectin as electrode material for super-capacitors and support material for Pt nanowires towards electrocatalytic methanol and ethanol oxidation *Electrochim. Acta* **163** 140–8

[41] Ormsby R, Kastner J R and Miller J 2012 Hemicellulose hydrolysis using solid acid catalysts generated from biochar *Catal. Today* **190** 89–97

[42] Li S, Gu Z, Bjornson B E and Muthukumarappan A 2013 Biochar based solid acid catalyst hydrolyze biomass *J. Environ. Chem. Eng.* **1** 1174–81

[43] Nguyen H K D, Pham V V and Do H T 2016 Preparation of Ni/biochar catalyst for hydrotreating of bio-oil from microalgae biomass *Catal. Lett.* **146** 2381–91

www.ingramcontent.com/pod-product-compliance
Ingram Content Group UK Ltd.
Pitfield, Milton Keynes, MK11 3LW, UK
UKHW051857100926
8556IPUK00005B/668

9 780750 326582